Handbook of Astronomy, Astrophysics and Geophysics
Volume I

The Earth:1
The Upper Atmosphere, Ionosphere and Magnetosphere

D0087306

HANDBOOK OF ASTRONOMY, ASTROPHYSICS AND GEOPHYSICS
Edited by
CHARLOTTE W. GORDON, *University of Paris VI*
V. CANUTO, *Institute for Space Studies, New York*
W. IAN AXFORD, *Max Planck Institut für Aeronomie, W. Germany*

Associate Editors

J. L. Anderson, *Stevens Institute of Technology, New Jersey*
Orson L. Anderson, *University of California, Los Angeles*
Peter G. Bergmann, *Syracuse University, New York*
Geoffrey R. Burbidge, *University of California, San Diego*
Raynor L. Duncombe, *University of Texas, Austin, Texas*
D. Lamb, *University of Illinois, Urbana*
Richard S. Lindzen, *Harvard University, Cambridge, Mass.*
Robert W. Noyes, *Smithsonian Astrophysical Observatory, Cambridge, Mass.*
Tobias Owen, *State University of New York at Stony Brook*
R. Smoluchowski, *Princeton University, New Jersey*
Karl K. Turekian, *Yale University, Connecticut*
L. Woltjer, *Columbia University, New York*

ISSN 0141-0326

Handbook of Astronomy, Astrophysics and Geophysics Volume I

The Earth: 1
The Upper Atmosphere, Ionosphere and Magnetosphere

Edited by
CHARLOTTE W. GORDON
University of Paris VI

V. CANUTO
Institute for Space Studies, New York
and
W. IAN AXFORD
Max Planck Institut für Aeronomie, W. Germany

Gordon and Breach

Copyright © 1978 by Gordon and Breach, Science Publishers, Inc.

Gordon and Breach, Science Publishers, Inc.
One Park Avenue
New York, N.Y. 10016

Gordon and Breach, Science Publishers Ltd.
42 William IV Street
London WC2N 4DE

Gordon & Breach
7-9 rue Emile Dubois
Paris 75014

Library of Congress Cataloging in Publication Data
Main entry under title:

The Earth.

(Handbook of astronomy, astrophysics, and geophysics; 1)
1.   Atmosphere, Upper—Addresses, essays, lectures.
2.   Ionosphere—Addresses, essays, lectures. 3. Magnetosphere—
Addresses, essays, lectures. I. Gordon, Charlotte W. II. Canuto,
Vittorio.   III.   Axford,   W. Ian.   IV.   Series.
QB43.2.H35 vol. 1 [QC879.2]   520'.8s [551.5'14]
ISBN 0-677-16100-X   77-19187

Library of Congress catalog card number 77-19187
ISBN 0 677 16100 X

INTRODUCTION

In recent years scientists have been forced to work in ever-narrowing specialities at a time when interdisciplinary information is more important than ever before. It is increasingly necessary to publish articles which form basic references in the subject they treat and can be used by a non-specialist. Such articles are intended to bring out the foundations of knowledge in each field whether they are theoretical, experimental or observational. Many of the articles are first published in the journal *Fundamentals of Cosmic Physics* to ensure speed of publication. These will be updated or revised where necessary for compilation in the "Handbook of Astronomy, Astrophysics and Geophysics".

CHARLOTTE W. GORDON
V. CANUTO
W. IAN AXFORD

CONTENTS

Hydrogen in the Upper Atmosphere

BRIAN A. TINSLEY
The University of Texas at Dallas, Texas 75230, U.S.A.

1 HISTORICAL INTRODUCTION: IDENTIFYING THE GEOCORONA

The existence of hydrogen in the upper atmosphere and its thermal escape was of interest in the mid 19th century (Waterson 1846, Stoney 1868) and a theory of its height distribution and escape rate was well developed by the early 20th century.

The atmosphere above the level of 75-100 km was then thought to consist almost entirely of molecular hydrogen (Hann 1903, Jeans 1904, 1910, 1916, Humphreys 1910, Wegener 1911a, b). The reason for this was that balloon measurements showed isothermal conditions at about $220°K$ and a lack of winds above 10 km. Thus it was thought the atmosphere was very stable above that level and the various constituents would separate out by diffusion. A condition of diffusive equilibrium at a temperature of $220°K$ would be set up in the stratosphere and above.

The distribution with height of particles in diffusive equilibrium at constant temperature in the presence of a vertical potential field is determined by the equation (see Section 2.1 or Chapman and Cowling 1952)

$$\frac{dp_i}{dh} + \rho_i \frac{d\psi}{dh} = 0 \tag{1}$$

where p_i is the partial pressure of the ith constituent, ψ is the potential energy per unit mass at height h, and ρ_i is the density.

For neutral constituents we take $\psi = gh$, $p_i = n_i kT$, and $\rho_i = m_i n_i$, where k is Boltzman constant, n_i is the number density of the ith constituent of mass m_i, g is the acceleration of gravity, and T is the absolute temperature.

Eq. (1) thus simplifies to

$$\frac{1}{n_i} \frac{dn_i}{dh} = -\frac{m_i g}{kT} \tag{2}$$

and can be integrated to give, for constant T and g,

$$n_i = n_{io} e^{-h/H_i} \tag{3}$$

where n_{io} is the number density at some reference level and $H_i = kT/m_i g$ is the scale height for the ith constituent. H_i is the height interval in which n_i decreases by a factor $1/e$. For hydrogen and nitrogen at $220°K$ the scale heights are 93 km and 6.6 km respectively. Consequently, although hydrogen was thought to be only one part to 30,000 of N_2 at 10 km, at

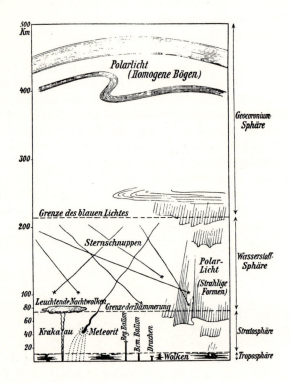

FIGURE 1   Atmospheric regions as depicted by Wegener (1911a)

100 km it would be ten times as abundant as N_2 with diffusive equilibrium in the stratosphere and above.   Figure 1 illustrates the situation according to Wegener (1911a), with hydrogen dominant in the "Wasserstoff-Sphäre". Wegener included 0.00058% of the hypothetical element geocoronium with molecular weight 0.4 in his tropospheric abundances, and it became dominant above 210 km, forming the outermost terrestrial atmosphere, or "Geocoronium-Sphäre".   Geocoronium was hypothesized by Wegener to account for the 5577Å line in the aurora and was stated to be required by Mendelejeff's periodic table of the elements.   But Chapman and Milne (1920) pointed out that an element with such a low molecular weight could not be retained by the atmosphere, and would be dissipated by thermal escape in a time com-

parable with 1 year.

They also argued that there was less than one part in 10^6 of free hydrogen in the atmosphere, and considered it possible that there was none at all at high altitudes. They discussed the assumption of the lack of mixing in the stratosphere. While suggesting a height of 20 km as a better approximation to the level (the turbopause) at which molecular diffusion became more important than mixing, they worked out model atmospheres for turbopause heights between 12 km and 50 km.

Maris (1928, 1929) whose work was later extended by Epstein (1932) and Mitra and Rakshit (1938) emphasised the long time taken to reach diffusive equilibrium, should mixing stop. At the end of a thousand years, separation would be far from complete for the lower altitudes. Observations of the movements of persistent meteor trains gave information on the mixing and indicated that only about 5×10^5 sec (6 days) could be allowed for diffusion to proceed, and using this time it was possible to calculate down to what level diffusive separation should occur. Depending on the constituent concerned, Maris obtained heights around 110 km for the turbopause. Thus even if hydrogen were 0.01% of the atmosphere at the turbopause, its relative abundance would not be appreciable for another 100 km or so.

Interest in hydrogen as an upper atmospheric constituent waned until Meinel (1950a, b) identified OH emission in the night airglow, and the chemistry of oxy-hydrogen compounds in the D and lower E region was analysed by Bates and Nicolet (1950). The altitude region (70-100 km) is also known as the mesosphere and lower thermosphere, and the chemistry down to 60 km, or even 40 km is sometimes included in models. Instead of hydrogen in the upper atmosphere being in molecular form as a continuation of the ground level abundances, the source was now the photodissociation of water vapor at 70-90 km, with a contribution from methane dissociation at lower altitudes. The production of atomic and molecular hydrogen, with peak concentrations near 80 km, would sustain an upward flow of up to $\sim 10^8$ hydrogen atoms $cm^{-2} sec^{-1}$. This hydrogen would be mainly in atomic form above 100 km, where the H_2 would be dissociated, and the H would be lost from the earth by upward diffusion and thermal escape at higher altitudes. The supply was maintained in the D and E region by water vapor and the products of methane dissociation being carried up by gentle mixing winds.

Bates and Nicolet also pointed out that the amount of oxygen liberated into the atmosphere, assuming $10^8 cm^{-2} sec^{-1}$ escape over geological time, is comparable with the present content of oxygen in the atmosphere.

The stimulus for a great deal of subsequent theoretical and experimental work on hydrogen in the upper atmosphere and interplanetary space came from the discovery of an intense glow in Lyman α at night at altitudes above 75 km. The glow was first observed by instruments on a rocket launched by the Naval Research Laboratory in 1955 (Byram *et al.* 1957), and in 1957 a detailed sky map of the emission was obtained when the sun was 44° below the horizon (Kupperian *et al.* 1958, 1959). Figure 2 shows the directional intensity contours obtained. The contours showed a minimum intensity close to the antisolar direction, and an emission was seen looking downward when the rocket was above 85 km, indicating an albedo of 42%. It was thought that emission from above was from scattering of solar Lyman α on atomic hydrogen in the solar system external to the orbit of the earth, and the emission from below from back scattering on D and E region hydrogen. Shklovsky (1960, 1959) pointed out that the narrow Balmer α (Hα) emission that had been observed by Prokudina (1958), probably arose from Lyman β scattering on the same hydrogen as for the Lyman α, but in this case the scattering would be non-conservative with some 12% resulting in fluorescence and production of the Balmer α emission. Shklovsky considered that the hydrogen was probably interplanetary but did not "exclude entirely" terrestrial hydrogen as the major scattering medium, i.e. a "geocorona" of atomic hydrogen "extending over several thousands or tens of thousands of kilometers".

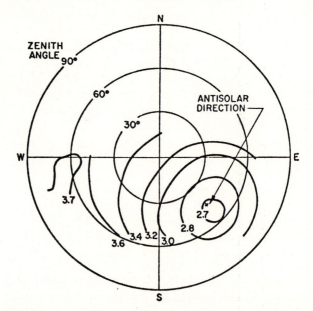

FIGURE 2   Directional intensity contours of Lyman α obtained above 120 km with the sun 44° below the horizon. The units are 10^{-10} watt cm^{-2} $ster^{-1}$. From Johnson and Fish (1961), *Astrophysical Journal*, © University of Chicago Press.

Brandt and Chamberlain (1959) discussed the scattering process, including polarization, in some detail, and the question of the interplanetary versus the terrestrial source for the 1957 Lyman α results. From the lack of change in intensity with height between 120 km and 146 km they reuled out an optically thick geocorona necessary to fill in the earth's shadow by multiple scattering. To explain the albedo as high as 42% they considered an extra source of Lyman α emission was probably necessary between 85 and 120 km, since the interplanetary source would be doppler shifted mostly outside the wavelength range for efficient scattering. A later development of the inter-planetary theory is due to Brandt (1961a).

Johnson (1959a, b) and Johnson and Fish (1960) argued strongly for the terrestrial source, pointing out that the nighttime albedo was quite naturally explained when exospheric hydrogen, with a temperature less than 2000°K, was scattering solar Lyman α and was the source of the photons incident on the E and D regions. The exosphere is the region above the "exobase" level (about 500 km) at which the probability of a fast upward moving particle avoiding further collisions becomes $1/e$. In the exosphere particles travel in ballistic orbits, mostly up from the exobase and down again, but some are escaping, and some may be in satellite orbits (Spitzer 1949).

Taking the temperature of the exosphere as determined by orbital decay of the Vanguard 1 satellite to be 1250°K, Johnson and Fish (1960) used the shape of the Lyman α intensity contours near the minimum to obtain an estimate of the vertical optical thickness above the E region. They obtained a value of 1.1, or 7×10^{12} atoms cm^{-2}. From a theoretical altitude profile normalized to this column abundance an escape rate of 10^8 cm^{-2} sec^{-1} was found, which agrees with the escape rate suggested by Bates and Nicolet (1950). The "constant" Lyman α intensity between 120 and 146 km was explained when a calculation of the expected change gave only 5%, as compared to 15% being the minimum detectable in the 1957 experiment. The position of the minimum in the intensity contours was 8° below the antisolar direction and this was accountable in terms of a parallax effect of the earth's shadow on terrestrial hydrogen, but not on interplanetary hydrogen. Johnson (1959a) predicted that the terrestrial hydrogen column abundance estimated as above would produce a narrow (0.03A) absorption core, 70% complete, in the center of the solar Lyman α line as seen from the E region. The several high resolution solar spectra obtained in July 1959 by Purcell and Tousey (1960, 1961) showed a narrow absorption core in Lyman α roughly consistent with this prediction. The profile of the solar line with the central absorption core is shown in Figure 3.

Brandt (1961b, c; 1962a, b) argued that the Johnson and Fish model, involving multiple scattering to fill in the shadow on the nightside, would have too low a hydrogen density when normalised to the Purcell and Tousey absorption core measurement, to produce the observed intensity. He proposed that the observed Lyman α originated from single scattering in the outer part of a terrestrial hydrogen cloud, at distances greater than 5-10 earth radii. The cloud was considered to be spherically symmetric, and to have a characteristic

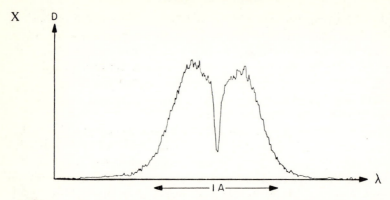

FIGURE 3   Microphotometer tracing of the solar Lyman α line profile, obtained above 134 km, showing the narrow centered absorption core due to terrestrial hydrogen. From Purcell and Tousey (1960), *Space Research,* © North Holland Publ. Co.

radius of $10^2 R_E$, and was called a "geocoma". However, quantitative difficulties with the geocoma hypothesis were pointed out by Donahue (1962).

The problem of the hydrogen column abundance measured by Purcell and Tousey and on other more recent rocket shots being low compared to that inferred from scattering measurements has still not been completely resolved, as is discussed in Section 9.1. The results seem best interpreted as due to low daytime abundances in the 100-300 km region, since other measurements at higher altitudes have established that there is sufficient hydrogen for radiative transfer to the nightside. If so, the reason for the low daytime abundances in the 100-300 km region has not been established.

Since 1962 the state of theory and observation of hydrogen in the upper atmosphere has been discussed in a number of reviews. Some of the more comprehensive are due to Kockarts and Nicolet (1962, 1963), Chamberlain (1963b), Kurt (1963), Donahue (1966), Krassovsky *et al.* (1966), Donahue (1968a), Rossbach (1969), Tinsley (1969c), Kockarts (1971) and Mange (1972).

2 THEORETICAL ALTITUDE DISTRIBUTIONS

2.1 General

The vertical distribution of atomic hydrogen in the D and E region of the atmosphere is determined by the chemistry of production and the amount of mixing between different regions resulting from transport processes.

These transport processes include the effects of internal gravity waves, tidal oscillations, and large scale circulations. Horizontal as well as vertical motions in the circulation may be important on account of latitudinal variations in composition. At present little accurate information is available on the relative importance and combined magnitude of the transport processes, and their altitude, latitude, and seasonal variations.

The combined transport effects are usually lumped together and represented as "eddy diffusion" by an "eddy diffusion coefficient" in models of the production and vertical distribution of constituents in the mesosphere and lower thermosphere.

At the turbopause molecular diffusion becomes more important than eddy diffusion, but because of the upward flow diffusive equilibrium is not achieved until 150 or 200 km, depending on the exospheric temperature. The flow results in a distribution tending to follow the atmospheric scale height, as for a mixed distribution. Where the temperature is varying with height, the effects of thermal diffusion are present. This applies below a level (the thermopause) at about 300 km. The region between the thermopause and exobase is nearly isothermal. The concept of "temperature" for the neutrals in ballistic and satellite orbits in the exosphere needs to be applied with some caution. In that region, as will be discussed later, the velocity distributions are incomplete Maxwell-Boltzman distributions. The extent to which various groups of particles are present in the actual distributions at the exobase and above affects the escape rate and the vertical profile.

The temperature controls not only the scale height above the turbopause but also the rate of escape, and hence the overall abundance. Consequently, the temperature distributions are of overriding importance to the altitude distributions above the turbopause.

The influence on vertical profiles of mixing, upward flow, and molecular diffusion including thermal diffusion was discussed by Mange (1955), Bates (1959), Johnson and Fish (1960) and Mange (1961). The steady state diffusion equation in the absence of eddy diffusion is as given in Chapman and Cowling (1952, p.244) for a two constituent atmosphere;

$$\bar{c}_1 - \bar{c}_2 = -\frac{n^2}{n_1 n_2} D_{12} \left\{ \frac{1}{p}\left(\frac{\partial p_1}{\partial r} - \rho_1 F\right) + k_T \frac{1}{T}\frac{\partial T}{\partial r} \right\} \tag{4}$$

$\bar{c}_1$ and $\bar{c}_2$ are the mean velocities of the two constituents with number densities n_1 and n_2. We take the hydrogen as having a number density n_1, partial pressure $p_1 = n_1 kT$, and partial density $\rho_1 = m_1 n_1$. With hydrogen being a minor constituent the total number density n can be taken equal to n_2. $p = nkT$ is the total pressure. Taking $\bar{c}_2 = 0$, then $S_{1,m} = \bar{c}_1 n_1$ is the upward flow of hydrogen by molecular diffusion in the direction r (which can be replaced by h), and $\rho_1 F = -m_1 n_1 g$ is the gravitational force per unit volume on the hydrogen. $\partial T/\partial r$ is the vertical temperature gradient, and D_{12} and k_T are the molecular diffusion coefficient and the thermal diffusion ratio respectively. Thus (4) becomes

$$S_{1,m} = -nD_{12}\left\{\frac{1}{p}\left(\frac{\partial p_1}{\partial h} + n_1 m_1 g\right) + k_T \frac{1}{T}\frac{\partial T}{\partial h}\right\} \tag{5}$$

This equation must be modified to include the effects of mixing for treatments of the region where it is important. This has been done by Colegrove et al. (1965, 1966) following the work of Lettau (1951). The result is the formalism in terms of the eddy diffusion coefficient K which is used as an adjustable parameter in models until a fit to experimental data determined by the degree of mixing is obtained.

This procedure has been relatively successful in that mesospheric composition models can then be constructed that agree reasonably well with measurements, but it tends to divert attention from the nature of the transport processes themselves and their spatial and temporal variations.

The eddy diffusion coefficient may be introduced into the formalism as follows:

If there are random motions in an atmosphere which is not of uniform composition then a volume element of fractional composition n_i/n will move relative to neighboring volume elements of slightly different composition.

The volume elements may be considered to be moving in a random walk (though the velocities and scale lengths may not necessarily be the same horizontally as vertically) and in so moving produce a net transport down the concentration gradient. The vertical eddy diffusion flux $S_{i,E}$ in the ith constituent will be proportional to the concentration gradient $\partial/\partial h(n_i/n)$ and to a coefficient which is a measure of the vigor of the mixing, and so we may define

$$S_{i,E} = -nK \frac{\partial}{\partial h}\left(\frac{n_i}{n}\right) \tag{6}$$

where K is the vertical eddy diffusion coefficient. Now n_i/n can be replaced by p_i/p in (6) and expanded to give

$$S_{i,E} = -nK\left\{\frac{1}{p}\left(\frac{\partial p_i}{\partial h}\right) - \frac{p_i}{p^2}\left(\frac{\partial p}{\partial h}\right)\right\} \tag{7}$$

and hydrostatic equilibrium requires

$$\frac{\partial p}{\partial h} = -nmg \tag{8}$$

where m is the mean molecular mass, and only gravitational forces are acting.

Hence the vertical flux due to the sum of molecular diffusion plus eddy diffusion for the constituent $i = 1$ is

$$S_1 = -nD_{12}\left[\frac{1}{p}\left(\frac{\partial p_1}{\partial h} + n_1 m_1 g\right) + \Lambda\frac{1}{p}\left(\frac{\partial p_1}{\partial h} + n_1 mg\right) + k_T\frac{1}{T}\frac{\partial T}{\partial h}\right] \quad (9)$$

where Λ is the eddy diffusion ratio, such that the eddy diffusion coefficient $K = \Lambda D_{12}$.

It can be seen that in an isothermal atmosphere where $D_{12} \gg K$, and $S_1 = 0$, (9) reduces to the relation $\partial p_1/\partial h = -n_1 m_1 g$, and leads to Eq. (3) with a scale height for n_1 of $kT/m_1 g$. But for the atmosphere where $D_{12} \ll K$ and $S_1 = 0$, (9) reduces to the relation $\partial p_1/\partial h = -n_1 mg$, and leads to a scale height kT/mg which is the same for all constituents, as expected for complete mixing.

To the extent that the mixing due to internal gravity waves and tides is the major process represented by the eddy diffusion coefficient, Lindzen (1971) has shown how these processes are related to the numerical value of the coefficient, and has been able to speculatively suggest a latitudinal, altitudinal, and seasonal structure for the coefficient which is not implausible.

We have introduced K as a coefficient measuring the vigor of the mixing. It has the dimensions of $[\text{Length}]^2 [\text{Time}]^{-1}$, and can be thought of as a weighted product of a characteristic vertical velocity and a characteristic vertical scale for the motion under consideration. Lindzen (1971) has shown that for internal gravity waves or tidal oscillations one can associate the vertical velocity with that present in the wave oscillation, and vertical scale with the vertical wavelength. The weighting factor could be considered zero for a pure adiabatic, sinusoidally oscillating coherent internal gravity wave, but might reach 0.5 for very rapid dissipation of the wave. These considerations lead to values of K decreasing inversely as the square root of the density below about 90 km, where near the equator the value might be about 3×10^5 cm^2 s^{-1}. Values up to 4×10^6 cm^2 s^{-1} might be found near the equator between 90 km and 108 km, where the diurnal tide is strong and unstable.

In the summer high latitude region the small fluctuations in temperature measured by rockets imply a coefficient less than ~4×10^5 cm^2 s^{-1}. Above 80 km in the very disturbed winter high latitude region K might reach 1.5×10^7 cm^2 s^{-1}, and may be ~3×10^6 cm^2 s^{-1} at 80 km, decreasing inversely as the square root of the density below that altitude. Gravity-wave and tidal effects are considered further by Lindzen (1970a, b) and Lindzen and Blake (1970).

We may note that Eq. (6), defines a velocity $V_{i,E}$ such that

$$V_{i,E} = \frac{S_{i,E}}{n_i} = -K\left\{\frac{n}{n_i} \cdot \frac{\partial}{\partial h}\left(\frac{n_i}{n}\right)\right\} = -\frac{K}{C_i} \qquad (10)$$

The term in curly brackets is the reciprocal of the characteristic length C_i, defining the concentration gradient, which is typically about an average scale height, or 10^6 cm in the mesosphere. Regarding $V_{i,E}$ as an effective "turbulent transport velocity", which would be ~1 cm sec^{-1} for $K \sim 10^6$ cm^2 s^{-1}, we can see what vertical bulk velocities due to large scale circulation would be needed in order for circulation to outweigh tides and gravity waves as transport processes. Vertical velocities up to 10 or 100 cm s^{-1} may exist at times in high latitude regions, hence circulation processes are clearly important.

The molecular diffusion coefficient D_{12} is given (Chapman and Cowling 1952, p.245) for a two constituent atmosphere by

$$D_{12} = \frac{3}{8n\sigma_{12}^2}\left[\frac{kT(m_1 + m_2)}{2\pi m_1 m_2}\right]^{1/2} \qquad (11)$$

where σ_{12} is the effective radius of the sphere of interaction for diffusion. For rigid elastic spheres of diameters σ_1 and σ_2,

$$\sigma_{12} = \tfrac{1}{2}(\sigma_1 + \sigma_2) . \qquad (12)$$

We may approximate the diffusion of hydrogen in the real atmosphere (consisting of N_2, O_2 and O) by the use of this binary diffusion coefficient where D_{12} and σ_{12} represent coefficients for interaction of species 1 with air of appropriate composition.

Eq. (11) can be transformed to give

$$D_{12} = \frac{3}{8\sqrt{2\pi}\sigma_{12}^2}\left(1 + \frac{m}{m_1}\right)^{1/2}\frac{(gH)^{1/2}}{n} \qquad (13)$$

showing that D_{12} increases rapidly with altitude as n decreases and H increases above about 80 km. The turbopause is the level at which $D_{12} = K$, and from measurements of the ratio $n(\text{He})/n(\text{N}_2)$ or $n(\text{A})/n(\text{N}_2)$, which will vary as diffusive separation sets in above the turbopause, it is found that the turbopause lies between 100 and 110 km. This implies that K lies between 10^6 cm^2 s^{-1} and 10^7 cm^2 s^{-1} in the range 100-110 km. As mentioned earlier the transport varies with geographical position, altitude, and time, and the turbopause height varies correspondingly. A wide range of vertical profiles of eddy diffusion coefficient have been assumed in model mesosphere calculations.

Eq. (9) can be transformed into a form suitable for numerically computing an altitude distribution i.e.

$$\frac{\partial n_1}{\partial h} = -\frac{n_1}{1+\Lambda}\left\{(1+\Lambda+\alpha_T)\frac{1}{T}\frac{\partial T}{\partial h} + \left(1+\frac{\Lambda m}{m_1}\right)\frac{m_1 g}{kT}\right\} - \frac{S_1}{D_{1z}(1+\Lambda)}. \quad (14)$$

The term $(n/n_1)k_T$ has been replaced by the thermal diffusion factor α_T, which is nearly independent of concentration, but depends on composition and temperature. The value of α_T for hydrogen in the range of composition and temperature in the upper atmosphere is not well known. Patterson (1966a) used $\alpha_T = -0.25$ for atomic and -0.33 (presumably) for molecular hydrogen (cf. Grew and Ibbs (1952) and McElroy and Hunten (1969a)).

The effects of sources and sinks and time dependency in the flow can be accounted for by having S_i vary with height, as given by the one-dimensional continuity equation

$$\frac{\partial n_i}{\partial t} = P_i - L_i n_i - \frac{\partial S_i}{\partial h} \quad (15)$$

where P_i is the production rate and L_i is the loss rate per unit volume.

2.2 D and E region density distributions

Vertical distributions of hydrogen, oxygen, and oxyhydrogen compounds in D and E regions have been worked out in increasing detail, starting with Bates and Nicolet (1950). Treatments of the oxyhydrogen chemistry have been made by Hesstvedt (1964, 1965a, b), Hampson (1964), Nicolet (1964), and Hunt (1966). The effects of eddy diffusion on the oxygen chemistry have been considered by Colegrove et al. (1965, 1966), Keneshea and Zimmerman (1970) and Shimizaki (1971). The effects of eddy diffusion on the oxyhydrogen chemistry were considered in an analytical model by Konashenok (1967). Detailed numerical calculations which include 38 reactions, the effects of eddy diffusion, and diurnal and latitude variations of the oxyhydrogen chemistry have been made by Hesstvedt (1968, 1969).

Similar sets of calculations with changes in rate coefficients, eddy diffusion profiles, and/or boundary conditions have been made by Shimizaki and Laird (1970, 1972), Bowman et al. (1970), Anderson (1971), and George et al. (1971). Nicolet (1970, 1971) has discussed many aspects of the photochemistry and transport from the stratosphere to 100 km, and Strobel (1972) has given a critical review of the subject.

The large number of chemical reactions that must be included in the model, the uncertainties in the input data, and the mathematical compromises necessary for computational feasibility lead to a great deal of uncertainty about the causes of certain features that emerge from the computations. It is helpful to note (Konashenok 1967, Hesstvedt 1968, Nicolet 1971, Anderson 1971) that the hydrogen species may be classified into two groups. The first includes OH, HO_2 and H, for which the photochemical reactions are sufficiently fast compared to the characteristic turbulent transport time ($C_i/V_{i,E}$, i.e. C_i^2/K) that photochemical equilibrium can be taken to apply below about 80 km. The second group includes H_2 and H_2O, for which the time to achieve photochemical equilibrium is slow compared to the characteristic turbulent transport time. For the first group under steady state conditions the continuity Eq. (15) reduces to

$$n_i = \frac{P_i}{L_i}. \tag{16}$$

For the second group Eq. (15) is used in conjunction with Eq. (6). Writing w_i for the mixing ratio n_i/n we have

$$\frac{\partial n_i}{\partial t} = P_i - L_i n_i - \frac{\partial}{\partial h}\left[-nK\frac{\partial w_i}{\partial h}\right] \tag{17}$$

and for steady state conditions $\partial n_i/\partial t = 0$, so

$$\frac{\partial^2 w_i}{\partial h^2} = \frac{w_i}{K}\left(L_i - \frac{P_i}{n_i}\right) - \frac{\partial w_i}{\partial h}\left(\frac{1}{K}\frac{\partial K}{\partial h} + \frac{1}{n}\frac{\partial n}{\partial h}\right). \tag{18}$$

The production rate of water is so slow compared to the loss rate and the transport time that P_i can be neglected in (18). The important loss processes are photodissociation

$$H_2O + h\nu \rightarrow OH + H \tag{19}$$

(chiefly due to Lyman α and below 70 km to the Shumann-Runge bands) and dissociation in the reaction

$$H_2O + O(^1D) \rightarrow OH + OH \tag{20}$$

which is important below 60 km.

Thus there is an equilibrium between the rate at which H_2O is transported into the mesosphere and the rate at which it is converted into other oxyhydrogen compounds. There will be a greater or lesser abundance of water vapor in the mesosphere depending on the magnitude of the actual eddy or circulation transport processes (Nicolet 1970, 1971; Anderson 1971).

The upward transport of H_2O from the troposphere is augmented in the stratosphere by CH_4 and H_2, which are broken down in that region with production of H_2O, so that the value of the mixing ratio for H_2O at the strato-pause is $\sim 6 \times 10^{-6}$, as compared to $\sim 3 \times 10^{-6}$ at the tropopause (Strobel 1972). The chemistry and transport of CH_4, CO_2 and CO in the atmosphere have been reviewed by Wofsy et al. (1972). In the mesosphere and lower thermo-sphere the breakdown of H_2O and reactions between the consequent OH and H and the ambient O_2, O and O_3 from the oxygen photochemistry are the source of the OH, H, HO_2, H_2O_2, H_2, $O(^1D)$, $O(^3P)$, and O_3 in that region. The chemical loss rate of H_2 is slow compared to transport times in the meso-sphere (Anderson 1971), and ultimately the oxyhydrogen compounds are converted to H_2 and H, which are transported upward with the remaining H_2O, to be broken down into H in the 90-150 km region in the processes

$$H_2O + h\nu \rightarrow H + OH \tag{21}$$

$$H_2 + O(^3P) \rightarrow H + OH \tag{22}$$

$$H_2 + O(^1D) \rightarrow H + OH \tag{23}$$

$$OH + O(^3P) \rightarrow O_2 + H \tag{24}$$

Reaction (24) is rapid compared to (21) through (23) so that the OH con-centration is small compared to the sum of those of H_2O and H_2.

Table I taken from Strobel (1972) illustrates schematically the transport processes and important photochemical reactions of oxyhydrogen compounds from the ground to the exobase. The total upward flux of hydrogen in all forms must be constant on the average, though the amounts in different forms change greatly. Where the effects of diffusion are small compared to mixing, the mixing ratio of total hydrogen in all forms should remain approximately constant. The flux of H_2O may in fact be downward at the Tropopause, on account of its production from CH_4 and H_2 in the stratosphere.

The various numerical calculations that have been made to determine altitude profiles of the oxyhydrogen compounds show a broad similarity but none can be considered reliable on account of the large uncertainties in the appropriate profile of eddy diffusion coefficient (not to mention its time and spatial variations). Great care must be exercised in setting up the boundary conditions (Strobel 1972). Figure 4 shows the vertical distribution of H_2O, H_2 and H calculated by Hesstvedt (1968) for both daytime condi-tions, and nighttime conditions where the H_2O and H_2 were determined by

BRIAN A. TINSLEY

TABLE I

Photochemistry and transport of hydrogen constituents, from Strobel (1972), *Radio Science,* © American Geophysical Union.

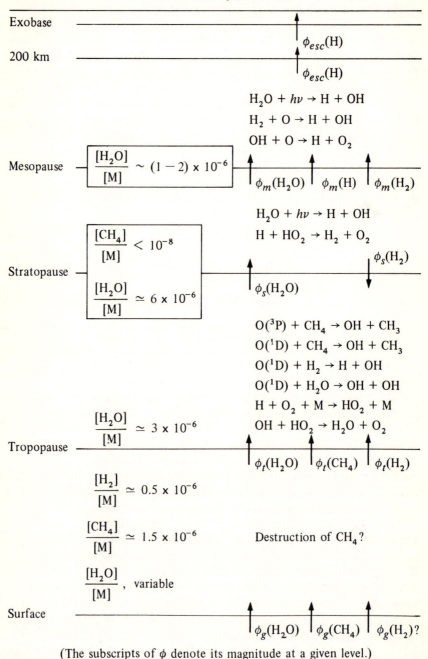

(The subscripts of ϕ denote its magnitude at a given level.)

production earlier in the daytime, but the H had rapidly decayed as the photo-chemical equilibrium conditions changed in the absence of sunlight.

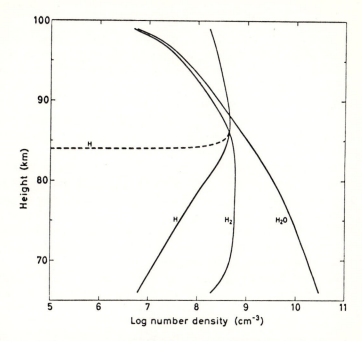

FIGURE 4   Vertical distribution of H, H_2 and H_2O in a steady state atmosphere with photochemical reactions and vertical eddy diffusion. Solid curves refer to a calculation for daytime conditions, the dashed curve is for nighttime conditions. From Hesstvedt (1968), *Geofysiske Publikasjoner*, © Universitets Forlaget, Oslo.

The upper boundary of the calculation shown in Figure 4 is 100 km, and the ratios n_{H_2}/n_H and n_{H_2O}/n_H given there are the photochemical equilibrium values assumed by Hesstvedt for the boundary. The n_{H_2}/n_H and n_{H_2O}/n_H ratios at 100 km, if of the order unity, are of considerable importance for the H altitude profile from 100 to 150 km. The photochemical equilibrium values at 100 km are very much less than one, but this is not necessarily the ratio in the upper atmosphere at that level. Bowman *et al.* (1970) found $n_{H_2O}/n_H = 0.5$ at 100 km, using a somewhat more realistic boundary condition. Enhanced turbulence, or upward flow at 10 cm s^{-1} could increase these ratios.

Bates and Nicolet (1965) advanced some arguments favouring the possibility that the n_{H_2}/n_H ratio might be greater than unity. This would keep the mixing ratio of total hydrogen more constant with height. A calculation by Patterson (1966a) later reinterpreted by Tinsley (1969a) indicates that data from the Purcell and Tousey (1960) Lyman α profile experiment is consistent with

n_{H_2}/n_H (or n_{H_2O}/n_H) being about 1.5:1. This would affect the hydrogen altitude profile significantly up to about 150 km, with the dissociating H_2 or H_2O acting as a source for H in the 100-150 km region. If indeed there is a flow of H_2 or H_2O through the 100 km level which averaged globally is significant compared to the H flow, then this might explain much of the problem of low altitude daytime abundances (Section 9.1).

The 100-150 km region can be considered as dividing the lower chemical source region from the upper region where the height profile is determined by thermal and molecular diffusion, upward flow and the effects of thermal escape. The upward flow of H through the 100 km level, when averaged over the globe and over several days, will equal the loss rate by thermal escape averaged similarly, provided other sources and sinks for hydrogen are not important. An upward flow of H_2 and/or H_2O is one possible source, and a diurnally averaged loss in the form of H^+ ions to the plasmasphere, and as a polar wind out the open field lines at the poles, may be important. Even with no other sources and sinks, the two distributions of upward flow at 100 km and at the exobase would be different spatially and temporally. This is a consequence of latitude and nightside-dayside movements of hydrogen, as lateral flow in the exosphere.

We shall next consider the theory of thermal escape and of exospheric altitude distributions, and then return to the region between .100 km and the exobase at about 500 km in Section 2.4.

2.3 Escape and exospheric density distributions

2.3.1 General   The theory of density distributions with altitude in the exosphere is closely associated with the theory of escape. The loss into space of sufficiently energetic atoms, above the level where the probability of collisions of upward moving particles becomes small, was originally discussed by Waterson (1846). He treated it in terms of uniform speeds for all atoms of a single species, as a velocity dispersion had not yet been introduced in kinetic theory. The theory was developed by Stoney (1868, 1898), who realized the importance of a distribution of velocities in allowing escape for a gas whose mean thermal velocity is less than the velocity of escape from the planet. Cook (1900) and Bryan (1901) improved the application of kinetic theory to the calculation of the escape flux.

Jeans (1904, 1910, 1916) refined the theory of escape, and found a negligible escape rate for hydrogen on earth at 220°K (in the 1910 paper an arithmetic error of a factor of 10^{55} and another smaller error did not invalidate this conclusion). While escape was found to be negligible at 220°K, he noted that only 10^7 years would be required to remove all the hydrogen in the atmosphere if the temperature of the isothermal region (supposed above 10 km) was 550°K instead at 220°K.

The theory of escape and of density distribution in the exosphere, including determination of the height of the exobase, developed in contributions from Emden (1907), Milne (1923), Jones (1923), Hulbert (1937), Mitra and Banerjee (1939), Spitzer (1949), and Gilvarry (1961a, b).

2.3.2 Components of the particle distribution   The hydrostatic distribution determined by (1) with the temperature constant with height may be considered to apply to neutral particles in the exosphere, where ψ is taken as the pure gravitational potential. Referred to a radius r_0 from the earth's center, we have at radius r ,

$$\psi = GM\left(\frac{1}{r_0} - \frac{1}{r}\right) \tag{25}$$

where G is the universal gravitational constant and M is the mass of the earth. Hence (1) can be integrated to give

$$n = n_0 e^{-\frac{mGM(r-r_0)}{kTr_0 r}} \tag{26}$$

where the subscripts i have been deleted. This equation, however, leads to a constant density as r tends to infinity, and is the situation which would apply if hydrogen travelling outward from the earth was able to fill up interplanetary space, and collisions established the Boltzmann distribution at the temperature T there. The necessary time for this is very much longer than the lifetime against the loss process of photoionization and charge exchange, about 10^7 sec, in interplanetary space. Thus a component of the Boltzmann distribution is missing, and as will be shown, an increasing fraction of the higher energy atoms are missing at increasing distances from the earth.

In order to calculate the altitude profile, we may consider that a fictitious Boltzmann distribution exists whose number density is given by (26), and that it is divided into components as follows:

1) The ballistic re-entry component, consisting of atoms on elliptic trajectories (less than escape energy) which travel up from the exobase and return to it.

2) The orbiting component, consisting of atoms on elliptic trajectories which lie wholly above the exobase.

3) The escape and capture components, consisting of atoms on hyperbolic trajectories (more than escape energy) whose perigees lie below the exobase. Half of these atoms (group 3a) are moving upward and escaping, while half (group 3b) are moving downward and will be captured.

4) The flyby component, consisting of atoms on hyperbolic trajectories whose perigees lie above the exobase.

The orbiting component is absent to an extent varying with height, as discussed in Section 2.3.3, and by Brandt and Chamberlain (1960) and Opik and Singer (1960).

The flyby component (group 4) and the capture component (group 3b) are the groups originating almost entirely in interplanetary space, and can be taken as of negligible density.

The fractional abundance in the different components at a given level may be evaluated by following particle trajectories up from the exobase (Opik and Singer 1959, 1960, 1961) or by integrating the Maxwellian distribution function over appropriate limits (Chamberlain 1960, Johnson and Fish 1960, Johnson 1961).

The general form of the Maxwellian distribution function in polar coordinates ϕ and ω can be written

$$n(v,\phi,\omega)d\omega\,d\phi\,dv = n\left(\frac{m}{2\pi kT}\right)^{3/2} e^{-\frac{mv^2}{2kT}} v^2 \sin\omega\,d\omega\,d\phi\,dv \qquad (27)$$

where $n(v,\phi,\omega)d\omega\,d\phi\,dv$ is the number of particles with velocities between v and $v + dv$, moving in directions between ϕ and $\phi + d\phi$ and ω and $\omega + d\omega$. We choose ω as the angle between the velocity vector and the local vertical, and ϕ as the azimuthal angle of the velocity vector.

The four groups of particles are separated by their velocities v being either less or greater than the escape velocity v_e (this separates groups (1) and (2) from groups (3) and (4)), and by the perigee of their trajectory being either inside or outside the exobase (this separates groups (1) and (3) from (2) and (4)). This latter mode of separation is made in terms of a limiting value of ω for a given v and r, defined by the condition that the perigee be at the exobase. We will now find this limit.

The motion of a particle in the earth's gravitational field can be specified by the constancy of the angular momentum

$$J = mrv \sin\omega \qquad (28)$$

and the constancy of the Hamiltonian

$$H = \frac{mv^2}{2} + GMm\left(\frac{1}{r_0} - \frac{1}{r}\right). \qquad (29)$$

The first term in the Hamiltonian is the kinetic energy at a point on the trajectory, and the second is the corresponding gravitational potential energy referred to a height r_0 (Eq. 25). The electrostatic potential energy is taken as zero. We are interested in the limiting case of the trajectory whose perigee is at the exobase, so we take r_0 as the exobase altitude. At the perigee, with velocity v_0,

$$J = mr_0v_0 \tag{30}$$

and we see

$$H = \frac{mv_0^2}{2}. \tag{31}$$

Thus anywhere on this limiting trajectory where a particle passes through r with a velocity v, from (30) and (31) we see

$$H = \frac{J^2}{2mr_0^2} = \tfrac{1}{2}mv^2 + GMm\left(\frac{1}{r_0} - \frac{1}{r}\right) \tag{32}$$

and substituting (28) in (32) the limiting ω_c is found from

$$\sin^2\omega_c = \frac{r_0^2}{r^2}\left[1 + \frac{2GM(r - r_0)}{v^2rr_0}\right]. \tag{33}$$

(This equation also applies to apogees at r_0, when $r < r_0$). For a given $r > r_0$, (33) also leads to a minimum velocity v_m for a perigee at the exobase, such that $\sin^2\omega_c = 1$ (apogee at r) and

$$v_m = +\left[\frac{2GMr_0}{(r + r_0)r}\right]^{1/2}. \tag{34}$$

For upward moving particles the limiting value ω_c of ω is given by

$$\cos\omega_c = +(1 - \sin^2\omega_c)^{1/2} = +\cos\omega_d \tag{35}$$

while for downward moving particles

$$\cos\omega_c = -(1 - \sin^2\omega_c)^{1/2} = -\cos\omega_d \tag{36}$$

where ω_d is always positive.

The trajectory which is an escape trajectory with velocity v_e at r is the limiting trajectory where the kinetic energy is zero at $r = \infty$, thus from (29)

$$H = \frac{GMm}{r_0} \tag{37}$$

and

$$\tfrac{1}{2}mv_e^2 = \frac{GMm}{r} \tag{38}$$

$$v_e = \sqrt{\frac{2GM}{r}} . \tag{39}$$

The fraction X_2 in velocity space of the total ficticious distribution that corresponds to group (2) is found by integrating the Maxwellian distribution function (27) over the limits

$$v_m < v < v_e$$

and $$-\cos \omega_d < \cos \omega < +\cos \omega_d .$$

The fraction X_1 corresponding to group (1) is the integral over the limits

$$0 < v < v_e$$

and $$-1 < \cos \omega < -\cos \omega_d \quad \text{and} \quad \cos \omega_d < \cos \omega < 1 .$$

The fraction X_3 corresponding to group (3) is the integral over the limits

$$v_e < v < \infty$$

and $$-1 < \cos \omega < -\cos \omega_d \quad \text{and} \quad \cos \omega_d < \cos \omega < 1$$

(group 3(a) has $\cos \omega$ positive, while group 3(b) has $\cos \omega$ negative). The fraction X_4 corresponding to group (4) is the integral over the limits

$$v_e < v < \infty$$

$$-\cos \omega_d < \cos \omega < +\cos \omega_d$$

Thus

$$X_2 = 2\pi \left(\frac{m}{2\pi kT}\right)^{3/2} \int_{v_m}^{v_e} e^{-\frac{mv^2}{2kT}} v^2 \left[-\cos \omega\right]_{+\cos \omega_d}^{-\cos \omega_d} dv \tag{40}$$

$$= 4\pi \left(\frac{m}{2\pi kT}\right)^{3/2} \int_{v_m}^{v_e} e^{-\frac{mv^2}{2kT}} v^2 \left\{1 - \frac{r_0^2}{r^2}\left[1 + \frac{2GM}{v^2}\left(\frac{r-r_0}{rr_0}\right)\right]\right\}^{1/2} dv \quad (41)$$

and

$$X_1 = 4\pi \left(\frac{m}{2\pi kT}\right)^{3/2} \int_0^{v_e} e^{-\frac{mv^2}{2kT}} v^2 \, dv - X_2 \quad (42)$$

$$X_4 = 4\pi \left(\frac{m}{2\pi kT}\right)^{3/2} \int_{v_e}^{\infty} e^{-\frac{mv^2}{2kT}} v^2 \left\{1 - \frac{r_0^2}{r^2}\left[1 + \frac{2GM}{v^2}\left(\frac{r-r_0}{rr_0}\right)\right]\right\}^{1/2} dv \quad (43)$$

and

$$X_3 = 4\pi \left(\frac{m}{2\pi kT}\right)^{3/2} \int_{v_e}^{\infty} e^{-\frac{mv^2}{2kT}} v^2 \, dv - X_4 . \quad (44)$$

Figure 5 is adapted from Opik and Singer (1961). Essentially it illustrates the variation of the integrands in Eqs. (41) to (44) with v^2 and y, where $y = r_e/r$, and the four values of y of 0.8, 0.6, 0.4 and 0.2 are illustrated.

The abscissa is u^2 where $u = v/v_{eo}$ and v_{eo} is the escape velocity at the exobase. The ordinate is

$$\frac{dn}{du^2} e^{\frac{mGM(r-r_0)}{kTr_0r}}$$

which removes the altitude variation in n.

Also shown are $u_e^2 = (v_e/v_{eo})^2 = y$, and $u_m = (v_m/v_{eo})^2$. The ratio $mv_{eo}^2/2kT = E$ was taken as 4.65, appropriate to hydrogen at 1500°K in the terrestrial exosphere.

Figure 6 is taken from Johnson (1961) and shows the different altitude profiles obtained for (a) hydrostatic equilibrium, i.e. n as in Eq. (21); (b) ballistic re-entry plus escaping plus orbiting components, i.e. $n(X_1 + X_2 + \frac{1}{2}X_3)$; (c) ballistic re-entry plus escaping components, i.e. $n(X_1 + \frac{1}{2}X_3)$. The temperature was again 1500°K.

Further contributions to the theory of escape were made by Herring and Kyle (1961), Aamodt and Case (1962), Shen (1963).

Chamberlain (1963a), in a comprehensive treatment of exospheric and escape theory, worked out a mathematical formalism with the actual distribution of velocities at any altitude obtained by using Liouvilles theorem to connect the distributions at that point with those at the exobase. The term

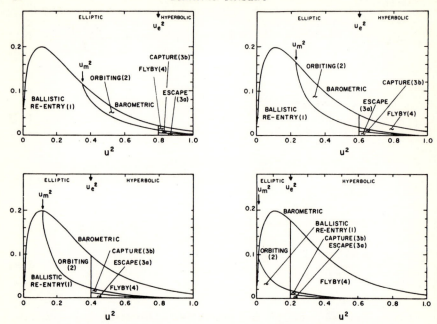

FIGURE 5   Figure illustrating the division of the Boltzmann distribution into groups on the basis of energy being greater or less than escape energy at chosen altitude, and perigee of the trajectory being inside or outside the exobase. See text for explanation of symbols. Adapted from Opik and Singer (1961), *J. Phys. Fluids,* © American Physical Society.

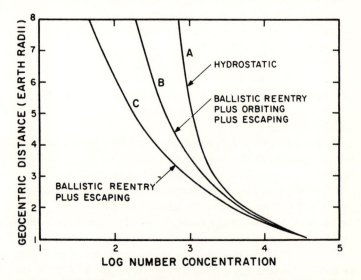

FIGURE 6   The vertical profiles at 1500°K for the hydrostatic distribution (A); the ballistic re-entry plus escaping plus orbiting components (B); and the ballistic re-entry plus escaping components (C). Adapted from Johnson (1961), *Astrophysical Journal,* © University of Chicago Press.

"critical level" is used as an alternative to "exobase", and the mathematical approach is an alternative way of obtaining the same height profiles and velocity distributions as those based on the work of Johnson and Fish (discussed above) and Opik and Singer.

As before, the atmosphere is considered spherically symmetric and non-rotating, and for the calculation of the density and velocity distributions as a function of altitude a Maxwell-Boltzmann velocity distribution is assumed at the exobase. Limits equivalent to v_e and ω_c which divide the particle groups are established. The results of computer calculations of the appropriate integrals are tabulated, and from these the fractions, or "partition functions" corresponding to X_1, X_2 and $\frac{1}{2}X_3$ can be readily evaluated for different heights and exospheric temperatures.

There is an interesting point to be noted in Chamberlains treatment of the oribiting particles, and the determination of X_2. These particles are not on orbits which intersect the exobase, and therefore one cannot apply Liouvilles theorem moving along a dynamic trajectory from the exobase to obtain the density distribution of satellite particles. Some assumption must be made about the population of orbiting particles, in both velocity distribution and density variation with height.

With regard to the velocity distribution Chamberlain has assumed that it is always part of the same Maxwellian function (Eq. 27) characterized by the same temperature T as for the ballistic and escape components. At increasing heights the orbiting component is absent to an increasing degree; Chamberlain has treated this by specifying a satellite critical level R_{sc}, such that all possible orbits are filled in the Maxwellian function below R_{sc}, and above it only those particles are present which have perigees below R_{sc}. The exospheric density distributions for heights above R_{sc} then correspond to escape and ballistic components for a fictitious exobase at R_{sc}.

As discussed in Section 2.3.3 this may not be an adequate model for the orbiting component. So far however, it is the best available.

Tables are given by Chamberlain (1963a) for which the integrated density in a column above a given height, for a given exospheric temperature can be obtained, depending on the assumed value of R_{sc}. Also expressions are given from which doppler profiles of particles in a column along a specified direction can be evaluated numerically if needed.

The treatments of the exosphere so far have been assuming spherical symmetry, which is not the case in the real atmosphere, since there is a diurnal temperature variation and necessairly a resulting density variation at the exobase (Section 3). Immediately above the exobase one would expect that the density and altitude profile would be determined by the density and temperature at the exobase, but at altitudes an appreciable fraction of an earth radius higher, this would be affected by particles which had come from exobase regions of different density and temperature.

Since higher temperatures are associated with lower densities at the exobase, and higher temperatures are associated with larger scale heights in the exosphere, it is clear that at increasing heights the trend is to smooth out the diurnal variation, so that at some height the density might remain constant during 24 hours. At even greater heights the diurnal variation could be reversed, with the scale height effect resulting in higher densities above the higher temperature exobase.

Numerical calculations by Vidal-Madjar and Bertaux (1972) have shown the above general description to be correct. They have made accurate calculations for a model with ballistic and escape components (no orbiting particles) with the following assumptions about temperature and density variation over the exobase. The temperature distribution is axially symmetric, and varies sinusoidally from the maximum to the minimum. The ratio of maxima to minimum temperature was taken as 1.5 for the three levels of solar activity considered. (The Jacchia (1971) ratio is 1.28.) The density ratios at the exobase were those given by the zero net ballistic flux condition (see Section 3.2 and Quessette 1972).

Vidal-Madjar and Bertaux found that below 1.5 earth radii vertical profiles defined by the local temperature and density at the exobase (e.g. as given by Chamberlain 1963a) fitted their accurate model to within 5%. Between 1.5 and 2 earth radii, the density was uniform around the earth, and a vertical profile defined by the temperature and density at the exobase 90° from the maximum or minimum fitted the accurate model to within 10%. From 2 to 10 earth radii the accurate model exhibited a higher density above the higher temperature exobase. Vertical profiles defined by a constant temperature (that for 90° from the maximum) and an exobase density n varying as

$$n = 0.95 \left(\bar{n}_0 + \frac{\bar{n}_0 - n_0(\alpha)}{3.5} \right),$$

where $\bar{n}_0$ is the mean exobase density and $n_0(\alpha)$ is the density at angle α from the maximum, were found to fit the accurate model to within 5%.

The effect of the rotation of the exobase (whether or not at the same angular velocity as the solid planet) on the altitude profile has been considered by Bryan (1901), Hagenbuch and Hartle (1969) and Hartle (1971). There is a small effect, which for the terrestrial atmosphere assuming uniform exobase density and temperature, results in about 2% greater density at the equator than at the poles (Hagenbuch and Hartle 1969) at an altitude of two exobase radii. With non-uniform exobase density and temperature (Hartle 1971) the enhancements due to rotation are greatest above regions of exobase or density minima, amounting to 15%–17% at altitudes 10-20 earth radii.

2.3.3 Interactions in the exosphere Interactions in the exosphere would have small effect on the altitude profile of the ballistic component, since the time constants for interaction are generally long compared to ballistic flight times. But there is a large effect on the orbiting component, since the abundance of these atoms depends on the relative rates of population and loss for the satellite orbits.

At all altitudes above the exobase, population can occur by rare collisions between neutrals, and by charge exchange collisions between protons and neutral hydrogen or oxygen atoms. Loss can be by photoionization, ionization by fast electrons, collisions between neutrals, and between neutral hydrogen and oxygen ions or protons. The population processes will not necessarily produce an orbiting component that will be characterized by the same temperature as the ballistic component, as the exospheric temperature varies on a time scale short compared to the population and loss process. Also the proton ion temperature can be considerably above the neutral temperature and the discrepancy increases with height. Charge exchange produces neutrals with a temperature characteristic of the protons, and at proton temperatures above about 5700°K (mean proton velocity ~11 km/sec) most of the neutrals produced will be on hyperbolic trajectories. The charge exchange reactions (Eqs. 77 and 90) are discussed in Section 4. The reaction involving oxygen neutrals and ions is important only in the lower part of the exosphere (upper F region) whereas the reaction involving neutral and ionized hydrogen is important at all levels, including the radiation belts and beyond the magnetosphere. Solar wind protons produce 400 km/sec neutrals, effectively removing them from the distribution, and radiation belt protons are even more energetic. The effect of solar radiation pressure on perturbing the trajectories of hydrogen atoms is important for particles at several tens of earth radii, as pointed out by Brandt (1961b). He suggested that a "geotail" existed, which was a "tenuous downward extension of the terrestrial hydrogen cloud". His abundances are not reliable, partly because he considered the solar wind also assisted solar radiation pressure by "dragging" the neutral hydrogen into a tail, instead of producing 400 km/sec neutrals by charge exchange which would no longer play a part in the scattering. He considered the geotail to be probably not important in the Lyman α scattering observed at low altitudes, but with a length $\sim 10^3 R_E$ and a width $\sim 10^2 R_E$ to be observable from interplanetary space.

There is evidence of a "geotail" produced by solar Lyman α radiation pressure on geocoronal hydrogen, but much narrower and closer to the earth. From measurements at $10 R_E$ Thomas and Bohlin (1972) concluded that such a concentration in the antisolar direction existed, however it was sufficiently close to the earth that the terrestrial shadow was clearly discernable in it.

As discussed qualitatively by Thomas and Bohlin, and quantitatively by Bertaux and Blamont (1973) the Lyman α radiation pressure acts on atoms

in both ballistic and orbiting trajectories, and by perturbing them can
effectively both populate and depopulate the orbiting component. The net
effect of the pressure on both components is to concentrate their densities
toward the antisolar direction. In addition it is possible that removal of
atoms by solar wind charge exchange, when trajectories pass outside the
magnetosphere, leaves mostly those with apogees in the antisolar direction,
i.e. with the atoms "protected" by the magnetosphere.

No adequate calculation of the various interactions in the exosphere, which
will determine the abundance and variations of the orbiting component, has
yet been made. Chamberlain's (1963a) treatment provides density distributions if the concept of a satellite critical level is appropriate, the temperature
characterizing the orbiting component is the same as for the ballistic, and the
value of R_{sc} is known. Chamberlain obtained $R_{sc} \sim 2.5$ Earth radii
from an "illustrative numerical example" of a method of calculating R_{sc}
from the equilibrium of population by ion-neutral charge exchange collisions
and loss by ion-neutral charge exchange collisions plus photoionization.
There is considerable uncertainty in this value, and in how well the concept
of a satellite critical level in general can represent the real distribution.

It should be noted that a distribution of satellite particles for which a
satellite critical level R_{sc} applies (i.e. no orbits with perigees above R_{sc})
is only one of a number of ways in which satellite orbits could be partially
populated, e.g. removal by charge exchange with solar wind might better
correspond to removal of orbits with apogees above a certain level with a
certain orientation of major axes. Wallace et al. (1970) suggested that the
depletion of long period satellite particles by photoionization, and by charge
exchange with solar wind protons, might explain the shape of the variation of
Lyman α intensity with geocentric distance observed by Mariner 5. Bertaux
and Blamont (1973) support the suggestion of significant loss rates by charge
exchange with the solar wind for particles whose orbits extend outside the
dayside magnetosphere. They calculate that this effect, and/or perturbations
due to solar Lyman α radiation pressure could account for the Mariner 5
and Ogo 5 observations (Section 6.5).

2.3.4 Classical escape rate   The number of particles crossing a reference
surface of unit horizontal area in unit time, moving upward with a velocity
greater than v_e is given by the integral of the right-hand side of (27)
multiplied by the upward component of velocity $v \cos \omega$

$$S_e = n \left(\frac{m}{2\pi kT} \right)^{3/2} \int_{\phi=0}^{2\pi} \int_{\omega=0}^{\pi/2} \int_{v_e}^{\infty} e^{-\frac{mv^2}{2kT}} \, v^3 \sin \omega \cos \omega \, d\omega \, d\phi \, dv \quad (45)$$

$$S_e = n \left(\frac{kT}{2\pi m} \right)^{1/2} e^{-\frac{mv_e^2}{2kT}} \left(1 + \frac{mv_e^2}{2kT} \right) \quad (46)$$

Substituting for v_e^2 from (38) and using (26)

$$S_e = n_0 \left(\frac{kT}{2\pi m}\right)^{1/2} e^{-\frac{GMm}{r_0 kT}} \left(1 + \frac{GMm}{r_0 kT}\right) \quad (47)$$

where the escape flux at a height r is now referred to the constant number density n_0 at the exobase of height r_0. This is equivalent to the expression derived by Jeans (1904), for which the global flux through a spherical surface at radius r is an increasing function of r. The reason, as he noted, is that a Maxwellian distribution is assumed at the radius r, which implies the presence of the flyby component, which is counted as escape flux when crossing the reference surface. One avoids the problem of the absence of the flyby component by evaluating (47) at the exobase with $r = r_0$, and little error is present in practical terrestrial situations if (47) is evaluated at 500 km.

To obtain the true escape flux at any altitude, i.e. to obtain an expression for any altitude which does not include the fictitious flyby component, we simply evaluate (47) at the exobase, where the flyby component is zero, and apply an inverse square decrease in escape flux with r, i.e.

$$S_e = n_0 \left(\frac{kT}{2\pi m}\right)^{1/2} e^{-\frac{GMm}{r_0 kT}} \left(1 + \frac{GMm}{r_0 kT}\right)\left(\frac{r_0}{r}\right)^2. \quad (48)$$

This is equivalent to the expression given by Chamberlain (1963a, Eq. 40).

2.3.5 Departure from classical escape rate Chamberlain (1963a) also considered the effect on the escape rate calculated from the exobase due to the departure from the Maxwellian velocity distribution in that region caused by the escape of the high energy component. Because of the finite time needed for collisions to replace the component X_{3a} there will be a reduction in the escape rate, and a small effect of the altitude profile. Chamberlain found a reduction of ~20% in the escape rate, where atomic hydrogen is a minor constituent in an oxygen atmosphere. Other calculations of this effect have been made by Biutner (1958, 1959a) and Hays and Liu (1965).

Monte Carlo calculations can give a more precise evaluation of this reduction of escape flux due to the escape of the high energy component. A number of such calculations have been made (Lew and Venkateswaran 1965, Liwschitz and Singer 1966, Chamberlain and Campbell 1967, Brinkman 1970, also see Liwschitz 1966 and 1967, Lew and Venkateswaran 1966, Venkateswaran 1971, Chamberlain and Smith 1971, and Brinkman 1971). The two most recent calculations are consistent with each other, and Brinkman (1971) gives a discussion of the earlier work. Reliable values for the reduction in escape rate below the Jeans rate would appear to be 25-30% for hydrogen and 1-3% for helium in oxygen of the terrestrial atmosphere. There is a small increase of the effect with temperature.

The average number of collisions needed to change the velocity of an atom of mass m_1 depends very much on the ratio of m_1 to the mass m_2 of the atoms or molecules it is colliding with. For $m_2 \gg m_1$ the velocity change of m_1 on impact is on the average m_1/m_2 of the velocity before impact. Thus approximately three times as many collisions are needed to replace the component X_{3a} in a CO_2 atmosphere ($m_2 = 44$) as in the terrestrial O atmosphere ($m_2 = 16$) at the exobase.

Consequently the reduction in escape rate below the Jeans rate is more important in the CO_2 atmospheres of Mars and Venus. Chamberlain (1969) and Chamberlain and Smith (1971) show that the reduction is approximately (depending on temperature) 50% for hydrogen ($m_1 = 1$) on Mars. About the same ratio should apply on Venus.

The departure from the classical escape rate is accompanied by a departure in the altitude profile in the exobase region from that which would apply if the complete Maxwellian distribution were present there. The density increases slightly with altitude compared to what it would be with the complete distribution present. This is because the diffusion is affected by upward diffusing particles in the exobase region retaining more of the Maxwellian high velocity tail than the downward moving ones. The increase begins about three mean free paths below the exobase level, and continues up to about the exobase level, above which the densities remain higher than they would otherwise be.

Chamberlain and Smith (1971) find that for H on earth the density enhancement would be about 20% if the exospheric temperature were as high as $3570°K$. Typically it would be much smaller, about 2% for a $1500°K$ exospheric temperature.

2.4   Thermospheric distributions and changes with exospheric temperature

The calculation of the altitude profile in the region from about 100 km to the exobase may be made using an assumed number density n_0 and upward flux S_1 at the exobase as the upper boundary, and obtaining a self consistent solution to (14) and (15) between the lower and upper boundary.

If the upward flow S_1 is taken as constant with height, it corresponds to the escape flux S_e when possible sources and sinks and time dependence for atomic hydrogen in the thermosphere are neglected. Then Eq. (14) may be used alone, and steady state vertical n_1 profiles can be computed for different temperature profiles and values of S_1.

Taking S_1 as the escape flux S_e and constant with height, Bates and Patterson (1961a, b) and Kockarts and Nicolet (1962, 1963) computed steady state vertical profiles, using Eq. (5) and ignoring eddy diffusion since the scale height determined by eddy diffusion between 100 and 120 km was close to that determined by diffusive flow, as will be shown later. Given S_1, n_0 at

500 km can be calculated from (47). It was necessary to assume a model for the major atmospheric constituents and the temperature variation with height.

To the accuracy that these steady state vertical profiles correspond to the real atmosphere, production below 100 km and the upward flux through the 100 km level and hence through all levels may be considered constant over the solar cycle, and this was assumed. Figure 7, from Kockarts and Nicolet (1962) demonstrates the large changes in shape of these vertical profiles with changes in exospheric temperature as occur during the solar cycle. The profiles yield a density at 100 km that depends on S_1 but very little on n_0 at 500 km, and so the density at 100 km also remains constant over the solar cycle. The 100 km density was taken as 10^7 cm^{-3} corresponding approximately to $S_1 = 2.5 \times 10^7$ cm^{-2} sec^{-1}. At 500 km a factor of 10 increase in density occurs when the temperature is reduced from 1500°K to 1000°K. The increase in abundance at reduced temperature maintains a constant escape flux under the conditions of a reduced escape rate per atom. This corresponds to varying n_0 in Eq. (47) to keep S_e constant as T is varied. This variation of density with temperature is in direct contrast with the variation of non-escaping elements, which have lower densities at a given height, for lower temperatures, because of the reduction of scale height with reduced temperature. Only at altitudes of the order of 50,000 km does the effect of scale height changes for hydrogen become important, and there the abundances tend to remain more constant through the solar cycle than those in the lower exosphere. The dashed lines in Figure 7 show the diffusive equilibrium profiles which would occur if flow and eddy diffusion were negligible. The departure from diffusive equilibrium occurs at lower altitudes with decreasing exospheric temperature as n_1 increases relative to S_1.

Some insight into the form of the vertical profiles can be obtained by looking at approximate analytic solutions to Eq. (5) or Eq. (14) where eddy diffusion is ignored, i.e. solutions to

$$\frac{\partial n_1}{\partial h} = -n_1 \left[(1 + \alpha_T) \frac{1}{T} \frac{\partial T}{\partial h} + \frac{1}{H_1} \right] - \frac{S_1}{D_{12}} \tag{49}$$

where H_1 is the scale height for atomic hydrogen. The diffusive equilibrium solution for the number density, n_{d_1} is obtained by setting $S_1 = 0$, i.e.

$$n_{d_1} = n_t \left(\frac{T_t}{T} \right)^{1+\alpha_T} \exp\left(-\int \frac{dh}{H_1} \right) \tag{50}$$

where T_t and n_t are the turbopause temperature and density in the absence of flow. The same treatment as is being followed here applies also to

X

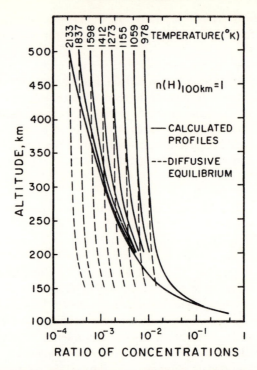

FIGURE 7　　Steady state vertical profiles for different exospheric temperatures. The dashed lines show the diffusive equilibrium profiles which would occur if flow and eddy diffusion were negligible. From Kockarts and Nicolet (1962), *Ann. Geophys.*, © Centre National de la Recherche Scientifique.

Deuterium, as discussed in Hunten (1971). The solution in the presence of flow can be considered a ratio to the diffusive equilibrium solution

$$r = n_1/n_{d_1} \tag{51}$$

and (49) transforms to

$$n_{d_1} \frac{\partial r}{\partial h} = -\frac{S_1}{D_{12}} \tag{52}$$

$$= -\frac{S_1 n_2}{b} \tag{53}$$

where the major constituents are represented by the index 2, and from (11)

$$b = \frac{3}{8\sigma_{12}^2} \left[\frac{kT(m_1 + m_2)}{2\pi m_1 m_2} \right]^{1/2} . \tag{54}$$

The density of the major constituents n_2 is the derivative of the column abundance N_2

$$n_2 = -\frac{dN_2}{dh} \tag{55}$$

and we can write

$$\frac{dr}{dh} = \frac{S_1}{n_{d_1} b} \cdot \frac{dN_2}{dh} . \tag{56}$$

Now both n_{d_1} and b vary slowly with height, but by assuming them constant, we obtain the analytic solutions which are useful for insight into the form of the profiles, i.e.

$$r = \frac{S_1 N_2}{n_{d_1} b} + \text{constant} \tag{57}$$

or $$n_1 = r n_{d_1} = \frac{S_1 N_2}{b} + r_\infty n_{d_1} . \tag{58}$$

The second term on the right of (58) can be normalized to n_0, the exobase density, provided the first term is negligible, which is true except for extremely high exospheric temperatures. The second term represents the diffusive equilibrium solution. The complete solution in the presence of flow is thus the sum of the diffusive equilibrium term and a term which varies rapidly with altitude representing the effects of flow, depending only on the upward flux and the neutral atmosphere parameters.

The two contributions can be seen in the curves of Figure 7. Note that the abscissa is a logarithmic scale. Given a constant upward flux and constant background atmosphere, the first term in (58) representing the effects of flow does not change with exospheric temperature, whereas the second term in (58), being normalized to n_0, as in Eq. (47) varies strongly with exospheric temperature.

The column abundance N_2 of the major constituents can be written approximately

$$N_2 = H_2 n_2 \tag{59}$$

(this is exactly true where g and T are constant, from Eq. (3)) and hence where the flow term in (58) dominates we have

$$n_1 = \frac{S_1 H_2 n_2}{b} \tag{60}$$

i.e. the atomic hydrogen density is proportional to the ambient density, and the scale height of the atomic hydrogen is the same as that of the major constituents, which is nearly the same as in the mixing region below the turbopause.

If the scale height of hydrogen in the 100-120 km region was determined only by flow, it would be close to that which would be obtained by complete mixing, and thus the effect of changing the height of the turbopause within that region is small (Kockarts and Nicolet 1963). With a change of turbopause height from 102.5 km to 110 km, there is a 20% reduction in density at 200 km and in the escape flux, provided the exospheric temperature is above about 900°K. At temperatures as low as 600°K the reduction is by 55%, since at that temperature the flow effects are giving way to diffusion at heights as low as 110 km.

Where the flow term dominates, the velocity of flow is, from (58)

$$V_f = \frac{S_1}{n_1} = \frac{b}{N_2} \tag{61}$$

which is a few cm sec^{-1} at turbopause heights. Eq. (61) represents a limiting or maximum diffusive flux for the given N_2, and the column of neutral atmosphere may be considered a "bottleneck" to the flow upwards to the region where there is rapid loss by thermal escape.

A velocity of flow at the exobase may also be defined for the thermal escape using (43)

$$V_{te} = \frac{S_e}{n_0} = \left(\frac{kT}{2\pi m}\right)^{1/2} e^{-\frac{GMm}{r_0 kT}} \left(1 + \frac{GMm}{r_0 kT}\right). \tag{62}$$

The value of V_{te} ranges from about 5 metres/sec at 1400°K to about 1 cm/sec at 500°K, for hydrogen on earth. If the exospheric temperature was below about 600°K on earth, or other parameters on a planet resulted in V_{te} for hydrogen (or some other constituent) being comparable with V_f, then the diffuse equilibrium solution would be dominant down to the turbopause.

Although exospheric temperatures as low as 600°K may occur at night at solar minimum, the diurnal effects, as we shall see in the next section, prevent the high densities and the diffusive equilibrium profile corresponding to 600°K from being attained.

Separate vertical profiles of both molecular and atomic hydrogen can be calculated using (14) and (15). The calculations of vertical profiles that were mentioned in Section 2.2, with a source of atomic hydrogen in the 100-150 km region due to the dissociation of molecular hydrogen, were made by Patterson (1966a) using this method. The reactions (22) and (24) simultaneously gave the loss rate of H_2 and the production rate of H, i.e.

$$\frac{dS_1}{dh} = -2 \frac{dS_2}{dh} = 2k_{22}n_3n_2 \tag{63}$$

where S_2 was the upward flux of H_2, and n_2 and n_3 were the number densities of H_2 and $O(^3P)$. The rate of dissociation is essentially determined by k_{22}, the reaction rate for Eq. (22), as reaction (24) is very much faster than (22) at the temperature of the lower thermosphere.

3 THEORETICAL DIURNAL VARIATIONS

3.1 Time dependent distributions

The steady state vertical profiles with $S_1 = S_e$ represent the gross changes in the mean diurnal altitude profiles over a solar cycle as the mean exospheric temperature changes, but not the diurnal changes with diurnal exospheric temperature changes. One important reason is the finite time required for the hydrogen distribution to adjust to a change in escape rate per atom caused by a change in exospheric temperature. A characteristic time constant

$$\tau(h) = N/S_e \tag{64}$$

was defined by Bates and Patterson (1961), which is the ratio of the column number density N above a given altitude to the escape flux. For an altitude of 300 km the time constants were 50 hours at $1000°K$ exospheric temperature, and 8 hours at $1500°K$, comparable to the period of diurnal variation. Thus the time dependence of the hydrogen density distributions is alone sufficient to smooth greatly the large diurnal variation which would otherwise result. Hanson and Patterson (1963) (HP for short) found that time dependence alone smoothed it to less than a factor of 2. One may evaluate the effect as follows. The escape rate per atom

$$B(T) = \frac{S_e}{n_0} \tag{65}$$

varies strongly with temperature as given in Eq. (47). The diurnal temperature variation is several hundred degrees K and may be obtained from a source such as Jacchia (1971), and expressed as

$$T = A + g(t) . \tag{66}$$

The upward flux S_1 of hydrogen from the source region is taken as constant up to the exobase, (say 500 km) and lateral flow and other sources and sinks are ignored.

Then if N is the abundance of hydrogen above the exobase, in a 1 sq cm base cone whose apex is the center of the earth, then approximately

$$N \propto n_0 H_0 \tag{67}$$

$$= e n_0 T \tag{68}$$

where the constant e includes the effect of the incomplete Boltzmann distribution above the exobase and

$$S_1 - S_e = \frac{\partial N}{\partial t} . \tag{69}$$

Differentiating (68) and using (69) and (65)

$$\frac{\partial N}{\partial t} = eT \frac{\partial n_0}{\partial t} + e n_0 \frac{\partial T}{\partial t} \tag{70}$$

i.e.

$$\frac{\partial n_0}{\partial t} = \frac{S_1}{eT} - \frac{n_0 B(T)}{eT} - \frac{n_0}{T} \frac{\partial T}{\partial t} . \tag{71}$$

An equation equivalent to (71) was integrated numerically by HP using a cosine function for $g(t)$ in (66), and e as 1.98 km $(°K)^{-1}$. The initial value of n_0 was varied until it was the same as the value at 24 hours. It is clear from (69) that with this cyclic solution the mean value of S_e is equal to S_1 .

Table II gives the maximum and minimum temperatures used by HP and the ratio $R_{time\ dep.}$ of the maximum to minimum values of n_0 . The fourth column gives the steady state ratios of maximum to minimum n_0 , calculated from (47).

TABLE II

Result of time dependent effect in smoothing the theoretical
steady state density distribution

Maximum temp.	Minimum temp.	$R_{time\ dep.}$	$R_{steady\ state}$
1150°K	750°K	1.57	21.9
1450°K	950°K	1.69	11.0
1750°K	1150°K	1.91	7.1

The time dependence of the density distribution has smoothed the diurnal variation to less than a factor of 2. The maximum and minimum concentrations were found to occur up to two hours after the minimum and maximum temperatures.

The diurnal temperature variation was taken as a factor of 1.53. A better factor for the equatorial region would be 1.28 (Jacchia, 1965) or 1.25 for mid-latitudes. At low temperatures S_1 or S_e integrated over 12 hours becomes negligible with respect to N, and the only n_0 variation is due to scale height changes in the exosphere. Equating (69) to zero yields N = constant and from Eq. (68)

$$n_0 \ \propto \ \frac{1}{T}. \tag{72}$$

Thus the value of $R_{time\ dep.}$ calculated by HP tends to 1.53 at low temperatures, and would be near 1.3 for the real atmosphere.

Providing lateral flow and loss processes other than thermal escape are negligible, the solution to Eq. (71) might be considered to be a good approximation to the diurnal variation of n_0 at the exobase. Wallace and Strobel (1972) have drawn attention to the region between about 120 km and the exobase, neglected by HP who considered that the region above the exobase accounted for most of the variation of hydrogen content.

Wallace and Strobel calculated the diurnal variation as if the column abundance between 120 km and the exobase alone smoothed the diurnal flow. They integrated Eq. (5) and Eq. (15) (with P_i and $L_i = 0$) between 120 km and the exobase, and used as boundary condition at the exobase the thermal escape flux relation between S_e and n_0.

Since there is less hydrogen available to smooth diurnal variations in the region below the exobase than in the region above it, they obtained greater diurnal variations than Hanson and Patterson, except for low temperatures where the HP asymptotic solution was greater. This is due to scale height changes in the exosphere and they did not consider the hydrogen content of the exosphere (Tinsley 1973).

A treatment which includes the effects of the region above the exobase as well as the region below would be one in which Eqs. (5) and (15) were integrated from 120 km to the exobase as before, but the upper boundary condition was not the flux S_e , but the flux S_u , where

$$S_u = S_e + \frac{\partial N}{\partial t} \tag{73}$$

and $\partial N/\partial t$ is given by (70). This assumes that there is negligible lag in the response of N to changes in n_0 and T , which is so since the ballistic flight time for one scale height is only about 15 minutes.

The treatment using (73) should yield a low temperature asymptotic solution slightly smaller than given by

$$R_{time\ dep.} = T_{max}/T_{min} \tag{74}$$

i.e. less than about 1.3.

At all temperatures the treatment considering the whole profile should yield a solution less than the solution considering only the region above the critical level. Referring to the HP solutions calculated by Wallace and Strobel (1972), $R_{time\ dep.}$ ranges from below 1.3 at solar minimum to below 1.6 at solar maximum, for the Jacchia (1965) temperature excursions.

3.2   Lateral flow

Lateral flow must also be considered when calculating accurate diurnal variations. Lateral flow is a net horizontal transport of atoms in ballistic trajectories in regions across the exobase where temperature and density gradients are present. The temperature effect drives a flux from hotter to colder regions along diurnal and latitudinal temperature gradients, while the density gradient thereby produced and enhanced by reduced escape rate in the colder regions produces a compensating return flux.

Lateral flow has been considered by Hanson and Patterson (1963), Donahue and McAfee (1964), Patterson (1966b), Joseph and Venkateswaran (1966), McAfee (1967) (see also Donahue (1966)), Patterson (1970), and Quessette (1972). Several approaches have been used. The first consists of numerically integrating the contributions to fluxes arriving at an element of area on the exobase from adjacent regions, taking into account the varying temperature, density, and range of the source. Flat earth, cylindrical geometry, and spherical geometry have been used. Joseph and Venkateswaran (1966) traced large number of individual paths of atoms in ballistic orbits.

Only Patterson (1966b) was able to incorporate time dependence in the density distribution taking into account both the escape flux and the lateral flow, and even so only for cylindrical geometry. A more powerful approach to lateral flow has been used by Patterson (1970), using numerical integration of expressions for the net lateral flux derived by Hodges and Johnson (1968). A spherical geometry was used, incorporating the equinoctial global temperature model of Jacchia (1965), and diurnal and latitudinal variations were calculated. For a temperature model with maximum and minimum temperatures of $1300°K$ and $1000°K$ the steady state ratio (as from Eq. (47)) was 4.5. The time dependent effect without lateral flow reduced this to 1.4. Then, the density distribution at the exobase which would produce zero lateral flux all over the exobase was evaluated, and the maximum to minimum density ratio found was 1.6, slightly higher than when lateral flux was not considered. Quessette (1972) has refined the calculations of McAfee (1967) using a spherical geometry but an axial symmetry of density and temperature about the minimum to maximum temperature axis, and a sinusoidal temperature dependence from temperature minimum to maximum. The density distribution which would produce zero net lateral flux was evaluated for a number of temperature models, and it was found that the maximum to minimum density ratio (R_{ZNLF}) varied linearly with the value of Y for the models considered, where

$$Y = 4 \frac{(T_{max} - T_{min})}{(T_{max} + T_{min})^2} \tag{75}$$

Since the value of R_{ZNLF} was not unity when $T_{max} - T_{min}$ was set equal to zero in the linear relation, as it must be for an isothermal atmosphere, a variation valid for smaller Y, which fits Quessette's results, is that given by Tinsley (1973)

$$R_{ZNLF} = 1 + aY^{1/2} + bY \tag{76}$$

where a and b are constants.

An even better relation is given by Hodges (1973), using analytic expressions for the lateral flow based on the expressions of Hodges and Johnson (1968).

Hodges' and Quessette's results agree closely with that of Patterson (1970). Hodges was able to treat non-zero lateral flow, exobase rotation, and diurnal tidal winds.

Taking the real atmosphere as corresponding to the Jacchia (1965) range of Y values, R_{ZNLF} ranges from 1.9 at solar minimum to about 1.3 at solar maximum.

3.3   Effects of ion-neutral interactions

A third effect on the diurnal variation is the diurnal variation in the flow of hydrogen ions into and out of the plasmasphere. The plasmasphere can be considered as an extension of the ionosphere along magnetic field lines which extend out to approximately 4 earth radii in the equatorial plane. Hydrogen atoms are converted to ions and vice versa in the upper F region by the charge exchange reaction

$$H + O^+ \rightleftarrows H^+ + O \tag{77}$$

In the daytime the ionospheric conditions are such that there is a flow of ions into the plasmasphere, and at night a flow out, thus increasing nightside neutral hydrogen concentrations at the expense of dayside concentrations.

The effect on the diurnal variation of exchange of hydrogen with the plasmasphere was first discussed by Hanson and Patterson (1963).

As discussed in more detail in Section 4, protons are formed in the charge exchange reaction (77), and then move into the plasmasphere under the influence of diffusion, electric fields and pressure gradients. There is a theoretical upper limit on the upward flux during the day, determined by the finite ability of diffusive transport in the lower exosphere to support a flow. There is no corresponding limit on downward flow, except the finite content of plasma in the field tube in the plasmasphere. To the extent that there is not a steady accumulation or depletion of plasma in the field tube the total quantity of plasma flowing in during the day will equal that flowing out at night. This is not true when electric fields convect plasma across field lines during magnetic storms.

Calculations give an upward limiting flux a few times the upward flux through the 100 km level, i.e. a few times the globally averaged escape flux. In addition, as will be discussed in Section 4.3, it is likely that diurnal variations in escape fluxes and heating of neutrals due to charge exchange with energetic hydrogen ions in the plasmasphere is important. Given that such fluxes occur for several hours each day and night, the time dependent effects, as discussed before, would be applied to the sum of the diurnally varying escape fluxes and the flux into the plasmasphere.

No calculations of the diurnal variation taking the ion-neutral exchange fluxes explicitly into account have yet been made. However it is unlikely that fluxes only a few times the escape flux would produce a diurnal variation more than a factor 2 as lateral fluxes would act to smooth out the density variation once it becomes of greater amplitude that the R_{ZNLF} density variation. Conversely, lateral fluxes would increase the density variation if it would otherwise have been smaller than R_{ZNLF}. This description is not strictly correct if the phases disagree. The zero lateral flow density maxima and minima correspond to the temperature minima and maxima, whereas lags of up to two hours may be present in the time dependent vatiations.

The density distribution in the polar regions at the solstices is a special case of the diurnal variation, where time dependent variations are very small, and the sustained sinks and sources due to thermal escape, the loss of ions to the polar wind, and the effects of large scale circulation would probably result in large lateral fluxes being generated.

4 ION-NEUTRAL INTERACTIONS

4.1 Theoretical H^+ altitude profiles

The charge exchange reaction (77) was pointed out by Dungey (1955) to be a source and sink of hydrogen ions in the thermosphere. Because of the almost exact energy balance the cross section ($3 \times 10^{-15}\,cm^2$ from the measurement of Stebbings and Rutherford, 1968) is high enough so that in the thermosphere the reaction is the most important source of H^+ ions, and charge transfer equilibrium is achieved over much of it.

For equal neutral temperature T_n and ion temperature T_i the equilibrium is expressed by

$$\frac{n(H)}{n(O)} = \frac{8}{9}\frac{n(H^+)}{n(O^+)} \tag{78}$$

while for a small difference in neutral and ion temperatures the expression due to Banks (1968)

$$\frac{n(H)}{n(O)} = \frac{8}{9}\frac{n(H^+)}{n(O^+)}\left[\frac{T_i + 16T_n}{T_n + 16T_i}\right]^{1/2}\frac{32T_i + 3T_n}{32T_n + 3T_i} \tag{79}$$

may be used.

In the absence of sources and sinks, the altitude profile of H^+ would be determined by the electrostatic equilibrium of the H^+ ions in the presence of O^+ ions. The situation has been discussed by Dungey (1955), Mange (1960, 1961) and Bates and Patterson (1961a).

The electrons, because of their small weight, tend to diffuse upwards until the electric field E set up by the very slight charge separation between the ions and electrons holds them down. In a mixture of ions the electric field determined by the major ion strongly affects the distribution of the minor ions. For example, where $n(H^+)$ is very small compared to $n(O^+)$ the electric field exerts an upward force equal to half the weight of the O^+ ions or eight times the weight of the hydrogen ion.

In accordance with Eq. (1), the altitude distribution of the ith ion species can be obtained by putting

$$\frac{d\psi}{dh} = g - \frac{eE}{m_i} \tag{80}$$

so that

$$\frac{1}{n_i} \frac{dn_i}{dh} = \frac{-m_i g + eE}{kT} \tag{81}$$

where e signifies here the electronic charge. Eq. (1), (80) and (81) apply at constant temperature, or at variable temperature if thermal diffusion is neglected. Thermal diffusion may be incorporated by deriving the relation from Eq. (4).

An analogous equation holds for the distribution of electron number density n_e at height h

$$\frac{1}{n_e} \frac{dn_e}{dh} = \frac{-m_e g - eE}{kT} \tag{82}$$

where m_e is the electron mass.

The condition of electrical neutrality can be expressed as

$$\Sigma n_i = n_e . \tag{83}$$

Eqs. (81) for each ion and (82) and (83) can be integrated numerically to yield the altitude distribution of the ions. Figure 8 from Mange (1960) is a calculated distribution for O^+ and H^+ as the only ions present, with the arbitrary concentration shown at 200 km. The four asymptotic scale heights for each constituent bear simple relations to the scale height $H_i = kT/m_i g$ for the neutral species corresponding to the ion species. This can be seen by transforming the equations to give

$$\frac{1}{n_i} \frac{dn_i}{dh} = -\frac{m_i g}{kT} \left[1 - \frac{m_+ - m_e}{2m_i} \right] \tag{84}$$

where m_+ is the mean ion mass. m_e can be neglected, and

$$\frac{1}{n_i} \frac{dn_i}{dh} = -\frac{1}{H_i} [1 - m_+/2m_i] \tag{85}$$

m_+ will be equal to the mass of the dominant constituent at the asymptotes, hence the scale height for the dominant constituent (which is O^+ at low F region altitudes, H^+ at high altitudes) is twice that of the corresponding neutral constituent. The scale height for O^+ at high altitudes where it is a minor constituent, will be 32/31 times that for neutral oxygen, and the scale height for H^+ at low altitudes will be negative ($n(H^+)$ increases with altitude) and one seventh that of neutral hydrogen. Bates and Patterson (1961) show that the ratio of H^+ to O^+ will be determined by the charge transfer equilibrium equation (78) in the lower part of the exosphere. Above that level the most likely distribution of O^+ ions is that controlled by diffusion (the source due to photoionization and the sink due to recombination will be small), and they demonstrate that the sources and sinks due to charge transfer will have little effect on the altitude profile of H^+ and O^+ calculated by the electrostatic equilibrium theory outlined above.

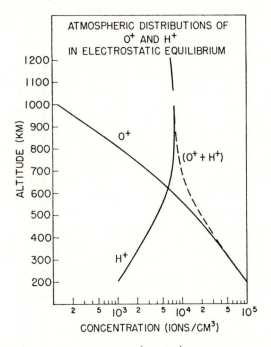

FIGURE 8   Calculated vertical profiles of O^+ and H^+ for only these ions present, starting from the arbitrary concentrations of 200 km shown. The temperature of $1500°K$ and the gravitational acceleration were taken as constant with altitude. From Mange (1960), *J. Geophys. Res.*, © American Geophysical Union.

The situation described above is an equilibrium situation where the asymptotic pressure for each constituent at high altitudes is that required to form an upper boundary to the solution. This is not true in general, and with total or partial asymptotic pressures which are less or greater than the equilibrium value, upward or downward flows will occur. Ions are accelerated upwards or downwards when the pressure gradients are not balanced by potential gradients. In the region where $n(H^+)$ is small compared to $n(O^+)$, and the H^+ partial pressure at high altitudes is lower than that required to form the upper boundary to the equilibrium solution, the electric field due to the O^+ ions can rapidly accelerate the H^+ ions so that they move upwards with velocities which may become supersonic, provided the altitude is such that drag from collisions with other constituents is not too great. If the O^+ partial pressure at high altitudes is lesser or greater than the equilibrium value then there will be upward or downward acceleration of O^+ also. The ions are constrained to move along magnetic field lines.

In the plasmasphere magnetic field lines are closed, so that upward or downward fluxes are only for the relatively short periods compared to a day necessary to restore equilibrium. The fluxes serve to balance both the pressure changes along the field line and the concentration changes in H^+ and He^+ in the ionospheres at either end as one or other ionosphere changes from day to night conditions or vice versa.

In the polar regions the magnetic field lines are open to the magnetospheric tail for part or all of each 24 hour period, and the tail pressures are apparently much below the necessary values for equilibrium, so that flows for periods which are long compared to 24 hours can occur, and supersonic velocities are easily attained. These flows are called the polar wind.

The energy required to drive these flows comes from the internal energy of the plasma, and as in the case of the evaporation of a liquid, or adiabatic expansion of a neutral gas, there is a cooling of the plasma as the flows are generated. This point is elaborated and supported by observations by Hanson et al. (1973).

The equations governing the flows as used for example by Banks and Holzer (1968) are as follows: the condition for electrical neutrality (Eq. 83); the condition for continuity for each ion species where they are constrained to flow through a given field tube of cross section A i.e.

$$\frac{1}{A} \frac{\partial}{\partial r} (n_i v_i A) = q_i - l_i \tag{86}$$

where v_i is the bulk flow velocity at a distance r along the field line, and q_i and l_i are the production and loss rates for the ith species; and the conditions for momentum balance for electrons and for the ith ion species. Note that l_i is defined differently to L_i in Eq. (15).

The momentum equation for electrons is

$$\frac{1}{n_e m_e} \frac{\partial P_e}{\partial r} + \frac{eE}{m_e} = 0 \tag{87}$$

and the momentum equation for the ith ions is

$$v_i \frac{\partial v_i}{\partial r} + \frac{1}{n_i m_i} \frac{\partial P_i}{\partial r} + g - \frac{eE}{m_i} = \frac{A_c}{m_i n_i} - \frac{q_i}{n_i} v_i \tag{88}$$

where P_e is the electron pressure, P_i is the partial pressure of the ith ion, g is the component of gravitational acceleration along the field line, and

$$A_c \simeq \sum_k \left[\frac{P_i}{D_{ik}} (v_i - v_k) \right]. \tag{89}$$

The term $A_c/m_i n_i$ in (88) represents the effect of frictional forces between ith species ions moving with velocity v_i colliding with kth species (ions or neutrals) moving with velocities v_k. D_{ik} is the coefficient of molecular diffusion between the ith and kth species. The last term in (88) is the momentum imparted ions that have been formed at rest. The first term accounts for inertia under accelerating flow, and the next three account for the pressure gradient, component of gravitational force, and electric field. Terms in (87) are equivalent to those in (88), with negligible terms omitted. Eqs. (87) and (88) are equivalent to Eq. (1) generalized for flow and acceleration.

Calculations of the magnitude of the upward and downward fluxes for various atmospheric temperatures and boundary conditions have been made by Hanson and Patterson (1963), Geisler (1967), Banks and Holzer (1968, 1969a, b, c) Dessler and Cloutier (1969) and Ho and Moorcroft (1971).

Banks et al. (1971), Fontheim and Banks (1972), Nagy and Banks (1972), Moffett and Murphy (1973) and Hanson et al. (1973) have discussed the flows at middle and low latitudes and the conditions under which the velocities might become supersonic for short periods.

An additional upward accelerating mechanism for the polar wind mentioned by Axford (1968) is the electric field produced by upward moving photoelectrons in the summer polar region. This mechanism is presumably effective at low latitudes also. These flows have a very large effect on the altitude profiles of H^+, He^+, and H in the polar regions, and at lower latitudes at certain times.

4.2   Limiting upward ion flux

Hanson and Ortenburger (1961) first pointed out that the flow of protons between the lower exosphere and the plasmasphere is impeded by the necessity for diffusion through the intervening oxygen ions.

For upward fluxes above the lower exosphere (say 1000 km) there is a finite limit to the flux because of the finite ability of diffusive transport in the region below that level to support a flow. There is no corresponding limit for downward fluxes. In their calculations based on a model atmosphere Hanson and Patterson (1963) found that at an exospheric temperature at 1000°K the upward limiting H^+ flux was about 2.5×10^7 cm^{-2}, about the same as the H flux from the source region below 100 km. This flux was less than the flux that was necessary to account for diurnal changes that had been observed in plasmasphere ion content via whistler observations. A similar result was obtained using a different mathematical approach by Geisler (1967). It seems however that the calculated fluxes were much too low because of the use of hydrogen abundances in the model atmospheres that were too low in comparison with the actual atmospheres at the time of the observations of protonosphere ion content.

For intercomparing hydrogen model atmospheres and measurements on the actual atmospheres it is convenient to use as a reference the models of Kockarts and Nicolet (1962, 1963) (KN models for short). These models were calculated for a wide range of exospheric temperatures (T_∞ for short) and altitudes. They are tabulated in even more detail above 500 km in the U.S. Standard Atmosphere Supplements, 1966, and by Jacchia (1971).

Since the KN models are for steady state conditions, and there is a diurnal variation of about a factor of two in the real atmosphere (Section 9.2) it is necessary to allow for diurnal variations when using the KN models as a reference for intercomparing various hydrogen density measurements. This is done most simply by evaluating the mean diurnal (or global) T_∞ value for the time of observation, and looking up the number density provided by the KN models at that temperature and the appropriate height. Then a diurnal variation by a factor of two with about two hours phase lag with respect to the temperature maxima and minima is applied to the model density, and a comparison made with the observations.

Hanson and Patterson (1963) and Geisler (1967) used KN abundances in their calculation of upward limiting H^+ fluxes, and the KN abundances are at least a factor of 3 lower than indicated by measurements (see Section 8) of mean diurnal abundance at mean diurnal T_∞. Since the daytime T_∞ values were actually used in the calculations to select a KN model abundance, the error was even greater, for as shown in Section 3.1 the steady state models overestimate the diurnal variation. This was probably the main reason for the inadequate limiting flux calculated.

The calculations of Banks and Holzer (1969c) were made to evaluate the upward flux of both hydrogen and helium ions escaping along open field lines in the polar wind. They obtained H^+ fluxes of between 3.8 and 6.6 x 10^8 cm^{-2} sec^{-1} at 1000 km for neutral temperatures of 1500°K to 750°K, for an ion temperature of 3000°K. The increasing flux with decreasing temperature partly reflects the increasing neutral hydrogen abundance in the model as the temperature is reduced. The model atmosphere used (Nicolet *et al.* 1969) has about 6 times the KN abundances. The calculations showed that lower temperatures led to larger atomic oxygen density gradients and lower O^+ densities which also increase the H^+ escape flux. Decreasing the ion and electron temperature however decreased the flux.

4.3 Other effects of ion-neutral interactions on neutral distribution

The effect of vertical ion fluxes on the diurnal variation of the hydrogen neutrals has already been considered. There are two other effects on the neutral distributions due to ion neutral interactions but so far little theoretical work has been done on them.

The first effect is the depletion of neutral hydrogen in the polar regions resulting from conversion to ions in charge exchange and loss via the polar wind. Fluxes of ions of the order of a few times 10^8 cm^{-2} sec^{-1} are generated by charge exchange in the locality from which they will escape, and even if the hydrogen abundances in the upper atmosphere are as large as the Nicolet *et al.* (1969) models, the loss flux through ions will be significant compared to the source flux from chemical production below 100 km. Consequently some depletion of the abundance of H in the thermosphere relative to low latitudes is to be expected. A crude model of the loss has been given by Meier (1970a). While lateral flow brings in hydrogen from low latitudes at a rate proportional to the difference from low latitude abundances (taking into account temperature differences) this flow occurs above the exobase level, near 500 km for T_∞ around 1000°K, and higher with higher T_∞. The charge exchange proceeds most rapidly in the 250-450 km region. Consequently the greatest depletion should occur below the exobase.

The effect on the global abundance (which is determined by the competition between source and loss processes) of the loss of hydrogen in the polar wind will probably not be large, since the area of the "open" field lines, beyond 75° geomagnetic latitude, is only about 3% of the area of the surface of the ionosphere, and the area of the region in which the field lines are sometimes open, say above 60° geomagnetic latitude, is 13%. Given that the polar wind fluxes at maximum are only a few times the chemical source fluxes and the thermal escape fluxes, the integrated polar wind flux will remain a small fraction of the integrated thermal escape flux.

During magnetic storms the plasmasphere boundary, which is normally not far inside the boundary of open and closed field lines, moves to geomagnetic latitudes which may be considerably less than $60°$. The depletion of the ion content of the field lines somewhat below $60°$ geomagnetic latitude was observed by Park (1970). The ions were subsequently replenished by upward flow. The process has been examined theoretically by Banks et al. (1971). During strong magnetic storms the area from which plasma can be convected to open field lines may become significant compared to the total surface area of the ionosphere, and significant temporary reductions of the global abundance could result.

The second ion-neutral interaction effect that has not been evaluated with any precision is the heating and enhanced thermal escape caused by energetic ions in the plasmasphere charge exchanging with cool neutrals in the reaction

$$H^+ + H \rightleftarrows H + H^+ . \tag{90}$$

Cole (1966) suggested that the loss of hydrogen by this process might be significant. If the hydrogen ion energy corresponds to the mean thermal velocity for a temperature of $5750°K$, equivalent to a mean velocity of 11 km/sec or energy of 0.63 eV, and a neutral is produced by charge exchange in the lower exosphere, it can escape from the earth. At higher altitudes less energy is required. Collisions with ions of less than this energy can still heat the hydrogen and perhaps alter the altitude profile and global distribution (Tinsley 1973).

Serbu and Maier (1970) measured ion temperatures and concentrations from $1.5R_E$ to $7R_E$. They found temperatures in the region within the plasmasphere ranging from as low as $5000°K$ at $1.5R_E$ on the nightside and $7000°K$ at $1.5R_E$ on the dayside to $30,000°K$ at $3.6R_E$ on the dayside. Ion densities were about 10^4 cm^{-3} at $1.5R_E$.

A very rough evaluation of the escape flux is based on the integral

$$F = \int MK \left(\frac{r}{r_0}\right)^2 n_h(r) n_h^+(r) \, dr \tag{91}$$

where K is the rate coefficient for Eq. (90) taken from the data of Dalgarno (1960); M is a geometric factor to take into account that some collisions will produce particles on trajectories which intersect the exobase. M varies from 0.5 at low altitudes to a limit of 1 at very high altitudes; $(r/r_0)^2$ refers the escape flux to the exobase level; $n_h(r)$ and $n_h^+(r)$ are the neutral and ion densities respectively. Using a $n_h(r)$ distribution given by three times KN abundances, and the ion densities and temperatures of Serbu and Maier (1970), the integral outwards from $1.5R_E$ on the dayside yields an escape flux of about 4×10^8 cm^{-2} sec^{-1}. This will be increased by fluxes from below $1.5R_E$, from the region of diminishing ion temperature but increasing neutral and ion density.

To evaluate the global effect of this loss and of the larger production of heated neutral atoms, it will be necessary to use a global distribution of ion density and temperature from heights of about 1000 km to 15,000 km. Such a distribution is not available at present. About all that can be said is that the heating and loss may have important effects on the neutral hydrogen abundance, altitude profile and global distribution (including solar cycle and diurnal variations).

5 RADIATIVE TRANSFER THEORY

Because a very large number of optical observations of hydrogen in the upper atmosphere have been made, some basic principles of the radiative transfer theory necessary to interpret them will be given here. Radiative transfer theory is also a necessary preliminary to evaluating the extent of transfer of Lyman α, Lyman β, and other resonance lines to the nightside and their effects as a source of nighttime D and E region ionization (Section 10). The analysis of optical observations is difficult, especially of the brightness distribution when the solar flux has been resonantly absorbed and then re-emitted as resonance re-radiation or fluorescence in Lyman α, Lyman β, or Balmer α. The optical thickness for Lyman α and Lyman β allow high orders of multiple scattering, and transfer of radiation to all points on the nightside.

The theoretical treatment of the radiative transfer process is rendered very complex by the spherical geometry involved, the relatively large optical depth for Lyman α (less so for Lyman β or for Lyman α at very high altitudes or at extreme sunspot maximum) the variable temperature with height below the thermopause, and the fact that the hydrogen distribution is not spherically symmetric nor even symmetric about the sun-earth line.

5.1 Absorption line profile

The simplest radiative transfer problem is to interpret the profile of the absorption core in the solar Lyman α line, due to terrestrial hydrogen above the point of observation.

The scattering cross section $\sigma(\nu)$ in the Doppler approximation to the profile (sufficient for most purposes) is given at a frequency ν by

$$\sigma(\nu) = \sigma_0 e^{-\left(\frac{\nu - \nu_0}{\Delta \nu_D}\right)^2} \tag{92}$$

where

$$\sigma_0 = \frac{\pi e^2}{m_e c} \frac{f_{12}}{\sqrt{\pi} \Delta\nu_D} \tag{93}$$

and

$$\Delta\nu_D = \frac{\nu_0}{c} \sqrt{\frac{2kT_n}{m_H}} \tag{94}$$

and ν_0 is the center frequency of the absorption line; f_{12} is the oscillator strength for Lyman α.

The optical depth τ at a height h is then

$$\tau_h(\nu) = \int_h^\infty \sigma(\nu, T_n)\rho(h) \sec\theta \, dh \tag{95}$$

provided $\sec\theta$ is not large. θ is the zenith distance of observation, and $\rho(h)$ is the density of atomic hydrogen in the ground state. The intensity remaining in the solar line profile at frequency ν is

$$I(\nu) = I_0(\nu) e^{-\tau_h(\nu)} . \tag{96}$$

The solar line shape $I_0(\nu)$ is inferred from the part of the profile outside the absorption core. It is usually easier to measure the equivalent width of the absorption core, which is insensitive to instrumental broadening, than the core profile itself. The equivalent width is given by

$$W(h) = \int_0^\infty (1 - e^{-\tau_h(\nu)}) \, d\nu . \tag{97}$$

Since both T_n in (94) and $\rho(h)$ in (95) are variable in height, it is necessary to assume altitude distributions of both the hydrogen density and the temperature, and then the profile of the absorption core or its equivalent width may be calculated for comparison with observations. Figure 9, from Meier and Prinz (1970) is an example of calculations of equivalent width as a function of altitude for various profiles in T_n and $\rho(h)$.

5.2　Brightness distribution calculation

The much more difficult calculation of the brightness distribution of scattered radiation, especially on the nightside of the earth, has been solved by Thomas (1963a), (see also Thomas 1963b, Donahue and Thomas 1963a, b) by a method of successive approximations.

X

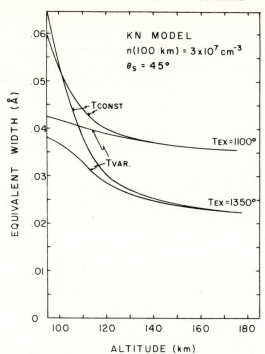

FIGURE 9   Calculated equivalent widths as a function of altitude for hydrogen altitude profiles determined by exospheric temperatures of 1100°K and 1350°K. The curves labelled T_{VAR} take into account the variation of the doppler width with the decreasing temperature below the exobase, whereas the curves labelled T_{CONST} are for constant doppler width. From Meier and Prinz (1970), *J. Geophys. Res.,* © American Geophysical Union.

The technique has been applied to many Lyman α observations, and is also discussed by Donahue (1962), Donahue (1966), Donahue and Meier (1967) where it is applied to sodium observations), and Meier and Mange (1970). The application to Lyman β and Balmer α scattering is discussed by Donahue (1964a) and Meier (1969). The equations are for flux scattered in the whole line profile. The integral equation relating the density distribution of excited atoms in the scattering medium is

$$N(r) = N_0(r) + B\sigma_0\rho(r)\ \frac{d\Omega}{4\pi} \int dl N(r')G(r,r')e^{-\tau O_2}(r,r') \qquad (98)$$

where $N(r)$ and $N(r')$ are the excited atom densities at r and r', $N_0(r)$ is the density of atoms excited directly by solar radiation, $G(r,r')\,dr\,dr'$ is the probability that a photon will be emitted from dr' at r' and absorbed in dr

at r, neglecting O_2 absorption. Depending on whether Lyman α or Lyman β is being considered, the value of B the branching probability is either 1.0 or 0.8844, and τ_{O_2} the optical depth for O_2 absorption has values appropriate to that emission and the concentration along the paths considered. The term $N(r')G(r,r')e^{-\tau O_2(r,r')}$ is proportional to the rate at which photons emitted from dr' at r will be absorbed in dr at r. The integration with respect to l where $l = |r - r'|$ and with respect to Ω takes account of the illumination of the ground state atom at r by atoms at all distances and directions included in the radiative transfer treatment. The complete second term on the right of (98) is the density of excited atoms produced by multiple scattering, which is added to the density excited directly by solar radiation, where the region considered is not in the shadow for direct sunlight.

Eq. (98) is usually solved in terms of a source function $S(\tau)$, i.e. normalized excitation rate per unit optical depth at the center of the doppler profile, such that

$$S(\tau) = \frac{\gamma}{(\pi F_0 \sqrt{\pi}\Delta\nu_D)} \frac{N(r)}{\sigma_0 \rho(r)} \tag{99}$$

where γ is the reciprocal of the natural lifetime, πF_0 is the Lyman α flux at the center of the solar line in photons cm^{-2} sec^{-1} Hz^{-1}, thus

$$S(\tau) = S_0(\tau) + B \int \frac{d\Omega}{4\pi} \int d\tau' S(\tau') G(\tau,\tau') e^{-\tau O_2(\tau,\tau')} \tag{100}$$

$S(\tau)$ is in units of flux per unit solar flux per unit absorption coefficient at the center of the doppler profile, i.e. it has dimensions $(Hz)^{-1}$.

Eq. (100) is solved for the source function distribution by iteration. An approximate source function distribution is used for $S(\tau')$ in (100) which gives $S(\tau)$ to a better approximation, which is then used for $S(\tau')$ for a further improvement, and so on. The initial approximation for $S(\tau')$ can be $S_0(\tau)$, given by

$$S_0(\tau) = T(\tau, \text{sun}) \tag{101}$$

where (τ, sun) is the optical depth at the line center between the point and the sun.

The quantity which is measured in the observations is the apparent emission rate which is usually expressed in terms of the unit the Rayleigh. (Strictly speaking the measured quantity is the intensity I, or the flux per unit area per unit solid angle at the detector.) The rayleigh is the apparent emission rate, or the rate, assuming no losses between the source and detector, of emission into 4π steradians of the whole column of unit cross-sectional area from the detector to infinity along the line of sight. The Rayleigh is defined as an emission rate of 10^6 photons per cm^2 column per second, and is related to I by

$$\text{Emission Rate in Rayleighs} = R = 10^{-6} 4\pi I \qquad (102)$$

where I is the observed intensity in photons cm^{-2} sec^{-1} $ster^{-1}$. (The terms "emission rate", "intensity", and the symbol "I" are often used loosely by workers in the field, as can be seen by examining some of the figures reproduced later in this paper.)

The integral of the source function along the direction of observation could be called the true emission rate (in normalized units) and for an emission for which the medium was optically thin the apparent emission rate would be equal to the true emission rate.

In the case of Lyman α or Lyman β the medium is not optically thin, and hence what is observed in the contribution of the source function from each point reduced by the losses along the path to the observer. The losses are by resonant scattering out of the beam and by absorption on O_2. Hence the apparent emission rate in terms of the normalized units is

$$E_N(\tau) = B \int d\tau' S(\tau') T(\tau, \tau') e^{-\tau O_2(\tau, \tau')} \qquad (103)$$

where $T(\tau, \tau') d\tau' d\tau$ is the probability that a photon will be emitted from the optical depth element $d\tau'$ at τ' and reach $d\tau$ at τ without being absorbed. The functions G (Eq. 98 and 100) and T (Eq. 103) are defined by Holstein (1947).

The emission rate in rayleighs is therefore

$$E(\tau) = 10^{-6}(\pi F_0 \sqrt{\pi \Delta \nu_D}) E_N(\tau) . \qquad (104)$$

The apparent emission rate for Balmer α is unaffected by absorption and is the integral along the direction of observation

$$E_N(\tau) = (1 - B) \int d\tau' S(\tau') \qquad (105)$$

where the B value for Lyman β is used.

The results of calculations of brightness distributions by the above method (the Thomas-Donahue-Meier method) are shown in the next few figures. One source function distribution corresponds to one hydrogen model atmosphere, but the apparent emission rate can be calculated and displayed in many different ways depending on whether one wishes to compare with observations of altitude profile (e.g. from rockets), solar zenith distance profiles (e.g. from observations made along a satellite orbit) or observation zenith distance profiles, etc. Solving for several model atmospheres can give the variation with various vertical optical thicknesses, necessary if one wishes to use the observations to evaluate the hydrogen abundance. The vertical optical thickness τ_V is defined by (95), evaluated for h as the height of the lower boundary and with θ taken as zero.

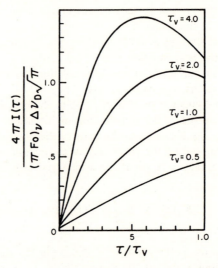

FIGURE 10   Calculations of normalized Lyman α apparent emission rate for vertical optical thickness τ_V of 0.5, 1.0, 2.0 and 4.0. The solar zenith distance was 60°. The scale τ/τ_V is the fraction of the optical depth measured down from the top of the medium. Adapted from Donahue (1966), *Ann. Geophys.*, © Centre National de la Recherche Scientifique.

An example of the variation of normalized zenith emission rate for optical thicknesses of 0.5, 2.0, 2.0 and 4.0 is shown in Figure 10. The solar zenith distance is 60°. In this and following calculations the molecular oxygen is assumed to produce negligible absorption above the lower boundary, and total absorption below it. The abscissa is the fraction of the optical depth, measured down from the top of the medium, and the normalized zenith emission rate $E_N(\tau)$ is given for the four different models as a function of depth in the medium. For optical thicknesses above about unity the zenith intensity maximizes above the lower boundary. Figure 11 is an extension of the same calculations to $\tau_V = 10$, transformed to an abscissa scale of actual altitude for hydrogen altitude distributions proportional to a KN model at a temperature of 850°K. There is little variation with altitude above 120 km in

Figure 11 because about half of the hydrogen is below 120 km, and the amount above a given altitude changes very little up to 1000 km.

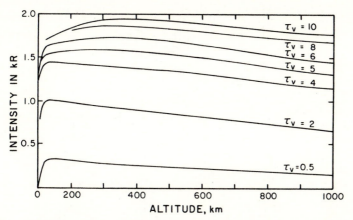

FIGURE 11   Extension of the calculations of Figure 10 to $\tau_V = 10$ transformed to an abscissa scale of actual altitude for hydrogen vertical profiles proportional to a KN model at a temperature of 850°K. The ordinate scale is apparent emission rate for an assumed solar flux. Adapted from Donahue (1966), *Ann. Geophys.*, © Centre National de la Recherche Scientifique.

Figure 12 is part of a set of results for Balmer α calculated by Meier (Tinsley and Meier 1971). Calculations were made for four model hydrogen atmospheres with differing temperatures, for a range of solar zenith distances 105° to 180°, i.e. solar depression angles 15° to 90°. The model abundances at 100 km were taken as 3×10^7 cm^{-3}, and the altitude profiles corresponded to three times the KN abundances up to 500 km joined to a Chamberlain (1963a) profile with satellite critical altitude $2.5R_c$, where R_c is the geocentric radius of the exobase. Since these models, and models with number densities at a given altitude a constant factor times these number densities have been used in many radiative transfer calculations, they are listed here in Table III.

The abscissae for the eight plots in Figure 12 are the azimuth of observation relative to the solar azimuth. The ordinate is emission rate in rayleighs for a solar flux which was arbitrarily adjusted so that the brightness at 30° solar depression in the azimuth of the sun remained constant. These results are for nighttime conditions, where only very high altitude hydrogen is directly illuminated, and Lyman β flux is transported by multiple scattering to the nightside. The intensity at 40° zenith distance shows less variation with azimuth relative to the sun than the 80° zenith distance results, where at 15° depression and 0° relative azimuth one is looking 25° away from the sun into much directly illuminated hydrogen, whereas at 180° relative azimuth one is looking 5° from the antisolar direction.

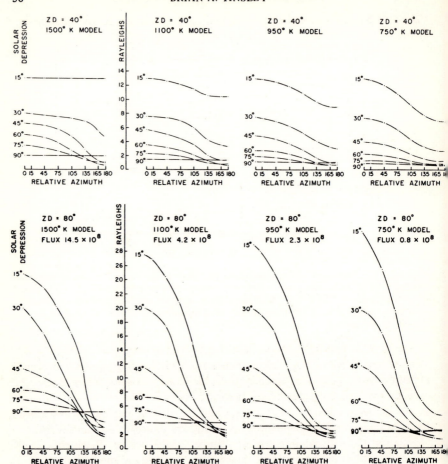

FIGURE 12   Calculations of Balmer α emission rate at the ground, for four model
vertical profiles of hydrogen determined by exospheric temperatures of 1500°K, 1100°K
950°K, and 750°K. Emission rates were evaluated for solar depression angles from 15°
to 90° and relative solar azimuths from 0° to 180° for each model, at zenith distance of
observation of 80° and 40°. The emission rate in Rayleighs is for a solar flux which was
adjusted for each model so that the emission rate at 30° solar depression in the azimuth
of the sun remained constant. From Tinsley and Meier (1971), *J. Geophys. Res.*, © American
Geophysical Union.

There is little change in the shape of the distributions for the 750°K to
1500°K exospheric temperature change, although there is a change in the
Lyman β vertical optical thickness above 650 km from 1.76 at 750°K to 0.057
at 1500°K. Concerning the changes that are present, the brightness at midnight (solar depression = 90°) in most directions is greater relative to the
twilight part for the lowest optical thickness (T_∞ = 1500°K) than it is for the
highest optical thickness (T_∞ = 750°K). The reason for this is that the

TABLE III

Hydrogen models with 3 times Kockarts–Nicolet (1962, 1963) abundances to 500 km, joined to Chamberlain models with $R_{sc} = 2.5 R_c$

Height km	$T = 750°K$			$T = 950°K$			$T = 1100°K$			$T = 1500°K$		
	$n(H)$	$\tau_{Ly\alpha}$	$\tau_{Ly\beta}$	$n(H)$	$\tau_{Ly\alpha}$	$\tau_{Ly\beta}$	$n(H)$	$\tau_{Ly\alpha}$	$\tau_{Ly\beta}$	$n(H)$	$\tau_{Ly\alpha}$	$\tau_{Ly\beta}$
100	3.01(7)	26.4	4.03	3.01(7)	8.84	1.35	3.01(7)	5.92	9.04(−1)	3.01(7)	4.29	6.55(−1)
103	2.20(7)	25.1	3.83	2.04(7)	7.69	1.17	2.02(7)	4.79	7.31(−1)	2.04(7)	3.33	5.08(−1)
105	1.62(7)	24.0	3.67	1.40(7)	6.85	1.04	1.38(7)	4.02	6.14(−1)	1.39(7)	2.67	4.08(−1)
110	8.98(6)	22.7	3.47	6.94(6)	5.89	8.99(−1)	6.67(6)	3.13	4.78(−1)	6.74(6)	1.91	2.92(−1)
120	3.18(6)	21.5	3.29	2.16(6)	5.10	7.79(−1)	1.95(6)	2.45	3.74(−1)	1.91(6)	1.33	2.03(−1)
135	1.73(6)	20.8	3.17	8.02(5)	4.73	7.22(−1)	6.39(5)	2.15	3.28(−1)	5.62(5)	1.09	1.66(−1)
160	1.17(6)	20.0	3.06	3.62(5)	4.48	6.83(−1)	2.37(5)	1.97	3.01(−1)	1.88(5)	9.65(−1)	1.47(−1)
215	9.28(5)	18.5	2.83	1.84(5)	3.91	5.97(−1)	9.75(4)	1.81	2.76(−1)	5.37(4)	7.04(−1)	1.08(−1)
400	7.04(5)	15.0	2.29	1.22(5)	3.15	4.82(−1)	5.81(4)	1.55	2.37(−1)	1.59(4)	4.55(−1)	6.94(−2)
650	4.98(5)	11.5	1.76	9.19(4)	2.62	3.99(−1)	4.08(4)	1.34	2.04(−1)	8.90(3)	3.74(−1)	5.71(−2)
1000	3.22(5)	8.30	1.27	6.50(4)	2.07	3.16(−1)	3.03(4)	1.11	1.70(−1)	7.11(3)	3.30(−1)	5.04(−2)
2500	7.34(4)	2.92	4.46(−1)	2.02(4)	9.81(−1)	1.50(−1)	1.09(4)	6.02(−1)	9.20(−2)	3.31(3)	2.16(−1)	3.30(−2)
5000	1.48(4)	1.02	1.55(−1)	5.63(3)	4.50(−1)	6.88(−2)	3.58(3)	3.11(−1)	4.74(−2)	1.40(3)	1.33(−1)	2.02(−2)
10,000	2.53(3)	3.28(−1)	5.01(−2)	1.34(3)	1.82(−1)	2.78(−2)	9.99(2)	1.37(−1)	2.08(−2)	4.99(2)	6.80(−2)	1.04(−2)
15,000	8.39(2)	1.69(−1)	2.57(−2)	5.06(2)	1.02(−1)	1.56(−2)	4.01(2)	7.76(−2)	1.18(−2)	2.19(2)	4.24(−2)	6.47(−3)
25,000	2.29(2)	7.18(−2)	1.10(−2)	1.56(2)	4.77(−2)	7.28(−3)	1.32(2)	3.43(−2)	5.23(−3)	7.94(1)	2.19(−2)	3.35(−3)
40,000	6.57(1)	3.01(−2)	4.60(−3)	4.95(1)	2.13(−2)	3.26(−3)	4.37(1)	1.35(−2)	2.07(−3)	2.84(1)	1.07(−2)	1.63(−3)
50,000	3.60(1)	1.90(−2)	2.90(−3)	2.82(1)	1.38(−2)	2.10(−3)	2.53(1)	7.44(−3)	1.14(−3)	1.70(1)	7.09(−3)	1.08(−3)
60,000	2.21(1)	1.25(−2)	1.91(−3)	1.78(1)	9.24(−3)	1.41(−3)	1.62(1)	3.73(−3)	5.70(−4)	1.12(1)	4.86(−3)	7.42(−4)
70,000	1.53(1)	8.68(−3)	1.32(−3)	1.25(1)	6.45(−3)	9.85(−4)	1.15(1)	1.24(−3)	1.90(−4)	8.01(0)	3.44(−3)	5.25(−4)

geometric effect of greater scale height for the higher temperature allows more directly scattered radiation from very high altitude hydrogen to be seen. In the exact antisolar direction however, the relative brightness is smaller for the lower hydrogen density, because of the reduced transport by multiple scattering.

Other calculated brightness distributions are given later in comparison with observations. An alternative approach to the solution of the radiative transfer problem has been given by Kaplan *et al.* (1965) and Kaplan and Kurt (1965). Their results appear to be more approximate but consistent with those obtained by the above method, for the dayglow and twilight range treated.

There are deficiencies in the foregoing treatments whose effect on the accuracy of the calculations is not at present clear, but probably not large. These are the neglect of the temperature dependence of the line width with altitude, and the assumption of an isotropic scattering cross section. The non-coherency of the scattering process has not been accurately treated, but Wallace (1971) found that the accurate treatment made little significant difference to the results of the calculation. The Monte-Carlo approach to calculation of the brightness distribution appears to be possible, and may be better able to handle the region from 90–150 km for Lyman α scattering where the use of the accurate resonance scattering formula, with the correct temperature dependence is desirable. Another advantage of such calculations is that they would provide an independent check of the Thomas-Donahue-Meier method for large solar zenith distances, where there is a high degree of multiple scattering for large optical thicknesses.

6 OPTICAL OBSERVATIONS

Optical observations have both the advantage and disadvantage that they are affected by hydrogen at large distances from the detector, in contrast to say mass spectrometer observations, which sample the nearby hydrogen only. Thus it is necessary to set up model hydrogen atmospheres, and calculate the integrated effect the atmosphere has had on the optical emission seen, to obtain information on the abundance, altitude variation, or diurnal variation. If the procedure is successful, then the advantage is obtained of obtaining information on regions not readily accessible, and with Balmer α data, from the ground. However there are so many variables in the optical work, e.g. the value of the solar flux at the line center which has been a particular problem, that observations by non-optical methods have proven desirable as an independent check on the results, and as the best method of measuring certain types of variations, e.g. the diurnal variation.

6.1 Rocket measurements of absorption core in solar line

The first high resolution spectrograms of solar Lyman α showing the terrestrial absorption core were obtained from a rocket measurement on July 20, 1959 by Purcell and Tousey (1960, 1961). The line profile for heights between 134 km and 163 km is shown in Figure 3. The absorption cores recorded near 100 km and 200 km were interpreted using a curve of growth analysis and a two temperature model hydrogen atmosphere by Bates and Patterson (1961a, b). From the ratio of the inferred column abundance above 100 km (5.7×10^{12} atoms cm^{-2}) to that above 200 km (1.8×10^{12} cm^{-2}) they also inferred an exospheric temperature of 1100°K. Further high resolution observations of Lyman α were made on April 19, 1960 (Tousey 1963) and April 14, 1966 (Bruner and Parker 1969) and October 19, 1967 (Bruner and Rense 1969, Jones *et al.* 1970). A discussion of the earlier approximate analyses, and detailed analyses of the four sets of data was made by Meier and Prinz (1970). They assumed exospheric temperatures given by the method of Jacchia (1966) and the measured equivalent widths were compared to calculated equivalent widths, for model altitude distributions of hydrogen based on the KN steady state profiles. The results were consistent with models that had abundances between 1.25 and 2.0 times the KN abundances evaluated for T_∞ values for the local time of flight. While these T_∞ values yield the correct doppler line profile, they give an incorrect altitude profile $\rho(h)$, since the steady state models overestimate the diurnal variation, which affects the upper part of the profile. Allowing for diurnal effects the results would be consistent with models which had mean diurnal abundances somewhat less than 1.25-2.0 times KN. The variation in equivalent width with height for the three flights where there was sufficient data, did not agree well with that calculated from the models, and allowances for diurnal effects would probably not help the agreement.

6.2 Emission line widths from absorption cell measurements

On April 17, 1961 an evening (solar depression 25°) rocket flight was made to 177 km with a hydrogen absorption cell in front of a Lyman α detector. The cell was filled with molecular hydrogen which dissociated when heated by a tungsten ribbon filament, absorbing out incident Lyman α to 0.04Å either side of the line center, i.e. the atomic hydrogen in the cell had an equivalent width (Eq. 97) of about 0.08Å. The absorption removed the entire signal from below the rocket, and 85% of that from above the rocket (Morton and Purcell 1962). From the same rocket, and also from another on October 31, 1961, photographic spectra showed Lyman α but no other line strong enough to provide the 15% residual (Morton 1962). A further measurement with an absorption cell was made on April 26, 1965 (Winter and Chubb 1967), with similar results.

An interpretation of the residual 15% signal from the upward hemisphere in the Morton and Purcell experiment as due to an extraterrestrial Lyman α component, in particular to scattering of solar Lyman α on interplanetary hydrogen, was made by Patterson et al. (1963). The extraterrestrial source for the emission has been confirmed by Lyman α measurements in interplanetary space, but the relative amounts of emission from several interplanetary components and from galactic sources is not well established.

Satellite measurements of Lyman α using a hydrogen absorption cell have been made by Metzger and Clark (1970) on Ogo 6. Absorption widths used were typically 0.020 to 0.025A from the line center, with different widths for the three different operating currents corresponding to different degrees of dissociation of the molecular hydrogen. These widths were sufficiently narrow that the doppler width and hence the temperature of the geocoronal hydrogen could be inferred from the ratio of intensities of the cell off and on in different current modes. The satellite was nearly in a dawn-dusk orbit, and the geocoronal intensities were high enough to make the extraterrestrial component small even with the cell on. A mean exospheric temperature of $900°K \pm 200°K$ was found, with the dusk temperatures greater than the dawn temperatures by about $200°K$. Further, the summer polar region was found to be about $75°K$ warmer than the winter polar region.

The small effect attributable to extraterrestrial Lyman α was strongest in the ecliptic and weak near the ecliptic poles, and may be partly due to interplanetary hydrogen streaming nearly parallel to the ecliptic plane. Observations and theory of extraterrestrial Lyman α have been reviewed by Tinsley (1971), and Axford (1972). The extraterrestrial emission must be taken into account in terrestrial hydrogen studies since it provides a background which must be subtracted to interpret nighttime observations (as in Donahue 1964b).

6.3   Rocket intensities, altitude profiles, and polarization

There have been a large number of rocket flights at various times during daytime, twilight, and nighttime conditions, that have measured the intensity and altitude variation of Lyman α in various directions, at heights up to 1200 km.

Seventeen experiments carried out up to June 1965 have been reviewed by Donahue (1966). Earlier reviews of some of these experiments are due to Donahue (1962, 1964b), Chubb and Bryam (1963), Donahue and Thomas (1963a, b), Donahue and Fastie (1964), Chamberlain (1963a, b, 1964), Kurt (1963), Kaplan et al. (1965), Kurt (1966).

Figure 13 shows typical results, obtained in three rocket flights in June 1962 and May 1963 (Fastie et al. 1964). In general the rocket flight results presented difficulties and inconsistencies for the interpretation of the data against the theoretical models, e.g. conflicts between the abundance derived from the altitude profile and that derived from the intensity. The latter always posed problems because of the difficulty of obtaining a reliable calibration in the UV and the lack of simultaneous measurement of the solar flux, either integrated flux or line center flux, both of which undergo relatively large short and long term variations.

General features of expected intensity distributions such as height and solar zenith distance variations were usually confirmed. A solar cycle effect on abundance was quite clear, with a marked increase in density at all altitudes above 120 km with the decreasing exospheric temperature from solar maximum to solar minimum, consistent with the theoretical predictions. The density normalizations at 100 km were usually consistent with daytime abundances one to three times KN abundances. The profile from one rocket shot to 1200 km was consistent with abundances about three times more than the low altitude determinations near that time. Further rocket experiments were conducted by Moos and Fastie (1967), Kondo and Kupperian (1967), Sheffer (1968), Fastie (1968), Young et al. (1968), Ingham (1969), Meier et al. (1970), Donahue and Kumer (1971), Buckley et al. (1971), Weller et al. (1971), and Paresce et al. (1972).

As before, problems have appeared in the data, e.g. Meier et al. (1971) again find a conflict between the abundances derived from the intensity and altitude profile, and Donahue and Kumer (1971) find differences between the upleg and downleg profiles that are much larger than the expected diurnal variation. As before, the results in general are compatible with scattering on hydrogen models with exospheric temperatures as given by the Jacchia (1966) method, with density normalizations between one and three times KN abundances.

A measurement of the polarization of scattered Lyman α in the dayglow was made by Heath (1967) and interpreted by Donahue (1967) to be consistent with three times KN abundances. However a more accurate analysis by Modali et al. (1972) gives the observed upper limit to the polarization, only if the hydrogen optical depth is greater than 7 above 200 km, which requires hydrogen abundances more than 4.4 times KN abundances. (This is allowing for a 20% increase in column abundance over the mean diurnal value, to account for the diurnal variation for the observation at 0820 LT.

6.4 Satellite solar zenith distance and altitude profiles

Lyman α intensities seen by satellites depend to a large extent on the orbit and altitude parameters applicable. Large solar zenith distance variations are usually seen. Altitude variations may be small or large, and latitude variations may occur. Observations in fixed directions and also angular scans, e.g. from zenith to nadir, have been made.

X

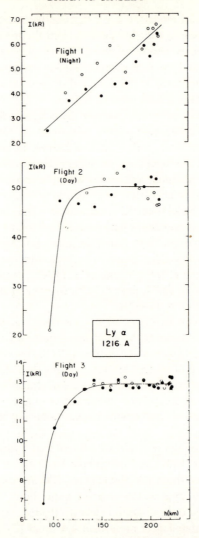

FIGURE 13   Altitude profiles of Lyman α measured on three flights in June 1962 and May 1963. The daytime emission rates fall of rapidly below 130 km on account of O_2 absorption. The nighttime emission rates fall of slowly below the peak altitude on account of the increasing H optical thickness between the observation point and the source region at high altitude above the shadow. (The optical thicknesses were considerably larger in 1962 than in 1957 on account of the lower exospheric temperature.) From Fastie *et al.* (1964), *J. Geophys. Res.,* © American Geophysical Union.

Lyman α measurements made from Venera 2, Venera 3 and Zond 1 have been discussed by Kurt (1967); from Ogo 3 by Mange and Meier (1970); from Ogo 3, Ogo 4 and Oso 4 by Meier (1970b); from Ogo 4 by Meier and Mange

(1970); from Mariner 5 by Wallace *et al.* (1970); from Ogo 4 and Ogo 5 by Thomas (1970); from Ogo 6 by Metzger and Clark (1970); from Ogo 4 and Oso 4 by Meier and Mange (1973) and from Oso 5 by Vidal-Madjar *et al.* (1973).

Figure 14 illustrates measurements made with filter photometers on Ogo 4 and Oso 4 (Meier and Mange 1973). The intensity looking 15° from the zenith is plotted as a function of solar zenith angle for one pass, and compared to theoretical curves. Useful data was obtained only for solar zenith distances greater than about 60° on Ogo 4, but the Oso 4 experiment provided 0° to 180° data points. The spherically symmetric hydrogen model atmosphere used for the theoretical brightness calculation had a T_∞ value of 1100°K with three times KN abundances up to 500 km, joined to a Chamberlain (1963a) profile above 500 km. The value of the satellite critical altitude R_{sc} used was $R_{sc} = 2.5R_c$ where R_c is the geocentric radius of the exobase.

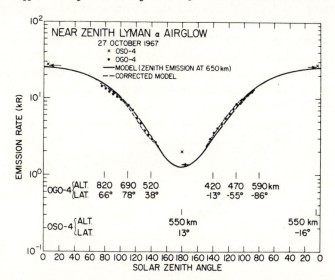

FIGURE 14   Comparison of measured and calculated Lyman α emission rates for Ogo 4 and Oso 4 observations, as a function of solar zenith angle. The solid line is the theoretical curve for intensities at 650 km, with $R_{sc} = 2.5R_c$. The dashed line includes corrections for the altitude and direction of the Ogo 4 observations. The arrows indicate calculated values for Oso 4 observations. From Meier and Mange (1973), *Planet. Space Sci.*, © Pergamon Press.

The observations were not actually made in the zenith, but 15° away from it. A correction to the theoretical zenith model for this results in the dashed lines in Figure 14. Apart from the Ogo 4 point at 180°, the data and corrected model agree quite well.

The $R_{sc} = 2.5R_c$ model is given in Table III. There is a $R_{sc} = 1$ model (no satellite orbits included) which has about 40% less density at

10,000 km, and an $R_{sc} = \infty$ model (all satellite orbits included) which has about 40% more density at 10,000 km. Brightness distribution calculations based on these three models have been made by Meier and Mange (1970). The nighttime emission rates are strongly dependent on the hydrogen concentrations above 10,000 km, determined by R_{sc}. This is principally a geometric effect due to the relative ease with which atoms between 10,000 and 70,000 km can scatter photons into the shadow. The R_{sc} value of $2.5R_c$ used in Figure 14 was based on Ogo 3 high altitude observations (Mange and Meier 1970).

Similar experimental results were obtained with a scanning spectrometer on Ogo 4 by Thomas (1970), and over a restricted solar zenith angle range on Ogo 6 by Metzger and Clark (1970) when the absorption cell was off.

The experiment on Oso 4 made zenith distance scans, and an example of these is given in Figure 15. Calculations of zenith distance variations were made for optical depths of 0.87, 1.3 and 3.5 above 650 km corresponding to 2, 3, and 8 times KN abundances for spherically symmetric models at a mean diurnal temperature of $1100°K$ with $R_{sc} = 2.5R_c$. The models were normalized to the observations at the closest approach to zenith. The best fit is for a vertical optical depth of 1.3. Thus 3 times KN is suggested for the density normalization at noon independent of the instrument calibration and choice of solar flux.

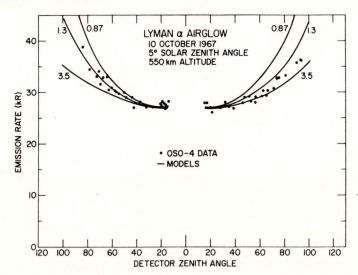

FIGURE 15   Variation of Lyman α emission rate with zenith distance of observation from Oso 4, near 500 km altitude, showing data obtained above the horizontal. Theoretical emission rates are shown for optical depths of 0.87, 1.3, and 2.5. From Meier and Mange (1973), *Planet. Space Sci.,* © Pergamon Press.

The changes of altitude of Ogo 4, Oso 4, and Ogo 6 were such as to produce changes in intensity which were noticeable but small compared to the change with solar zenith distance. Other satellites have had orbits which were highly elliptical, and the altitude variation becomes dominant, as illustrated in Figure 16 by Ogo 5 data from Thomas (1970). Data from two orbits are compared to Chamberlain profiles with $R_{sc} = 1.0R_c$, with temperatures chosen to fit the slope of the curves. However, such a large temperature change as from 800°K on April 2 to 1600°K May 11, 1968 is not consistent with the mean diurnal Jacchia (1966) exospheric temperatures for those days, which had the same value of 1100°K on both occasions. The assumption of an isotropic background (of $500R$ measured at the apogee of 152×10^3 km) was considered to be responsible for some error, since the background is anisotropic (Tinsley 1971, Thomas and Krassa 1971) and the direction of view changes over the orbit, but the main source of error is probably the seasonal change in the background emission (see Thomas and Bohlin 1972). The uncertainty in the temperature due to lack of knowledge of R_{sc} was about 200°K.

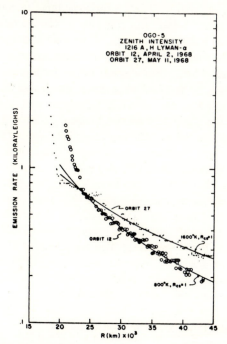

FIGURE 16   Variation of Lyman α emission rate with altitude observed on portions of Ogo 5 orbits on April 2 and May 11, 1968. Solid lines are theoretical models (Chamberlain, 1963) for $R_{sc} = R_c$ and temperatures of 800°K and 1600°K chosen to match the profiles. The inner, steeply rising region is due to radiation belt particles. From Thomas (1970), *Space Research,* ©North Holland Publ. Co.

Similar data out to 122×10^3 km and 102×10^3 km were obtained respectively on Ogo 3 by Mange and Meier (1970) and on Mariner 5 by Wallace *et al.* (1970). The data were compared to Chamberlain profiles at the Jacchia temperatures. Again, day to day variations were present in the slope of the curves in the Ogo 3 data during the operating period June-July 1966. The slopes of the curves corresponded to R_{sc} values from $2.5R_c$ to $1.0R_c$, with some slopes steeper than would correspond to $R_{sc} = 1.0R_c$.

The changes were interpreted as changes in the satellite population, with a possible mechanism being depletion of the neutral population by charge exchange with solar wind protons. It is difficult to compare observed absolute abundances with those derived from models, on account of uncertainty in the R_{sc} parameter. However both the Ogo 3 and Mariner 5 data at 50,000 km are consistent with about 3 x KN abundances if R_{sc} is taken as $2.5R_c$.

Data obtained on Ogo 5 by Bertaux and Blamont (1973) has allowed fairly accurate correction of the anisotropic background, and the derivation of H density distributions out to 100,000 km. They have determined the ratio of the density at $6R_E$ to that of $15R_E$, for their three periods of observation and also for the Mariner 5 data, and compared it for the ratios for Chamberlain (1963a) models for a range of exospheric temperature and R_{sc} values. Only for the nightside measurement could agreement be found, with R_{sc} between 1 and $2.5R_c$. The dayside ratios, including the value for Mariner 5, could only be matched by a model which was the same as the nightside to at least $6R_E$, but all satellite particles were absent at $15R_E$.

Calculations of the effect of solar wind charge exchange losses for orbits which extended outside the dayside magnetosphere showed that a combination of this loss mechanism, together with the perturbation of orbits away from the frontside by solar radiation pressure could explain the observations.

The earlier data from Zond 1, and from Venera 2 and Venera 3, which was launched through the dayside in November 1965, have been converted to abundances by Kurt (1967). The abundances are about a factor of four lower than those of Wallace *et al.*, and this is probably due to calibration differences and a different solar flux being used. The slope beyond 40×10^3 km is smaller than obtained in the Mariner 5 observations, corresponding better to a distribution with no depletion of satellite orbits ($R_{sc} = \infty$); however errors due to background subtraction may again be important.

6.5 Satellite measurements of temporal variations

Meier and Mange (1973) have also analysed the Oso 4 data to determine diurnal variations. These are evaluated by comparing the departures of the data from the brightness distribution of the $\tau_V = 1.3$ spherically symmetric model to the departures of brightnesses of other spherically symmetric models from that of the $\tau_V = 1.3$ model.

This procedure showed that there was a clear maximum in the density at 6-7 hr local time, and a minimum somewhat later than 16 hr. At 650 km the diurnal density ratio is about a factor of 1.7.

Day to day variations in the brightness of Lyman α under the same scattering geometry are due to the combined effects of changes in the solar Lyman α line center flux and the related changes in the terrestrial hydrogen abundance. Changes in the solar line center flux are associated with changes in the solar EUV flux which determines the exospheric temperature and hence the hydrogen abundance.

Meier and Mange (1973) corrected their measurements of day to day brightness changes measured on Ogo 4 for the effect of changes in exospheric temperature, and obtained a relationship between the solar Lyman α line center flux and the Zurich sunspot number R_z. The flux increases by about 50% when R_z increases from 40 to 220.

An experiment onboard Oso 5 has provided a great deal of data on temporal variations in the atomic hydrogen abundance. Measurements are made of the solar flux in a band 100A wide centered on Lyman α, and of the scattering of this flux by atomic hydrogen in a cell. The measurement of the scattered flux amounts to a measurement of the solar flux is a 10^{-2} A passband near the center of the solar Lyman α line, which scans across the absorption core due to terrestrial hydrogen between the satellite and the sun by means of the doppler shift corresponding to the satellite velocity. Scattering from atomic deuterium was also measured. Preliminary results were given by Blamont and Vidal-Madjar (1971) and results of a more sophisticated analysis by Vidal-Madjar et al. (1973).

A high correlation was found between the total Lyman α flux and the line center flux. A 30% variation in the total flux corresponds to a 47% variation in the central flux. The variation in the line center flux with R_z was the same as found by Meier and Mange (1973).

The diurnal variation at the exobase was found to be by a factor of 1.7, from a comparison of the data with an axially symmetric model exosphere, similar to those of Vidal-Madjar and Bertaux (1972), where the altitude profiles above any point on the exobase were determined only by the local exobase temperature and density. The best fit for the phase shift of the axis of symmetry was found to be about 4 hours from the sub-solar point, i.e. the temperature maximum and density minimum at 1600 hrs, or two hours later than the Jacchia temperature maximum.

Vidal-Madjar et al. (1973) pointed out that the diurnal variation of a factor of 1.7 is exactly that expected for the condition of zero net ballistic flux. They argue that this observed ratio is in fact due to the effects of ballistic fluxes which would quickly establish the equilibrium conditions on account of being much larger than other fluxes. This view differs from that of Tinsley (1973) who argues that the ballistic fluxes are not sufficiently large, but the agreement with the zero net ballistic flux condition is coincidental, on account of the effect of diurnal sources and sinks due to charge exchange with ions cancelling an opposite effect due to rotation.

Vidal-Madjar *et al.* (1973) find that the mean diurnal abundance of hydrogen varies with exospheric temperature as expected on the basis of Kockarts-Nicolet static models, with a density normalization that varied slowly with solar activity. For the 1969-1970 period the ratio varied from 2.3 x KN for low solar activity and 2.5 x KN for medium solar activity to 4 x KN for high solar activity.

6.6 Balmer α results

The first measurements of Balmer α began in January 1957 at Zvenigorod (Prokudina 1960, 1959, Shefov 1959), followed soon after by observations at Alma Ata (Gaynullina and Karagina 1960), at Abastumani (Fishkova and Markova 1958, 1960a, b), and at Ås (Kvifte 1959a, b). Further observations were made in 1959 and 1960 at Yakutsk (Yarin 1960) and at Haute Provence (Dufay and Dufay 1960). The spectrographic exposures usually extended over one or more nights. The emission was clearly not due to auroral emission, as the line was narrow (<2Å) and undisplaced. There was some question as to whether the emission was due to astronomical sources, i.e. gaseous nebulae as in Orion and Cygnus, especially since the nebular line [N II] 6583Å was seen to be present when Balmer α was enhanced. But careful observations showed that although very bright astronomical sources were present in restricted areas, there existed a faint emission of 5-20R extending over the whole sky. Fishkova and Markova (1960b) separated the sources on the basis of time variations, Dufay *et al.* (1961), with a spectrograph which could measure shifts as small as 0.2Å, showed that there were doppler shifts in the discrete bright sources which confirmed their galactic origin, while the fainter emission was undisplaced. Further discussion on the Abastumani observations has been given by Fishkova (1962, 1963a, b), Fishkova and Martzvaladze (1964, 1967a, b), Martzvaladze (1967, 1968). Figure 17 from Fishkova (1963a) shows variations at 67° zenith distance in the north for the exposures made at Abastumani. Observations near the antisolar point recorded intensities approximately 2.6 times smaller than at 67° z.d. in the north. The maximum in March (not present for pole observations) was due to galactic sources, while the remaining emission was attributed to resonant scattering of solar Lyman β in the geocorona.

The seasonal variation in intensity (excluding galactic emission) was found to be similar in shape to a plot of column abundance of hydrogen (from Johnson and Fish 1960) above the seasonally varying terrestrial shadow height in the midnight observing direction. An attempt to derive an absolute abundance was hampered by problems with the solar flux and the correction factor for multiple scattering.

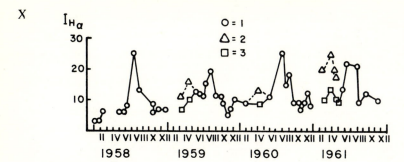

FIGURE 17   Variations of Balmer α emission rate at 67° zenith distance in the north from Abastumani, with exposures extending over most of each night. Triangles are observations contaminated by galactic emission, and squares are corrected intensities based on simultaneous observations of the celestial pole. From Fishkova (1963a), *Aurora and Airglow,* © Nauka Publishing House, Moscow.

Observations on Mt. Chacaltaya in Bolivia were made in 1961 by Ingham (1962), and results consistent with the previous work were obtained.

The use of a Fabry Perot etalon at Alma Ata in 1962-63 (Shcheglov 1963a, b, c, 1964) allowed exposures of about an hour to be made, and thus the large intensity variations during the night to be seen. Intensities ranged from $40R$ for small elongation (small angular distance from the sun), to $5R$ near the antisolar point at midnight. The doppler shift of the emission was less than 0.1Å, as expected for scattering on terrestrial hydrogen. For terrestrial hydrogen, the variations seen with elongation are the manifestations of varying shadow height (actually of the shadow of the O_2 absorbing region up to about 120 km) which can be specified uniquely by solar zenith distance, relative solar azimuth, and zenith distance of observation. Thus these coordinates are preferable for data analysis. Second order variations are due to the non-spherical hydrogen distribution and day to day changes in abundance.

Review of some of these observations was made by Donahue (1964a), with a comparison with radiative transfer calculations for hydrogen abundances deduced from Lyman α interpretations. The calculated intensities were lower than those observed by an order of magnitude. Part of this discrepancy was due to an error in the calculations (Tinsley 1969b, 1970) and to background effects present in the observations.

Discussions of Balmer α results up to 1965 are included in the reviews by Krassovsky *et al.* (1966) and Donahue (1966). Further observations with short integration times were made by Shcheglov (1967a, b), Armstrong (1967), Tinsley (1967, 1968, 1970) and Ingham (1968). A comparison of averaged distributions for three years was made with new radiative transfer calculations (Tinsley and Meier 1971). Observed data for 80° and 40° zenith distances of observation are shown in Figure 18, and they may be compared with the theoretical curves for the same observing parameters given previously in

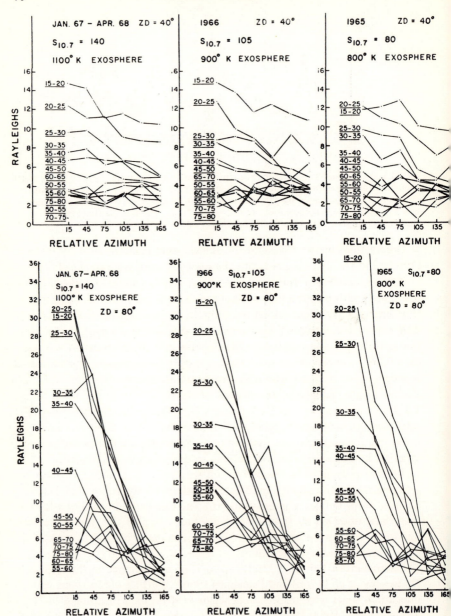

FIGURE 18 Variations of Balmer α emission rate observed from New Mexico averaged over a number of months and plotted in the same format as Figure 12. From Tinsley and Meier (1971), *J. Geophys. Res.*, © American Geophysical Union.

Figure 12. The agreement is fairly good considering experimental difficulties, which include the presence of the much brighter galactic sources in some areas of the sky. Correction was made for tropospheric scattering of the galactic sources, and tropospheric scattering and extinction of the non-uniform geocoronal source itself.

There is a clear diurnal variation in the Balmer α intensities noted by Shcheglov (1963a, b) and by other workers. The intensities just before dawn are about 20% greater than the evening intensities at the same scattering geometry. This is consistent with a diurnal variation in hydrogen abundance of about 2 (Tinsley 1970) and is consistent with the theory of diurnal variations (Section 3).

Seasonal variations of hydrogen abundance have been obtained by referring observed intensities to constant shadow height (Martzvaladze 1967, Fishkova and Martzvaladze 1972) or to an observed or model distribution of intensities with respect to solar and observing zenith distance and relative solar azimuth (Tinsley 1968, 1970). Figure 19 shows the seasonal variation obtained by Fishkova and Martzvaladze (1972) from 1962-65 observations, which is similar to that of Tinsley for 1965-69 observations. This seasonal variation is about 90° out of phase with the exospheric temperature variation invoked by Jacchia (1965) to explain semiannual density changes in the major constituents. Thus these data are an argument against the semiannual effect being a temperature variation, but make it more plausible that it results from composition changes in the lower thermosphere. The hydrogen abundance changes may also represent seasonal changes in mesospheric production.

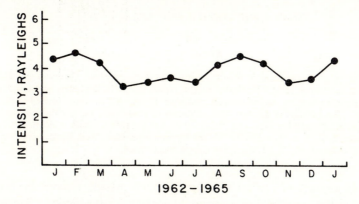

FIGURE 19   Seasonal variations of observed Balmer α emission rates normalized to constant shadow height. From Fishkova and Martzvaladze (1972), © Abastumani Astrophysical Observatory.

Solar cycle variations may also be obtained by normalizing observations (Fishkova and Martzvaladze 1972, Tinsley and Meier 1971). Figure 20 results from an analysis of solar cycle variations of the normalized intensity over the period 1957-1967, using data from several observers. Fluxes at the center of the solar Lyman β line were calculated, which, if scattered on a set of model atmospheres, gave the observed normalized intensities. The model atmospheres were effectively a continuous series, interpolated from Table III with respect to temperature. Also shown in Figure 20 is the best estimate of the variable flux at the center of the solar Lyman β line, based on a single solar Lyman β profile (Tousey 1963) and the variations in integrated Lyman β measured by Hall *et al.* (1969) and an assumed variation of profile with solar activity.

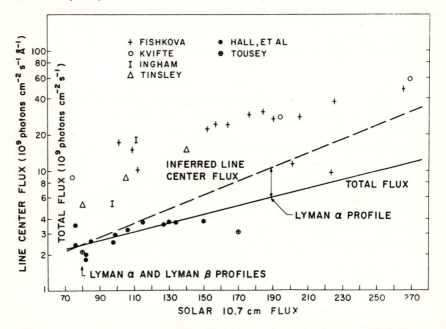

FIGURE 20   Comparison of required solar Lyman β line center fluxes to give Balmer α emission rates observed by Fishkova, Kvifte, Ingham, and Tinsley, after scattering on model hydrogen distributions, and measured total solar Lyman β fluxes due to Hall *et al.* and Tousey. The full line is the best fit to the data of Hall *et al.*, and the dashed line is the line center flux inferred from the measured Lyman β profile, which is assumed to vary with solar activity in the same way as the Lyman α profile. From Tinsley and Meier (1971), *J. Geophys. Res.,* © American Geophysical Union.

A discrepancy of a factor between 2 and 5 is present between line center Lyman β flux calculated as necessary to account for the observed Balmer α brightness after scattering on the model atmospheres, and the line center flux from the solar Lyman β measurements. This discrepancy could be due to

a)   the use of too low values of solar flux at the line center, owing to problems with the total intensity or line profile measurements.

b)   the use of too low values (3 times KN) for mean diurnal hydrogen abundances.

c)   the neglect of diurnal variations in the hydrogen models.

In general the trend of the variations found is consistent with constant production rate and thermal escape being the dominant loss process, but a difference between intensities is found on the ascending and descending parts of the solar cycle. This may be related to the geomagnetic activity lag with respect to solar EUV variations during the solar cycle, with geomagnetic activity increasing the hydrogen loss rate via the plasmasphere, and via charge exchange with hot ions (Section 4.3).

Short period variations of Balmer α intensity have been seen in a number of observations (Shcheglov 1967a, b; 1968, Karjagina et al. 1969a, b, c, d, Fishkova and Shcheglov 1969a, b, Fishkova et al. 1970, Martzvaladze et al. 1971). These were more pronounced for observations made at higher geomagnetic latitudes, and looking north. The amplitude has reached 50%, with time scales of the order of half an hour. Figure 21, from Fishkova and Martzvaladze (1972), shows fluctuations seen from Abastumani in 1968. They are possibly connected with short term variations of hydrogen content as a result of penetration of hydrogen ions in the auroral zone, or loss via the polar wind, or heating effects associated with auroral activity (Shefov 1970).

Krassovsky (1971) has discussed a correlation between Balmer α intensity variations and OH intensity and vibrational temperature variations, and suggests that these are related to very uneven vertical mixing of the upper mesosphere.

6.7   Rocket Lyman β results

Two rocket measurements of scattered Lyman β in the night sky have been made. The first was on August 10, 1967 (Young et al. 1968) and the second on October 13, 1969 (Weller et al. 1971), and with both flights simultaneous Lyman α and ground based Balmer α measurements were made. The 1967 results have been analysed for compatibility with Balmer α results and radiative transfer calculations by Meier (1969), and only approximate agreement is found using a solar Lyman β flux several times higher than the Tousey (1963) value, and with a model with higher abundances at high altitudes than have been found compatible with Lyman α results (Meier and Mange 1970). The 1969 results have been analysed by Weller et al. (1971). The intensity of Lyman β was within 50% of $27R$ at 217 km in the zenith, for a solar zenith angle of $134.3°$. When compared to radiative transfer calculations on a model atmosphere with three times KN abundances continued with a Chamberlain (1963a) model above 500 km with $R_{sc} = 2.5R_c$, the observations fitted the calculated spatial variations well from 100 to 130 km, but they did not fit well at higher altitudes. The presence of an additional emission in the wavelength range 740-1050A was necessary to explain the

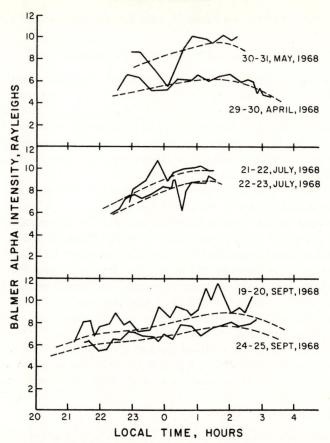

FIGURE 21   Short term fluctuations in Balmer α emission rate seen from Abastumani in 1968, for observations in the direction of the north celestial pole. From Fishkova and Martzvaladze (1972), © Abastumani Astrophysical Observatory.

discrepancy. The intensity of the line center solar Lyman β flux required to explain the observed intensities was about a factor of 5 higher than that obtained using the total solar Lyman β flux inferred from the measurements of Hall *et al.* (1969), and the line profile measurement of Tousey (1963). This fivefold increase was also necessary to account for the Balmer α intensities, which otherwise fitted the radiative transfer calculation well. Thus the same discrepancy as discussed in the previous section is found with the Lyman β measurements also. The fit between the observed Lyman α intensities and the calculated Lyman α intensities was considered to limit the possibility of higher hydrogen abundances to less than a factor of 2.

Calculations of the solar Lyman β emission by Avrett (1970) give a line shape that is narrower than the Tousey (1963) value by about a factor 2.

Use of this line shape would reduce the discrepancy correspondingly. Further measurements of the solar Lyman β profile would be very desirable. However, as pointed out by Weller *et al.* (1971) if simultaneous Lyman β and Balmer α measurements can be made with good absolute accuracies in each, then the ratio of intensities can give the optical depth of the hydrogen independent of a knowledge of the solar flux. The absolute accuracies in the 1969 measurements were not quite good enough to give useful results for this.

6.8 Observations of polar cap depletion

The polar wind loss of hydrogen ions is maintained by the creation of H^+ through the charge exchange of neutrals with oxygen ions (Eq. 77), as described earlier in Section 4.3.

The depletion of neutral hydrogen will affect the radiative transport through hydrogen, in such a way that intensities measured are less than in regions at the same solar zenith angle where there is no depletion. This effect has been seen in Lyman α nadir intensities by Meier (1970a), for Ogo 4 passes near the north and south magnetic poles as shown in Figure 22. It was present on many, but by no means all passes. In general the position coincided with the region of open magnetic field lines which is inside the evening (18 hr) side of the auroral oval, but extends beyond the morning (6 hr) side. No change was seen in the zenith observations from the satellite, implying little depletion at altitudes above 1000 km where most of the zenith light originates. This may be a consequence of lateral flow in the exosphere being fast compared to the rate of flow down to the region of 250-450 km altitude where most of the charge exchange reaction and depletion was occuring. The Ogo 4 orbit had a perigee 425 km and apogee 950 km. The observed reduction in nadir intensity can be related by radiative transfer models to reduction in H abundance, and a rough estimate leads to depletions as high as 40% in the region below 450 km. Precise calculations are probably not warranted until more is known about seasonal variations of mesospheric hydrogen production in the polar region (Donahue 1969a) and horizontal flows of the atmosphere as a whole (Johnson and Gottlieb 1970) which would bring neutral hydrogen into the region.

7 NON-OPTICAL OBSERVATIONS

In contrast to optical observations, the non-optical methods sample the hydrogen in the immediate vicinity of the detector, except for incoherent scatter and whistler observations which sample a localized region of ions somewhat removed from the detector. The neutral mass spectrometer and the satellite drag methods directly sample neutral hydrogen, and other

X

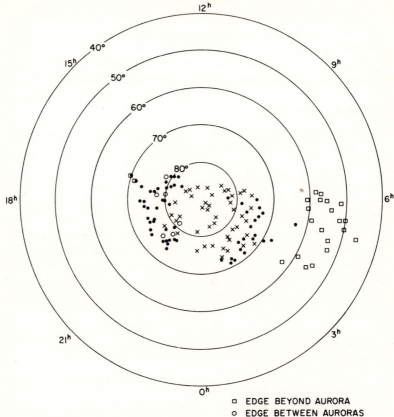

COMPOSITE OF OGO-4 NORTH POLAR PASSES
8/6/67 - 8/16/67

□ EDGE BEYOND AURORA
○ EDGE BETWEEN AURORAS
× DEPRESSION MAXIMUM
• EDGE OBSCURED BY AURORA

FIGURE 22 Composite of Ogo 4 north polar passes for the period 6 August – 16 August 1967. Crosses mark the point of maximum depression of the Lyman α nadir intensity, circles and squares indicate the edge of the depressed area. In the auroral oval itself the Lyman α emission from precipitating protons obscured the effect. From Meier (1970a), *J. Geophys. Res.*, © American Geophysical Union.

methods sample hydrogen ions, from which the neutral abundance can be evaluated by applying charge exchange equilibrium (Section 4.1).

7.1 Ion mass spectrometer measurements

From the 1961 midday rocket ion mass spectrometer measurement of Taylor *et al.* (1963), Hanson *et al.* (1963) obtained the neutral hydrogen concentration of 7×10^4 cm^{-3} at 500 km at an exospheric temperature T_∞ of 1235°K. Allowing for diurnal variations of T_∞ and $n(H)$ this is about 8 times the KN abundances.

From a 1964 midnight rocket ion mass spectrometer measurement Hoffman (1967) obtained the neutral hydrogen concentration of 5×10^6 cm^{-3} at 350 km at $T_\infty = 700 \pm 100°$K. A calibration revision (Hoffman 1970) reduces this to 1×10^6 cm^{-3}. Allowing for diurnal variations this is about 4 times the KN abundances.

From a 1966 rocket ion mass spectrometer measurements at 1300 hrs LT Brinton et al. (1969) obtained the neutral hydrogen concentration of 6×10^5 at 350 km at $T_\infty = 850°$K. Allowing for diurnal variations this is about 4 times the KN abundances.

An ion mass spectrometer on Explorer 32 has provided data for the period June 1966 to January 1967 (Brinton and Mayr 1971, 1972). The measurements were made in the altitude range 280-450 km. The values of $n(O)$ were obtained from the Jacchia (1965) model atmosphere, and were in agreement with density gauge measurements also made on Explorer 32.

Individual calculations of $n(H)$ from $n(H^+)$, $n(O^+)$ and $n(O)$ have been normalized to 350 km using a scale height of 600 km, which corresponds to $T_\infty = 900°$K (the average of diurnal and other variations of T_∞ for this period); The values of $n(H)$ show (1) a decrease due to the increase of T_∞ with solar activity; (2) a 27-day modulation due to variations of T_∞ with EUV variations associated with solar rotation; (3) a modulation due to the changing local time of the observations which were made at a restricted range of latitudes (23° to 47°) and longitudes (−65° to −124°).

A curve fitting technique has been used to separate the several variations into components, each associated with a primary component of the exospheric temperature. The results are shown in Figure 23. On the left side the variations in $n(H)$ associated with the solar cycle component (light line) and the sum of solar cycle and solar rotation components (heavy line) are shown. On the right side the heavy line denotes the diurnal variation.

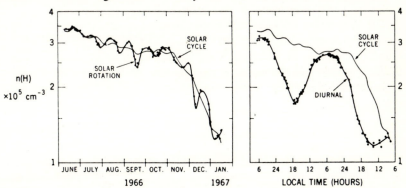

FIGURE 23   Separation into components of $n(H)$ variations. On the left are shown the variations in $n(H)$ associated with the solar cycle component (light line) and the sum of solar cycle and solar rotation components (heavy line). On the right the heavy line denotes the diurnal variation. From Brinton and Mayr (1971), *J. Geophys. Res.,* © American Geophysical Union.

The factor of 2.5 solar cycle decrease between June 1966 and January 1967 is in accordance with the theoretical effect of approximately a factor of 3 on the mean diurnal abundance, derived from the Jacchia (1965) model of mean T_∞ applied to the KN models.

The 27 day solar rotation variation is inversely correlated with 27 day exospheric temperature changes as expected. The variation can be as much as a factor of 1.3.

The diurnal variation is approximately a factor of 2, as predicted theoretically. The curve fitting technique revealed an increase in the magnitude of the diurnal variation with rising solar activity, consistent with the results of time dependant models (see Table II). The technique also revealed a phase lag between the Jacchia model T_∞ and the hydrogen density, which is two hours for the daytime density minimum and four hours for the nighttime maximum. The Patterson (1966) model found a smaller phase lag, but on the basis of cylindrical geometry. In any case the phase of the Jacchia (1965) model is a better representative of diurnal density than diurnal temperature variations.

The geomagnetic activity variations are not shown in Figure 23, but the expected negative correlation with geomagnetic T_∞ changes was found.

The analysis also revealed a small semiannual variation, which appeared to be directly correlated with the semiannual variation in T_∞ invoked by Jacchia (1965) to explain semiannual density changes in the major constituents. An anticorrelation with T_∞ is required by the Jacchia (1965) model. Thus, as pointed out in Section 6.6, the semiannual hydrogen abundance changes would appear to be related to some other effect such as composition changes in the lower thermosphere and/or mesosphere. The Balmer α and ion mass spectrometer results differ as to the exact phase of the semiannual variations but this is not surprising as the effect is small compared to other variations in each case, and the accuracy would not be high.

The abundances obtained have been compared to the KN models using mean diurnal temperatures derived from Jacchia (1965) models, and are found to be higher by an average (night and day) factor of about 3. The absolute accuracy of the ion mass spectrometer measurements and charge exchange equilibrium calculation should be good, and is estimated as $+35\% - 25\%$.

7.2   Neutral mass spectrometer measurements

Measurements have been made with a neutral mass spectrometer on Explorer 32. For observations at the end of May 1966 Reber et al. (1968) found densities ranging from 2.5×10^6 cm^{-3} at 300 km for mid-morning observations to 4×10^5 cm^{-3} at 900 km for afternoon observations. The mean diurnal T_∞ value from Jacchia (1965) for the period of observation was 916°K, and allowing for diurnal variations in the hydrogen the measured densities amount to 36 times KN abundances for the mid-morning observations, and 16 times KN abundances for the afternoon observations.

These densities are so much higher than densities evaluated by other methods that one suspects that instrumental problems such as slow outgassing of hydrogen from the instrument itself, have invalidated the results.

7.3 Electron cooling measurement

Brace et al. (1967) have arrived at an estimate of the neutral hydrogen concentration from Explorer 22 measurements of the rate of electron cooling after sunset, and an assumption as to the nature of the cooling. Near the magnetic equator the cooling of electrons by heat transfer to protons and then to neutrals should be more important than electron thermal conduction if the neutral density is high enough. The local loss term can be written

$$L = \Sigma_x k_x n(e^-) n(X) (T_e - T_n) \qquad (106)$$

where $n(X)$ is the concentration of the neutral particle coolant

T_n = neutral temperature

T_e = electron temperature, taken as equal to ion temperature T_i, i.e.

$$(T_e - T_i) \ll (T_i - T_n) \qquad (107)$$

k_x is a coefficient, which is 4×10^{-13} ev cm^3 sec^{-1} deg^{-1} for hydrogen and about ten times less for helium.

A value for L was deduced from the time rate of change of $n(e^-)$ and T_e from Explorer 22 measurements near the magnetic equator. Calculated electron thermal conduction was inadequate to account for it, as was cooling by neutral helium. Assuming that all the cooling was due to atomic hydrogen, and using Eq. (106), the required $n(H)$ at 1000 km in the post sunset period was about 5×10^5 cm^{-3} for early 1965. The Jacchia (1965) mean exospheric temperature was near 750°K for the period of observation. Allowing for $n(H)$ diurnal variations the measured abundance is about 6 times the appropriate KN abundance.

7.4 Satellite drag measurements

Satellite drag measurements can give hydrogen densities only when hydrogen provides the major component of the mass density, which occurs as low as 1000 km around solar minimum. The drag is comparable to the radiation pressure from the sun and the earth and electromagnetic drag effects.

Prior (1971) has made careful corrections for such effects for the orbital perturbations of two balloon satellites, Pageos and Dash II, for the period April-May 1967. The perigees were at about 2800 km and 2300 km respectively near 0600 local time at about $20°$ latitude.

It was found that after the corrections had been made, it was necessary to increase the abundance of hydrogen in the USSAS 1966 model atmosphere, for $900°K$ by a factor of 3 and then the drag of both satellites and the scale height of the atmosphere between them could be accounted for. Allowing for a $n(H)$ diurnal variation by a factor 2, and comparing to a KN model atmosphere for the applicable Jacchia (1965) mean T_∞ of $1076°K$ for the period, would require hydrogen abundances increased by a factor of about 4 instead of the factor of 3 quoted.

7.5   Incoherent scatter and whistler observations

With an incoherent scatter radar system it is possible to determine the O^+, H^+ and He^+ concentrations, and the electron and ion temperature, and using them deduce the neutral temperature, neutral hydrogen density, and vertical proton flux. Whistler measurements give electron density in the plasmasphere, and changes in this also yield the proton flux in or out.

Ho and Moorcroft (1971) have made an analysis of Arecibo data for the neutral parameters and proton flux. The data were obtained during a 30 hour period on December 7 and 8, 1965. The neutral hydrogen density was calculated from charge exchange equilibrium, using the measured values of $n(O^+)$ and $n(H^+)$, and $n(O)$ values from Jacchia (1965). The T_∞ values were obtained in an iterative solution for the cooling to neutral oxygen, and varied about $650°K$ to $950°K$. Applying charge exchange equilibrium, values for $n(H)$ were calculated, and varied from about 6×10^5 cm^{-3} to 12×10^5 cm^{-3} at 520 km. The mean value of 9×10^5 cm^{-3} is about 7 times the KN value for a mean T_∞ value of $800°K$.

The diurnal variation found by Ho and Moorcroft (1971) is rather irregular, with a maximum at 1900 hours as well as at 5 hours. There is a minimum at 14 hours as well as at 23 hours. While the maximum amplitude of the variation is a factor of 2, consistent with expectations, the irregular changes outside the error limits may reflect a lack of complete charge exchange equilibrium.

Data are given in a paper by Farley *et al.* (1967) which allow a similar analysis to be made for the Jicamarca observations of December 17, 1965. At noon and 550 km $n(O^+)/n(H^+)$ was 9.0, and T_n was close to $900°K$. Taking $n(O)$ for $900°K$ from Jacchia (1965) yields $n(H) = 4.5 \times 10^5$ cm^{-3}

For a noon T_∞ of $900°K$, the mean T_∞ is $816°K$. Allowing for the diurnal effect the hydrogen abundance is about 4 times the KN abundance. Possible errors include the assumption that complete charge exchange equilibrium was applicable at 550 km, and that the helium ion density was negligible.

Using the Millstone Hill incoherent radar facility Evans (1971) has obtained data for 1200-1600 hours on March 24, 1970 which have been interpreted by Schunk and Walker (1972) as consistent with a hydrogen model atmosphere with 5.9×10^4 cm^{-3} at 600 km. T_∞ was 1100°K, corresponding to a mean T_∞ of 1000°K. Allowing for diurnal variations in $n(H)$, this hydrogen abundance is 3 times KN abundances.

Measurements of vertical proton fluxes into and from the plasmasphere are important, as the magnitude of these fluxes affects the diurnal variation and possibly magnetic storm related losses.

Whistler measurements by Park (1970) of the electron density in the plasmasphere in June 1965 indicated a daytime upward flux averaging about 3×10^8 proton cm^{-2} sec^{-1}, and a downward nighttime flux of about 2×10^8 proton cm^{-2} sec^{-1}. (The net accumulation in the plasmasphere was lost from it at the time of magnetic storms as discussed in Section 4.3.)

The data of Evans (1971) has been interpreted by Schunk and Walker (1972) as implying an upward proton flux of 1.2×10^8 cm^{-2} sec^{-1} for a 1200-1600 hours LT time period in March, 1970.

An upward proton flux of 1.5×10^8 cm^{-2} sec^{-1} was calculated by Brinton et al. (1969) to be present at 1300 hours LT in March, 1966, from the results of the same rocket flight referred to in Section 7.1.

The observations of Ho and Moorcroft (1971) for December 7-8, 1965 at Arecibo yielded an average upward flux during daylight hours of 6×10^8 cm^{-2} sec^{-1}, with a peak value of 11×10^8 cm^{-2} sec^{-1} at 10 a.m. LT. The higher fluxes of the 1965 observations compared to the later ones may reflect the higher $n(H)$ values for low solar activity.

Ho and Moorcroft calculated the theoretical limiting upward flux using their derived $n(H)$ values, and other measured parameters, and obtained a value of 7.7×10^8 cm^{-2} sec^{-1} for T_∞ of 900°K, using Geisler's (1967) equation. Within the limits of error, this agrees with the flux obtained from the measurements.

Summarizing the results for the upward H$^+$ fluxes, it appears that they are a few times the Jeans escape fluxes corresponding to the neutral hydrogen abundances that are found to accompany them. (The upward flux corresponding to the KN models is 2.5×10^7 cm^{-2} sec^{-1}.) Fluxes close to the limiting value occur, but there is no clear evidence that the limit is exceeded.

7.6 Proton lifetimes at $2.5R_E$ to $4.8R_E$

As discussed in Section 2.3.3, hydrogen atoms in the exosphere can be removed by charge exchange collisions with protons if the proton energy is large compared to the thermal energy. A measure of the rate at which this process occurs, and in fact of the density of atomic hydrogen at heights corresponding to the outer part of the plasmasphere, can be obtained from the lifetimes of protons which have been injected into that region. Swisher and Frank (1968) obtained hydrogen densities for geocentric radial distances $2.5 - 4.8R_E$ from the decay of $\sim 200\,eV$ to $50\,keV$ proton fluxes injected during a magnetic storm in early July 1966. Densities ranged from $\sim 200\,cm^{-3}$ at 9500 km altitude to $30\,cm^{-3}$ at 24,200 km for a mean exospheric temperature near $850°K$. These are about a factor 5 lower than a 3 x KN model with $R_{sc} = 2.5R_c$, i.e. the model was found to be reasonably consistent with the Ogo 3 and Mariner 5 data (Section 6.4). The very fact that enhanced loss of protons was occurring implies some reduction in atomic hydrogen density, but it is not clear if this alone is responsible for the discrepancy.

7.7   Mesospheric molecular oxygen and hydroxyl emissions

Simultaneous measurements of the altitude profiles at the intensity of the Meinel hydroxyl (OH) bands and the 1.27μ emission from molecular oxygen $O_2(^1\Delta_g)$ in the mesosphere can be used to calculate the concentration of atomic hydrogen in the region. The relevant reactions are the production of ozone by association of O and O_2

$$O + O_2 + M \xrightarrow{k_{108}} O_3 + M \tag{108}$$

where $k_{108} = 1.5 \times 10^{-34} \exp(445/T)$, and its destruction by two competing processes. In the daytime photolysis occurs for $\lambda < 3100A$

$$O_3 + h\nu \xrightarrow{J_{109}} O_2(^1\Delta_g) + O(^1D) \tag{109}$$

where $J_{109} = 9.6 \times 10^{-3}\,s^{-1}$ at zero optical depth, and also reaction with H produces vibrationally excited OH

$$O_3 + H \xrightarrow{k_{110}} OH^*(\nu' < 9) + O_2 \tag{110}$$

where $k_{110} = 1.5 \times 10^{-2}\,T^{1/2}$.

The $O_2(^1\Delta_g)$ is quenched by atmospheric gases with a rate constant $k_D = 4.4 \times 10^{-19}\,cm^3\,s^{-1}$, but it is assumed by Evans and Llewllyn (1973) that the quenching of OH^* by O_2 and N_2 is negligible at mesospheric heights by comparison with the radiative lifetime, although there is some uncertainty on this point.

The O_3 concentration can be derived directly from the 1.27μ altitude profile:

$$n(O_3) = \frac{A + k_D n(M)}{A \cdot J_{109}} \left[\frac{dI(1.27\mu)}{dh} \right] \times 10^4 \qquad (111)$$

where $A = 2.6 \times 10^{-4}$ sec^{-1}, $dI(1.27\mu)/dh$ is the value of the 1.27μ emission in kilorayleighs per kilometer, and $n(M)$ is number density of the ambient atmosphere.

The assumptions of photochemical equilibrium and negligible OH* quenching leads to H concentrations from the OH emission profile:

$$\frac{\left[\dfrac{dI(OH^*)}{dh} \right] \times 10^4}{P k_{110} n(O_3)} \qquad (112)$$

where P is the number of photons produced per reaction. P depends on the particular OH band or bands passed by the photometer filter, and is evaluated by reference to a complete synthetic spectrum of the Meinel bands.

In a rocket measurement made by Evans and Llewellyn (1973) the filter passband extended from 1.65μ to 1.89μ, covering the 5-3 and 6-4 OH bands and part of the 7-5 band. For this spectral range, P was calculated to be 4.8.

The results of the analysis led to H concentrations of $2\text{-}5 \times 10^7$ cm^{-3} in the height range 77-97 km. This is an order of magnitude less than the model of Hesstvedt (Figure 4). At 100 km the value of $1 \pm 1 \times 10^7$ cm^{-3} was obtained, reasonably consistent with observations of the absorption core in the solar line.

8 DEUTERIUM

8.1 Theory

The thermal escape rate per atom is considerably less for deuterium (D) than for mass 1 hydrogen (H), owing to double the mass in the exponent of Eq. (47). Also the doubled mass means the scale height for D is only half that for H. Thus the altitude profiles are different, and the relative abundances are a function of altitude as well as temperature. The isotropic ratio is 1.44×10^{-4} by number at sea level, and it has been assumed that the ratio would be the same in the mesosphere. This would be so if thermal escape was sufficiently rapid so that the upward flux through the 100 km level was the limiting value allowed by diffusion. However it is apparent from the analysis of upward flow velocities of Section 2.4, that at temperatures below about 1000°K the thermal escape flux of D is so small that the vertical profile is determined by eddy mixing up to the turbopause and near diffusive equilibrium above. Thus there may be some enrichment of D relative to H in the mesosphere at low solar activity.

Profiles for D above 100 km have been calculated by Kockarts and Nicolet (1962) using Eq. (5) (eddy mixing neglected) and by McElroy and Hunten (1969a), and Kockarts (1972) using Eq. (13). Donahue (1969b) and Hunten (1971) have discussed general aspects of the profiles.

The enhancement of D relative to H reaches a maximum when the altitude is near 500 km and the exospheric temperature is near $1400°K$. At this temperature there is near diffusive equilibrium for D but appreciable flow effects for H.

At higher temperatures the escape of D is faster, and the trend is to a situation where the densities of both H and D are limited by upward flow, for which no enhancement would occur. At lower temperatures, the escape of H is slower, and the trend is to a situation where the densities of both H and D follow diffusive equilibrium curves above the turbopause, which reduces the enhancement. In fact the scale height differences would enhance the H at very low temperatures.

These results are calculated for static profiles and can be taken as referring to mean diurnal exospheric temperatures. Diurnal changes in the enhancement ratio are probably small.

Donahue (1969b) has discussed aspects of D in the terrestrial atmosphere in comparison to the situation on Venus (Section 12.2). Kockarts (1972) points out that the polar wind loss of D compared to thermal escape would be more important than for H at low exospheric temperatures.

8.2 Observations

Hoffman (1969) has measured the $n(D^+)/n(H^+)$ ratio at 700 km and obtains 1.1×10^{-3} for an exospheric temperature of $940°K$ at 1400 LT, which corresponds to a mean diurnal temperature of $840°K$. Provided charge exchange equilibrium applies as well for D^+ as H^+ the $n(D)/n(H)$ ratio will be the same, which is a factor of 7 enhancement over the sea level value, and presumably over the 100 km value for this exospheric temperature. McElroy and Hunten (1969a) find the ratio is 5.5 at 500 km for $840°K$, and allowing for scale height effects the ratio is about 5 at 700 km, in good agreement with the observations.

Bruner and Wilson (1969) observed a feature on the violet wing of a solar Lyman α profile obtained near 160 km. The wavelength of the dip corresponded to that to be expected from D absorption in the terrestrial atmosphere above the rocket, and from the amplitude an enhancement of a factor of 180 over the sea level ratio was determined. The Jacchia (1965) mean exospheric

temperature was $1040°K$ for the time of observation. The theoretical enhancement ratio at 160 km for $1040°K$ is 15, from Kockarts and Nicolet (1962).
The column enhancement ratio is not more than 30. Since the observation
was made at the limit of detectability, the dip was probably a noise feature.

9 DISCUSSION OF OBSERVATIONAL RESULTS AND OUTSTANDING PROBLEMS

9.1 Altitude profiles and absolute abundances

The observations of absorption core in the solar line are sensitive to hydrogen
densities in the 100-300 km altitude region, and yield daytime abundances
somewhat less than between 1.25 and 2.0 times the KN abundances, as do the
mesospheric OH and $O_2(^1\Delta_g)$ results. The rocket results for scattered Lyman α
altitude profiles also imply low altitude daytime abundances between 1 and 3
times KN values.

The satellite scattered Lyman α results are mainly sensitive to hydrogen
densities above 1000 km, since that is where most of the radiative transport
occurs, and on the nightside to hydrogen extending to above 10,000 km.
These measurements yield abundances about 3 times KN. It is difficult to
reconcile Balmer α and Lyman β results with abundances as low as 3 times
KN, but this can be done if a small allowance is made for the diurnal
variation and the solar Lyman β flux used is several times too low. Even so
the Lyman β height variation at low altitudes did not fit the calculation well.
The Balmer α and Lyman β results are also sensitive to hydrogen well above
1000 km.

The ion mass spectrometer, electron cooling, and incoherent scatter results
are sensitive to hydrogen at intermediate altitudes, between 300 and 1000 km,
and yield abundances between 3 and 8 times KN. The satellite drag results
for 2000-3000 km are about 4 times KN.

It is difficult to know how much of the discrepancies mentioned above are
due to inadequacies of the experimental techniques.

The observations have been evaluated using T_∞ values either measured
directly or from the Jacchia (1965) models. Since the seasonal variations do
not follow predictions based on the semiannual component of T_∞ in the
Jacchia (1965) models, it is better to use T_∞ values from the Jacchia (1971)
models which do not have a semiannual T_∞ variation. They also have better
representations of T_∞ solar activity changes, and improved $n(O)$ values
which are 15% or so higher. A reinterpretation of observational data in terms
of Jacchia (1971) values is probably not warranted, however, although it
might reduce some of the disagreements.

Considering all the results, it does seem that the daytime abundances from
100-300 km (1-3 times KN) are systematically lower than the abundances

between about 300-1000 km (3-8 times KN). The reason for this is not well understood. An appreciable concentration of H_2 (or H_2O) relative to H at 100 km would help to resolve the discrepancy (Section 2.2, and Tinsley 1969a). Note added in proof: An improved model of the mesospheric production and flow (Hunten and Strobel, *J. Atmos. Sci.*, 31, 305-317, 1974) gives $n(H_2)/n(H) = 1.2$ at 100 km, just as required.

Observations of polarization of Lyman α show promise of yielding values for abundances independent of the value of the solar flux. Observations of polarization of Lyman α at higher altitudes than 200 km, or of polarization of Lyman β or Balmer α would probably be even better, with a more rapid variation of polarization with abundance.

The altitude profiles above 10,000 km correspond in general to 3 times KN abundances, with $R_{sc} = 2.5R_c$, but show a number of unexplained variations.

9.2 Diurnal variation

A diurnal variation by about a factor of 1.7 at 350 km, with a few hours phase lag with respect to the temperature maxima and minima is consistent with the measurements and the approximate theoretical treatments for the exobase level. However, as discussed in Section 3, the theoretical treatments leave much to be desired. It is desirable to have calculations of time dependent effects based on solutions to the continuity equation for altitudes both above and below the exobase, with production and loss terms for charge exchange with oxygen ions included. A model for the diurnal variation of the vertical flow of H^+ would have to be assumed or calculated.

Some effort to include the escape of neutral hydrogen by charge exchange with hot hydrogen ions as a loss term should be included, and perhaps also the related effects of heating without escape (Section 4.3).

If the result of this calculation came near to the density distribution at the exobase for zero net lateral flow, then the correction for lateral flow effects would be small. The effects of lateral flow may be qualitatively evaluated by comparing the solution for zero net lateral flow (ZNLF) with the solution obtained without it (WLF). If the ZNLF solution has more diurnal variation than the WLF solution, and the phases approximately agree, then the effect of lateral flow is to increase the diurnal variation. If the ZNLF solution has less diurnal variation than the WLF solution, and the phases approximately agree, then the effect of lateral flow is to decrease the diurnal variation.

9.3 Sources and sinks

Further experimental and theoretical work is desirable on the mesospheric sources of atomic and molecular hydrogen, and on the seasonal variations in this. Donahue (1969a) has suggested an enhancement in production might occur in springtime in the polar regions. Johnson and Gottleib (1970) discuss horizontal flows in the atmosphere, which would also affect the chemistry The loss of hydrogen ions from the plasmasphere at the times of magnetic storms may be partly outward into the tail, and may be significant during periods of high magnetic activity.

9.4 Suggested hydrogen models

The discussion of altitude profiles given above is relevant to the choice of the best estimate of the absolute abundance of hydrogen in the upper atmosphere, for the purpose of construction of hydrogen model atmospheres. It would seem that for an altitude above 300 km about 3 times KN abundances would be a good weighted average of experimental results for the mean abundance at moderate solar activity. The T_∞ value for the KN model should be that of the mean diurnal T_∞ for the atmosphere, and a diurnal variation by a factor of 1.7 should be applied to this KN model. A phase lag of about 2 hours in the day and 4 hours at night with respect to the temperature maxima and minima would be a further refinement. Above 500 km the models could be continued with a Chamberlain (1963) profile with R_{sc} say $2.5R_c$, although the difference of this profile from the hydrostatic profile, or profiles with other R_{sc} , is not significant for several thousand kilometers.

The above model is very approximate. For the region for 60 km to about 200 km, the uncertainties appear too great to warrant even an approximate model at present.

10 IONIZATION OF THE NIGHTTIME D AND E REGIONS

Hydrogen in the upper atmosphere transports Lyman α and Lyman β by resonant scattering to the nightside, where they are important sources of ionization in the D and E regions. Of course the production rates for ionization are very much less than in the daytime, when the direct solar flux is much larger, but recombination quickly removes the daytime ionization, and the hydrogen emissions sustain ionization at a reduced level during the night.

Nicolet (1965) suggested that the nighttime E region could be maintained with electron density above 10^3 cm^{-3} by the ionization of nitric oxide NO by Lyman α. Swider (1965) discussed this and also the possibility of ionization of O_2 by Lyman β. Ogawa and Tohmatsu (1966) calculated that with isotropic fluxes of 4 kR of Lyman α and 10R of Lyman β at night the Lyman α ionization of NO was important only below the 100 km level, while the Lyman β ionization of O_2 was important above 100 km. The calculated electron densities and altitude profiles were in approximate agreement with those observed by ionosondes and rockets, with a peak value of about 3×10^3 cm^{-3} at 105 km.

Tohmatsu (1970) showed the diurnal variation of the observed electron density was also in approximate agreement with the diurnal variations of the Lyman α and Lyman β fluxes.

Improved calculations could now be made with improved values of the NO altitude profile and the absolute values of the Lyman α and Lyman β fluxes and their diurnal variation. The non-isotropic intensity distribution, especially for Lyman β, could also be taken into account by the use of the results of radiative transfer calculations.

Radiations of He II 304 (the He$^+$ Lyman α) and He I 584 are also found as a result of resonant scattering to the nightside, and contribute to the ionization at altitudes above 130 km.

11 EFFECT OF HYDROGEN LOSS ON OXYGEN EVOLUTION

The dissociation of water vapor in the mesosphere, loss of H to space, and consequent build up of oxygen in the earth's atmosphere was probably the main source of atmospheric oxygen in the period following the formation of the earth until significant production by plant life occurred.

The present loss rate may be taken as about 10^8 H atoms cm^{-2} sec^{-1}. This is actually the thermal escape rate corresponding to 4 times KN abundances. Other loss processes would increase this a little. If this loss rate had occurred for 4×10^9 years (say up to the Cambrian Era, 6×10^8 years ago) then 20 gm cm^{-2} of hydrogen would have been lost leaving 180 gm cm^{-2} of oxygen or 75% of the present atmosphere abundance. Little of this probably remained in the atmosphere however, since most of it would combine with reduced volcanic gases and reduced surface minerals.

If it is of considerable interest, with respect to theories of the evolution of life, to know what the oxygen abundance in the atmosphere was prior to the Cambrian Era. The earliest atmosphere had a much smaller abundance than the present one, which is considered to have almost entirely arisen since the formation of the earth by outgassing of the crustal material through volcanoes, hot springs, fumaroles, etc.

The study of this subject has a long history, and it has recently been reviewed by Johnson (1969). The composition of the material outgassed, is and probably always has been mildly reducing, thus no free oxygen is released, and the oxygen abundance would initially have been very small.

Molecular oxygen is a strong absorber of solar radiation in the wavelength range capable of dissociating water vapor (mostly $\lambda < 1840$A, but weakly out to 2000A). Thus the production of oxygen by water vapor dissociation is a self limiting process (Urey 1959), since as the oxygen builds up in the atmosphere it shields the water vapor from further photodissociation. There is some question as to how effective a self limiting process this is. Berkner and Marshall (1964, 1965) have assumed that after the oxygen abundance reaches one thousandth of the present atmospheric level there is no further significant production. Brinkman (1969) on the other hand has calculated that a production rate of 1.6×10^9 oxygen atoms cm^{-2} sec^{-1} should occur with the *present* level of oxygen shielding, which would allow a relatively rapid increase, doubling the present level in 5×10^8 years. He assumed however that no more than half of the H_2O dissociated then recombined, which is unrealistic.

A more reliable approach is to refer to the measurements on upper atmosphere hydrogen, which yield the escape rate of about 10^8 cm^{-2} with the present shielding. It can be seen that Brinkman's rates are too high, but nevertheless self limitation is not effective with even the present abundances.

To evaluate the rate of build up of oxygen before the development of widespread photosynthetic organisms it is necessary to estimate the rate of production by photodissociation in the early atmosphere, and the rate of loss due to oxidation of surface minerals and oxidation of reduced gases emanating from volcanoes. Brinkman (1969) calculated models for the build up on the assumption that the rate of oxidation of surface minerals was proportional to the partial pressure of oxygen. This may also be a very inaccurate assumption. He neglected the oxidation of volcanic gases which as discussed by Van Valen (1971) would appear to be the most important source of reduced material. Furthermore, unless the oxygen abundance is very small, most of the volcanic gases get completely oxidized in a relatively short time, so the atmospheric content of oxygen is less stable against this loss process than it is against loss by surface oxidation.

The whole problem of the stability of the atmosphere content of O_2 is a difficult one, and as Van Valen (1971) has pointed out, it is surprising that at some time since the origin of advanced forms of life the O_2 content has not dropped so low that they have nearly all been extinguished. The regulation of O_2 abundance provided by the increase in mesospheric photodissociation of H_2O as the O_2 abundance decreases may be more important than considered by Van Valen. In his model calculations Brinkman (1969) has used a photodissociation rate for H_2O which increases 10 times as the O_2 abundance drops to 1/16 of the present atmospheric level. While the recombination of the photodissociation products would proceed faster with increased rate of dissociation, there would be a net increase in abundance of H in the source region, and the upward flow and escape would increase in direct proportion.

There are many other variables that would affect the mesospheric production, including the limitation provided by the rate at which water vapor can pass through the stratospheric "cold trap" with temperatures around $200°K$. This in turn depends on the vigor of vertical mixing and large scale circulation processes, and the temperature of the "cold trap".

When the oxygen abundance is so low that there is no ozone layer in the mesosphere, and the composition is otherwise the same, the "cold trap" would be colder. McElroy and Hunten (1969a) have argued that an average escape rate of 10^{11} cm^{-2} sec^{-1} over geological time might have occurred for the Venus atmosphere. This atmosphere consists almost entirely of CO_2 and has no cold trap. Possibly the terrestrial atmosphere was similar, at one period in its history.

The effect of the cold trap may be bypassed by the upward transport of methane through it, as suggested by Bates and Nicolet (1965). If the source of the CH_4 is effectively in the reaction

$$CO_2 + 2H_2O \rightarrow CH_4 + 2O_2 \qquad (113)$$

such as occurs in some primitive organic processes, then the reactions of CH_4 above the cold trap and eventual loss of 4H constitute a net gain of O_2 to the atmosphere. If however, the source of CH_4 is in volcanic emanations, there is a net loss of O_2, since the reactions above the cold trap are effectively

$$CH_4 + O_2 \rightarrow CO_2 + 4H \qquad (114)$$

Yčas (1972) has pointed out that the methane and accompanying H_2 produced by the primitive anaerobic ecological system, that is considered to have existed prior to the presence of widespread photosynthetic organisms, could have been an important source of atmospheric O_2 as discussed above.

Further studies of the effectiveness of this and the photodissociative source, for various possible conditions in the early history of the earth, would be valuable.

12 H AND D AROUND OTHER PLANETS

12.1 General

Hydrogen atmospheres extending to high altitudes above Venus and Mars have been found by measurements of scattered Lyman α, as was originally the case for the terrestrial atmosphere. For Venus measurements were made by Barth et al. (1967, 1968) from Mariner 5, and by Kurt et al. (1968) from Venera 4. Also rocket observations have been made by Moos et al. (1969) and Moos and Rottman (1971). For Mars observations have been made from Mariners 6, 7 and 9 (Barth et al. 1971, 1972), and from Mars 2 and Mars 3 (Dementyeva et al. 1972). For both planets altitude profiles of Lyman α were obtained. The same basic theory as has been applied to the terrestrial atmosphere is used to calculate abundances and altitude profiles for H, D, H_2, and other hydrogen compounds around Venus and Mars. One important difference is that the dominant neutral constituent is CO_2 in the atmospheres of both planets, instead of the N_2, O_2, and O of the terrestrial upper atmosphere.

Scattered Lyman α has been observed from Jupiter by Moos et al. (1969), with a detector on a rocket which integrated the emission from the whole dayside of the planet. For Jupiter the exospheric temperature is so low, and the gravitational potential barrier so great that thermal escape is negligible.

12.2 Venus

The altitude profile of the Lyman α emission from Venus measured by Barth et al. (1967) was puzzling in that the data could not be fitted to scattering on an H distribution with a single scale height, but appeared to necessitate a second source at lower altitudes, characterized by a scale height which was a factor of 2 or 3 lower than expected for scattering on H (Barth et al. 1968, Barth 1968, Wallace 1969). Various suggestions have been put forward to explain this. Barth (1968) suggested that the smaller scale height component was Lyman α from the photodissociation of molecular hydrogen by solar radiation <850Å, but this has been seriously questioned by Donahue (1968b) and McElroy and Hunten (1969a, b). The presence of deuterium in greater abundance than mass 1 hydrogen at lower exospheric altitudes has been shown to be possible by Donahue (1969b) and McElroy and Hunten (1969a), but a high spectral resolution rocket measurement by Wallace et al. (1971) showed only the H component, and inadequate emission at the D wavelength for an explanation of the Mariner 5 scale height measurements. Possible sources of a "two temperature" distribution of H atoms were discussed by Barth (1968) but were considered improbable by McElroy and Hunten (1969a).

The effect of the large temperature differences that should exist between the dayside and nightside exobase regions and the lateral flow of hydrogen between the dayside and nightside is probably quite important. A large temperature difference should exist whether the upper atmosphere rotates with a 4 day period or a 117 day period (Dickinson 1971). The number of collisions needed for a H atom to approach thermal equilibrium with the CO_2 atmosphere is of the order of 10, on account of the 44 to 1 mass ration between CO_2 and H. This effect would allow lateral flow more effectively to bring in nightside "low temperature" H atoms and remove "high temperature" atoms from the dayside. There would be little loss by thermal escape from the nightside. But if the rotation period is indeed 117 days then the nightside probably also has very little production of H, and with low H abundances near the turbopause would act as a sink for H by downward flow. Thus, contrary to the situation on earth, the nightside may have a lower exobase

density than the dayside, as seems necessary to fit the Mariner 5 observations
(Wallace 1969). At any rate there could well be large gradients in hydrogen
temperature and abundance in the Venus exosphere, and it is possible that
these would have significant effects on the altitude profile. It should be
noted that it is not necessary to have just two sources with two temperatures
to obtain the sort of profile observed by Mariner 5, since many sources with
many temperatures would do as well.

12.3 Mars

The Mariner 6 and 7 and Mars 2 and 3 measurements showed that an atomic
hydrogen corona was present around Mars, and in contrast to the observations
on Venus, showed an altitude profile that was consistent with a single scale
height for hydrogen. The Mariner data have been analysed by Anderson and
Hord (1971), and the "Mars" data by Dementyeva et al. (1972). An
exospheric temperature of $350 \pm 100°K$ and a number density at 350 km of
$3 \pm 1 \times 10^4$ cm^{-3} were derived from the time of the Mariner 6 and 7 flyby.
These yield a thermal escape flux of 1.8×10^8 cm^{-2} sec^{-1}. Two years later at
the time of the Mars 2 and 3 flybys the temperature was again 350°K but the
number density at 250 km was apparently 6×10^3 cm^{-3}.

The photochemistry of water vapor and the production and escape of
atomic hydrogen on Mars was first considered by Biutner (1959b). More
comprehensive treatments have been made by McElroy and Hunten (1969b)
and Hunten and McElroy (1970). Spectroscopic observations have shown
water vapor to be present, and a photodissociation rate of perhaps
10^{10} cm^{-2} sec^{-1} was calculated to apply as a planetary average. H_2 was
found to dominate over H up to about 100 km but the escape flux, consistent
with the Mariner 6 and 7 observations, was almost entirely H. In the absence
of widespread photosynthesis as a source of oxygen on Mars, the major source
of free oxygen is from photodissociation of water vapor. Over 4×10^9 years,
production at the present rate would be 320 gm cm^{-2}, which is more than
10^4 times the present content of the Martian atmosphere.

An important loss mechanism for the oxygen is escape following dissociative
recombination of O_2^+, or CO_2^+, i.e.

$$O_2^+ + e \rightarrow O(\text{fast}) + O(\text{fast}) \tag{115}$$

and
$$CO_2^+ + e \rightarrow CO(\text{fast}) + O(\text{fast}) \tag{116}$$

The oxygen atoms produced in (115) and (116) have enough energy to escape
from Mars (but not from the Earth or Venus).

As discussed by McElroy (1972) and McElroy and Donahue (1972), the
calculated rate of escape of O is close to half the escape rate of H inferred by
Anderson and Hoord (1971). The constancy of this ratio may be maintained

by feedback effects on processes involving CO recombination and oxyhydrogen photochemistry, which tie the O loss rate to the H loss rate, preventing a buildup of O_2 on Mars.

12.4 Jupiter

The dominant atmospheric constituent in the atmosphere of Jupiter is probably H_2, although the relative abundance of He has not yet been determined. Above the Jovian turbopause the abundance of H_2 will increase relative to He, and there is a plentiful supply of H_2 for photodissociation and production of H. But since thermal escape is negligible on account of the large gravitational potential barrier and low exospheric temperature, the H produced must diffuse downward to recombine at lower levels. The composition and structure of the upper atmosphere of Jupiter has been reviewed by Gross and Rasool (1964), and Hunten (1969). While the H from photodissociation of H_2 dominates in the region higher than about 100 km above turbopause, the photolysis of methane (Strobel 1969) produces comparable concentrations at a slightly lower altitude. The methane absorbs Lyman α up to about 110 km, and the atomic hydrogen above that altitude resonantly scatters solar Lyman α.

Lyman α is also produced by photodissociation of H_2, as suggested by Barth (1968) for Venus. Carlson and Judge (1971) calculate that about $300R$ of Lyman α would be produced in this way.

The Lyman α emission from Jupiter measured by Moos et al. (1969) was equivalent to a uniform disc of brightness $4kR$. Hunten (1969) derived an H column abundance above a level taken as 110 km above the turbopause of 8×10^{15} cm^{-2}, by correcting the estimate of Moos et al. (1969). Essentially the same value was obtained by Carlson and Judge (1971). From this column abundance, in comparison with model altitude profiles for different diffusion coefficients, Hunten (1969) deduced an approximate value for the eddy diffusion coefficient K of 5×10^6 cm^2 sec^{-1}, about the same value as is thought to apply at comparable levels for the Earth.

Deuterium has been seen on Jupiter as CH_3D in observations made with a Michelson interferometer (Beer et al. 1972). A D/H ratio by number of 2.9 to 7.5×10^{-5} has been evaluated; the range of uncertainty reflecting a range of possible models for the Jovian atmosphere. The possible D/H values are all lower than the terrestrial value of 1.5×10^{-4}, presumably because of enrichment of deuterium in the inner solar system at the time of its formation (Geiss and Reeves 1972).

Acknowledgements

I wish to thank Dr F.S. Johnson for reading the manuscript and for helpful comments, and together with Dr W.B. Hanson, for many useful discussions on the subject.

This work was supported by NSF grants GA 18767 and GA 33262X and NASA Institutional Grant MGL 44-004-001.

References

Aamodt, R.E., and K.M. Case (1962). *Phys. Fluids*, **5**, 1019-1021.
Anderson, D.E., Jr., and C.W. Hord (1971). *J. Geophys. Res.*, **76**, 6666-6673, 1971.
Armstrong, E.B. (1967). *Planet. Space Sci.*, **15**, 407-425.
Avrett, E.H. (1970). *Studies of the Upper Chromosphere and Lower Transition Region*, paper presented at 2nd Oso Workshop, Goddard Space Flight Center, Greenbelt, Maryland, December 2, 1970.
Axford, W.I. (1968). *J. Geophys. Res.*, **73**, 6855-6859.
Axford, W.I. (1972). *The Interaction of the Solar Wind with the Interstellar Medium*, in Proceedings of the Solar Wind Conference, Asilomar, Pacific Grove, California, March, 1971. NASA SP308.
Banks, P.M. (1968). *Planet. Space Sci.*, **16**, 759-774.
Banks, P.M., and T.W. Holzer (1968). *J. Geophys. Res.*, **73**, 6846-6854.
Banks, P.M., and T.E. Holzer (1969a). Reply (to A.J. Dessler and P.A. Cloutier), *J. Geophys. Res.*, **74**, 3734-3739.
Banks, P.M., and T.E. Holzer (1969b). *J. Geophys. Res.*, **74**, 6304-6316.
Banks, P.M., and T.E. Holzer (1969c). *J. Geophys. Res.*, **74**, 6317-6332.
Banks, P.M., A.F. Nagy, and W.I. Axford (1971). *Planet. Space Sci.*, **19**, 1053-1067.
Barth, C.A. (1968). *J. Atmos. Sci.*, **25**, 564-567.
Barth, C.A., J.B. Pearce, K. Kelly, L. Wallace, and W.G. Fastie (1967). *Science*, **158**, 1675-1678.
Barth, C.A., L. Wallace, and J.B. Pearce (1968). *J. Geophys. Res.*, **73**, 2541-2545.
Barth, C.A., C.W. Hord, J.B. Pearce, K.K. Kelly, G.P. Anderson, and A.I. Stewart (1971). *J. Geophys. Res.*, **76**, 2213-2227.
Barth, C.W., C.W. Hord, A.I. Stewart, and A.L. Lane (1972). *Science*, **175**, 309-312.
Bates, D.R. (1959). *Proc. Roy. Soc.*, **A253**, 451-463.
Bates, D.R., and M. Nicolet (1950). *J. Geophys. Res.*, **55**, 301-327.
Bates, D.R., and M. Nicolet (1965). *Planet. Space Sci.*, **13**, 905-909.
Bates, D.R., and T.N.L. Patterson (1961a). *Planet. Space Sci.*, **5**, 257-273.
Bates, D.R., and T.N.L. Patterson (1961b). *Planet. Space Sci.*, **5**, 328.
Beer, R., C.B. Farmer, R.H. Norton, J.V. Martonchik, and T.G. Barnes (1973). *Science*, **175**, 1360-1361.
Berkner, L.V., and L.C. Marshall (1964). *Disc. Faraday, Soc.*, **37**, 122-141.
Berkner, L.V., and L.C. Marshall (1965). *J. Atmos. Sci.*, **22**, 225-261.
Bertaux, J.L., and J.E. Blamont (1973). *J. Geophys. Res.*, **78**, 80-91.
Biutner, E.K. (1958). *Astron. J. USSR*, **35**, 572-582. (Eng. trans. *Soviet Astronomy* – **AJ, 2**, 528-537, 1958.)
Biutner, E.K. (1959a). *Astron. J. USSR*, **36**, 89-99. (Eng. trans. *Soviet Astronomy* – **AJ, 3**, 92-102, 1959.)
Biutner, E.K. (1959b). Proc. Acad. Sci. *USSR (Phys.)*, **124**, 53-56. (Eng. trans. *Soviet Phys.* – Doklady 4, 3-7, 1959.)
Blamont, J.E., and A. Vidal Madjar (1971). *J. Geophys. Res.*, **76**, 4311-4324.
Bowman, M.R., L. Thomas, and J.E. Geisler (1970). *J. Atmos. Terr. Phys.*, **32**, 1661-1674.
Brace, L.H., B.M. Reddy, and H.G. Mayr (1967). *J. Geophys. Res.*, **72**, 265-283.
Brandt, J.C. (1961a). *Astrophys. J.*, **133**, 688-700.
Brandt, J.C. (1961b). *Astrophys. J.*, **134**, 394-399.
Brandt, J.C. (1961c). *Space Res.*, **2**, 624-638.
Brandt, J.C. (1962a). *Planet. Space Sci.*, **9**, 67-78.
Brandt, J.C. (1962b). *Nature*, **195**, 894-895.
Brandt, J.C., and J.W. Chamberlain (1959). *Ap. J.*, **130**, 670-682.
Brandt, J.C., and J.W. Chamberlain (1960). *Phys. Fluids*, **3**, 485-486.
Brinkman, R.T. (1969). *J. Geophys. Res.*, **74**, 5355-5368.
Brinkman, R.T. (1970). *Planet. Space Sci.*, **18**, 449-478.
Brinkman, R.T. (1971). *Planet. Space Sci.*, **19**, 791-794.
Brinton, H.C., M.W. Pharo III, H.G. Mayr, and H.A. Taylor (1969). *J. Geophys. Res.*, **74**, 2941-2951.
Brinton, H.C., and H.G. Mayr (1971). *J. Geophys. Res.*, **76**, 6198-6201.
Brinton, H.C., and H.G. Mayr (1972). *Space Res.*, **12**, 751-764.

Bruner, E.C., Jr., and R.W. Parker (1969). *J. Geophys. Res.*, 74, 107-113.
Bruner, E.C., Jr., and W.A. Rense (1969). *Astrophys. J.*, 157, 417-424.
Bruner, E.C., and T.E. Wilson (1969). *J. Geophys. Res.*, 74, 6491-6493.
Bryan, G.H. (1901). *Phil. Trans. Roy. Soc. Lond.*, A196, 1-24.
Byram, E.T., T.A. Chubb, H. Friedman, and J. Kupperian (1957). *Far Ultraviolet Radiation in the Night Sky*, pp.203-210. In M. Zelikoff (Ed), *The Threshold of Space*, Pergamon Press.
Buckley, J.L., H.W. Moos, and R.R. Meier (1971). *J. Geophys. Res.*, 76, 2437-2440.
Carlson, R.W., and D.L. Judge (1971). *Planet. Space Sci.*, 19, 327-343.
Chamberlain, J.W. (1960). *Astrophys. J.*, 131, 47-56.
Chamberlain, J.W. (1963a). *Planet. Space Sci.*, 11, 901-960.
Chamberlain, J.W. (1963b). *Science*, 142, 921-924.
Chamberlain, J.W. (1964). *The Geocorona*. In H. Odishaw (Ed), *Research in Geophysics, 1. Sun, Upper Atmosphere and Space*, MIT Press.
Chamberlain, J.W. (1969). *Astrophys. J.*, 155, 711-714.
Chamberlain, J.W., and F.J. Campbell (1967). *Astrophys. J.*, 149, 687-705.
Chamberlain, J.W., and G.R. Smith (1971). *Planet. Space Sci.*, 19, 675-684.
Chapman, S., and T.G. Cowling (1952). *The Mathematical Theory of Non-Uniform Gases*, 2nd Ed., Cambridge Univ. Press, p.244.
Chapman, S., E.A. Milne (1920). *Quart. J. Roy. Meteor. Soc.*, 46, 357-398.
Chubb, T.A., and E.T. Byram (1963). *Space Res.*, 3, 1046-1060.
Cole, K.D. (1966). *Nature*, 211, 1385-1387.
Colegrove, F.D., F.S. Johnson, and W.B. Hanson (1965). *J. Geophys. Res.*, 70, 4931-4941.
Colegrove, F.D., F.S. Johnson, and W.B. Hanson (1966). *J. Geophys. Res.*, 71, 2227-2236.
Cook, S.R. (1900). *Astrophys. J.*, 11, 36-43.
Dalgarno, A. (1960). *Proc. Phys. Soc. Lond.*, 75, 374-377.
Dementyeva, N.N., V.G. Kurt, A.S. Smirnov, L.G. Titarchuk, and S.D. Chuvahin (1972). *Preliminary Results of Measurements of UV Emission Scattered in the Martian Upper Atmosphere*, Preprint, Institute for Space Research, Academy of Sciences of the USSR, April, 1972.
Dessler, A.J., and P.A. Cloutier (1969). Discussion of a letter by Peter M. Banks and Thomas E. Holzer "*The Polar Wind*", *J. Geophys. Res.*, 74, 3730-3733.
Dickinson, R.E. (1971). *J. Atmos. Sci.*, 28, 885-894.
Donahue, T.M. (1962). *Space Sci. Rev.*, 1, 135-153.
Donahue, T.M. (1964a). *Planet. Space Sci.*, 12, 149-159.
Donahue, T.M. (1964b). *J. Geophys. Res.*, 69, 1301-1306.
Donahue, T.M. (1966). *Ann. Geophys.*, 22, 175-188.
Donahue, T.M. (1967). *Planet. Space Sci.*, 15, 1531-1534.
Donahue, T.M. (1968a). In S.K. Runcorn (Ed), *International Dictionary of Geophysics*, Pergamon Press.
Donahue, T.M. (1968b). *J. Atmos. Sci.*, 25, 568-573.
Donahue, T.M. (1969a). *J. Geophys. Res.*, 74, 3717-3719.
Donahue, T.M. (1969b). *J. Geophys. Res.*, 74, 1128-1137.
Donahue, T.M., and W.G. Fastie (1964). *Space Res.*, 4, 304-324.
Donahue, T.M., and J.B. Kumer (1971). *J. Geophys. Res.*, 76, 145-162.
Donahue, T.M., and J.R. McAfee (1964). *Planet. Space Sci.*, 12, 1045-1054.
Donahue, T.M., and R.R. Meier (1967). *J. Geophys. Res.*, 72, 2803-2829.
Donahue, T.M., and G.E. Thomas (1963a). *Planet. Space Sci.*, 10, 65-72.
Donahue, T.M., and T.E. Thomas (1963b). *J. Geophys. Res.*, 68, 2661-2667.
Dufay, M., and J. Dufay (1960). *Comptes Rendus Acad. Sci.*, 250 (No.25), 4191-4193.
Dufay, J., M. Dufay, and Nguyen Huu-Doan (1961). *Comptes Rendus Acad. Sci.*, 253, 974-977.
Dungey, J.W. (1955). *The Physics of the Ionosphere*, p.229, Physical Society, London.
Emden, R. (1907). *Gaskugeln*, p.270, Teubner Leipzig.
Epstein, P.S. (1932). *Beitr. Geophys.*, 35, 153-165.
Evans, J.V. (1971). *Radio Sci.*, 6, 855-861.
Evans, W.T.J., and E.J. Llewellyn (1973). *J. Geophys. Res.*, 78, 323-326.
Farley, D.T., J.P. McClure, D.L. Sterling, and J.L. Green (1967), *J. Geophys. Res.*, 72, 5837-5851.

Fastie, W.G. (1968). *Planet. Space Sci.,* **16**, 929-935.
Fastie, W.G., H.M. Crosswhite, and D.F. Heath (1964). *J. Geophys. Res.,* **69**, 4129-4140.
Fishkova, L.M. (1962). *Bull. Abastumani Astrophys. Observatory,* **29**, 77-91.
Fishkova, L.M. (1963a). *Aurora and Airglow,* **10**, 35-39. (Results of IGY USSR Acad. Sci.)
Fishkova, L.M. (1963b). *Astron. Circular USSR,* **253**, 1-4.
Fishkova, L.M., and G.V. Markova (1958). *Astron. Circular USSR,* **196,** 8-9.
Fishkova, L.M., and G.V. Markova (1960a). *Astron. Circular USSR,* **208**, 14-15.
Fishkova, L.M., and G.V. Markova (1960b). *Doklady USSR Acad. Sci.,* **134** (No.4), 799-801. *Soviet Physics Doklady* (trans.), **5**, 1042-1044, 1961.
Fishkova, L.M., and N.M. Martzvaladze (1964). *Brit. Ast. Ass. J.,* **74**, 162-165.
Fishkova, L.M., and N.M. Martzvaladze (1967a). *Aurora and Airglow,* **13**, 69-72. (Results of IGY *USSR Acad. Sci.*)
Fishkova, L.M., and N.M. Martzvaladze (1967b). *Geomagnetism and Aeronomy,* **7**, 1021-1025. (Eng. Trans., **7**, 828-831, 1967.)
Fishkova, L.M., and N.M. Martzvaladze (1972). *On the variations and spatial distribution of the upper atmosphere Hα emission,* paper presented at International Symposium on Solar Terrestrial Physics, Leningrad 1970. (See also *Bull. Abastumani Astrophys. Obs.,* **42**, 29-38 and 131-181, 1972.)
Fishkova, L.M., and P.V. Shcheglov (1969a). *Astron. Circular USSR,* **505**, 5-8.
Fishkova, L.M., and P.V. Shcheglov (1969b). *Astron. Circular USSR,* **516**, 6-7.
Fishkova, L.M., N.M. Martzvaladze, and P.V. Shcheglov (1970). *Astron. Circular USSR,* **555**, 7-8.
Fontheim, E.G., and P.M. Banks (1972). *Planet. Space Sci.,* **20**, 73-80.
Gaynullina, R.H., and Z.V. Karagina (1960). Izv. Astrophys. Institute Kazakh. *S.S.R. Acad. Sci.,* **10**, 52-63.
George, J.D., S.P. Zimmerman, and T.J. Keneshea (1972). *Space Res.,* **12**, 695-709.
Geisler, J.E. (1967). *J. Geophys. Res.,* **72**, 81-85.
Geiss, J., and H. Reeves (1972). *Astron. Astrophys.,* **18**, 126-132.
Gilvarry, J.J. (1961a). *Phys. Fluids,* **4**, 2-7.
Gilvarry, J.J. (1961b). *Phys. Fluids,* **4**, 8-12.
Grew, K.E., and T.L. Ibbs (1952). *Thermal Diffusion in Gases,* Cambridge Univ. Press.
Gross, S.H., and S.I. Rasool (1964). *Icarus,* **3**, 311-322. (Also erratum, *Icarus,* **4**, 10, 1965.)
Hall, L.A., J.E. Higgins, C.W. Chagnon, and H.E. Hinteregger (1969). *J. Geophys. Res.,* **74**, 4181-4183.
Hagenbuch, K.M., and R.E. Hartle (1969). *Phys. Fluids,* **12**, 1551-1559.
Hampson, J. (1964). Canad. Armament Res. Develop. Estab. Tech. Note 1627/64, Carde, Quebec, Canada.
Hann, J. (1903). *Meteorolog Zeitschrift,* **20**, 122-126.
Hanson, W.B., and I.B. Ortenburger (1961). *J. Geophys. Res.,* **66**, 1425-1435.
Hanson, W.B., and T.N.L. Patterson (1963). *Planet. Space Sci.,* **11**, 1035-1052.
Hanson, W.B., T.N.L. Patterson, and S.S. Degaonkar (1963). *J. Geophys. Res.,* **68**, 6203-6205.
Hanson, W.B., A.F. Nagy, and R.J. Moffett (1973). *J. Geophys. Res.,* **78**, 751-756.
Hartle, R.E. (1971). *Phys. Fluids,* **14**, 2592-2598.
Hays, P.B., and Y.C. Liu (1965). *Planet. Space Sci.,* **13**, 1185-1212.
Heath, D.F. (1967). *Astrophys. J.,* **148**, L97-L100.
Herring, J., and L. Kyle (1961). *J. Geophys. Res.,* **66**, 1980-1982.
Hesstvedt, E. (1964). *Geofys. Publi. Oslo,* **25** (No.3), 1-18.
Hesstvedt, E. (1965a). *Tellus,* **17**, 341-349.
Hesstvedt, E. (1965b). *Geofys. Publi. Oslo,* **26** (No.1).
Hesstvedt, E. (1968). *Geofys. Publi. Oslo,* **27** (No.4), 1-35.
Hessvedt, E. (1969). *A Photochemical Atmospheric Model Containing Oxygen, Hydrogen, and Nitrogen,* University of Oslo, Institute of Geophysics, Technical Report, December 1969.
Ho, M.C., and D.R. Moorcroft (1971). *Planet. Space Sci.,* **19**, 1441-1455.
Hodges, R.R. (1973). *J. Geophys. Res.,* **78**, 7340-7346.
Hodges, R.R., and F.S. Johnson (1968). *J. Geophys. Res.,* **73**, 7307-7317.
Hoffman, J.H. (1967). *J. Geophys. Res.,* **72**, 1883-1888.
Hoffman, J.H. (1969). Proc. *IEEE,* **57**, 1063-1067.
Hoffman, J.H. (1970). Personal communication.

Holstein, T. (1947). *Phys. Rev.*, **72**, 1212-1233.
Hulbert, E.O. (1937). *Rev. Mod. Phys.*, **9**, 44-68.
Humphreys, W.J. (1910). *Mount Weather Obs. Bull.*, **2** (No.2), 66-69.
Hunt, B.G. (1966). *J. Geophys. Res.*, **71**, 1383-1398.
Hunten, D.M. (1969). *J. Atmos. Sci.*, **26**, 826-834.
Hunten, D.M. (1971). *Comm. on Astrophys. and Space Phys.*, **3**, 1-6.
Hunten, D.M., and M.B. McElroy (1970). *J. Geophys. Res.*, **75**, 5989-6001.
Ingham, M.F. (1962). *Mon. Not. Roy. Ast. Soc.*, **124**, 523-532.
Ingham, M.F. (1968). *Mon. Not. Roy. Ast. Soc.*, **140**, 155-172.
Ingham, M.F. (1969). *Mon. Not. Roy. Ast. Soc.*, **145**, 401-404.
Jacchia, L.G. (1965). *Smithsonian Contrib. Astrophys.*, **8**, 215-257.
Jacchia, L.G. (1966). In *U.S. Standard Atmosphere Supplements*, p.37, Superintendent of Documents, U.S. Government Printing Office, Washington, D.C.
Jacchia, L.G. (1971). Smithsonian Astrophysical Observatory Special Report, **332**.
Jeans, J.H. (1904). *The Dynamical Theory of Gases*, Cambridge Univ. Press. (First Ed., 1904; Second Ed., 1916; Third Ed., 1921; Fourth Ed., 1925.)
Jeans, J.H. (1910). *Mount Weather Obs. Bull.*, **2** (No.6), 347-356.
Johnson, F.S. (1959a). Technical Report LMSD-49719, Missiles and Space Division, Lockheed Aircraft Corp., April 1959.
Johnson, F.S. (1959b). *Nature*, **184**, 1787.
Johnson, F.S. (1961). *Astrophys. J.*, **133**, 701-705.
Johnson, F.S. (1969). *Space Science Rev.*, **9**, 303-324.
Johnson, F.S., and R.A. Fish (1960). *Astrophys. J.*, **131**, 502-515.
Johnson, F.S., and B. Gottleib (1970). *Planet. Space Sci.*, **18**, 1707-1718.
Jones, J.E. (1923). *Trans. Camb. Phil. Soc.*, **22** (No.28), 535-556.
Jones, R.A. (1970). *J. Geophys. Res.*, **75**, 6966-6988.
Jones, R.A., E.C. Bruner, Jr., and W.A. Rense (1970). *J. Geophys. Res.*, **75**, 1849-1853.
Joseph, J.H., and S.V. Venkateswaran (1966). *Annales de Geophysique*, **22**, 553-576.
Kaplan, S.A., and V.G. Kurt (1965). *Kosmicheskie Issledovaniya*, **3**, 251-256. *Cosmic Research* (Transl.), **3**, 174-178, 1965.
Kaplan, S.A., V.V. Katyashina, and V.G. Kurt (1965). *Space Res.*, **5**, 595-611.
Karjagina, Z.V., V.E. Mozjaeva, and P.V. Shcheglov (1969a). *Astron. Circular USSR*, **502**, 5-7.
Karjagina, Z.V., V.E. Mozjaeva, and P.V. Shcheglov (1969b). *Astron. Circular USSR*, **516**, 4-6.
Karjagina, Z.V., V.E. Mozjaeva, and L.M. Fishkova (1969c). *Astron. Circular USSR*, **516**, 7-8.
Karjagina, Z.V., V.E. Mozjaeva, and P.V. Shcheglov (1969d). *Astron. Circular USSR*, **532**, 7-8.
Keneshea, T.J., and S.P. Zimmerman (1970). *J. Atmos. Sci.*, **27**, 831-840.
Kockarts, G. (1971). In F. Vernani (Ed), *Physics of the Upper Atmosphere*, Editrice Compositori, Bologna, Italy.
Kockarts, G. (1972). *Space Res.*, **12**, 1047-1050.
Kockarts, G., et M. Nicolet (1962). *Ann. Geophys.*, **18**, 269-290.
Kockarts, G., et M. Nicolet (1963). *Ann. Geophys.*, **19**, 370-385.
Konashenok, V.N. (1967). *Izv. Atmos. and Oceanic Phys.*, **3**, 227-235. (Eng. trans., **3**, 129-133, 1967.)
Kondo, Y., and J.E. Kupperian, Jr. (1967). *J. Geophys. Res.*, **72**, 6091-6097.
Krassovsky, V.I. (1971). *Ann. Geophys.*, **27**, 211-221.
Krassovsky, V.I., N.N. Shefov, and O.L. Vaisberg (1966). *Ann. Geophys.*, **22**, 208-216.
Kurt, V.G. (1963). *Usp. Fiz. Nauk.*, **81**, 249-270. *Soviet Physics Uspekhi* (transl.), **6**, 701-714, 1964.
Kurt, V.G. (1966). *Kosmicheskie Issledovaniya*, **4**, 111-115. *Cosmic Res.* (transl.), **4**, 98-102, 1966.
Kurt, V.G. (1967). *Kosmicheskie Issledovaniya*, **5**, 911-920. *Cosmic Res.* (transl.), **5**, 770-777, 1967.
Kurt, V.G., S.B. Dostovalov, and E.K. Shefler (1968). *J. Atmos. Sci.*, **25**, 668-671.

Kupperian, J.E., Jr., E.T. Byram, T.A. Chubb, and H. Friedman (1958). *Ann. Geophys.*, **14**, 329-333.

Kupperian, J.E., Jr., E.T. Byram, T.A. Chubb, and H. Friedman (1959). *Planet. Space Sci.*, **1**, 3-6.

Kvifte, G. (1959a). *J. Atmos. Terr. Phys.*, **16**, 252-258.

Kvifte, G. (1959b). *Geophys. Publi. Oslo*, **20** (No.12), 1-15.

Lettau, H. (1951). In T.F. Malone (Ed), *Compendium of Meteorology*, American Meteorological Society.

Lew, S.K., and S.V. Venkateswaran (1965). *J. Atmos. Sci.*, **22**, 623-635.

Lew, S.K., and S.V. Venkateswaran (1966). Reply (to M. Liwshitz), *J. Atmos. Sci.*, **23**, 817-819.

Lindzen, R.S. (1970). *Geophysical Fluid Dynamics*, **1**, 303-355.

Lindzen, R.S. (1971). *Geophysical Fluid Dynamics*, **2**, 89-121.

Lindzen, R.S., and D. Blake (1970). *Geophysical Fluid Dynamics*, **2**, 31-61.

Lindzen, R.S. (1971). In G. Fiocco (Ed), *Mesopheric Models and Related Experiments*, Reidel Publ. Co. Holland.

Liwshitz, M. (1966). *J. Atmos. Sci.*, **23**, 816-817.

Liwshitz, M. (1967). *J. Geophys. Res.*, **72**, 285-293.

Liwshitz, M., and S.F. Singer (1966). *Planet. Space Sci.*, **14**, 541-561.

Mange, P. (1955). *Ann. Geophys.*, **11**, 153-168.

Mange, P. (1960). *J. Geophys. Res.*, **65**, 3833-3834.

Mange, P. (1961). *Ann. Geophys.*, **17**, 277-291.

Mange, P. (1972). In E.R. Dyer (Ed), *Solar-Terrestrial Physics/1970: Part IV*, Reidel Publ. Co. Holland.

Mange, P., and R.R. Meier (1970). *J. Geophys. Res.*, **75**, 1837-1842.

Martzvaladze, N.M. (1967). *Astron. Circular USSR*, **427**, 1-4.

Martzvaladze, N.M. (1968). *Bull. Abastumani Astrophys. Observatory*, **36**, 82-92.

Martzvaladze, N.M., L.M. Fishkova, and N.N. Shefov (1971). *Astron. Circular*, **619**, 5-6.

Maris, H.B. (1928). *Terr. Mag.*, **33**, 233-255.

Maris, H.B. (1929). *Terr. Mag.*, **34**, 45-53.

McAfee, J.R. (1967). *Planet. Space Sci.*, **15**, 599-610.

McElroy, M.B. (1972). *Science*, **175**, 443-445.

McElroy, M.B., and T.M. Donahue (1972). *Science*, **177**, 986-988.

McElroy, M.B., and D.M. Hunten (1969a). *J. Geophys. Res.*, **74**, 1720-1739.

McElroy, M.B., and D.M. Hunten (1969b). *J. Geophys. Res.*, **74**, 5807-5809.

Meier, R.R. (1969). *J. Geophys. Res.*, **74**, 3561-3574.

Meier, R.R. (1970a). *J. Geophys. Res.*, **75**, 6218-6232.

Meier, R.R. (1970b). *Space Research*, **10**, 572-581.

Meier, R.R., and P. Mange (1970). *Planet. Space Sci.*, **18**, 803-821.

Meier, R.R., and P. Mange (1973). *Planet. Space Sci.*, **21**, 309-327.

Meier, R.R., and D.K. Prinz (1970). *J. Geophys. Res.*, **75**, 6969-6979.

Meier, R.R., D.M. Weiss, and P. Mange (1970). *J. Geophys. Res.*, **75**, 4224-4229.

Meinel, A.B. (1950a). *Astrophys. J.*, **111**, 207.

Meinel, A.B. (1950b). *Astrophys. J.*, **111**, 433-434.

Metzger, P.H., and M.A. Clark (1970). *J. Geophys. Res.*, **75**, 5587-5591.

Milne, E.A. (1923). *Trans. Camb. Phil. Soc.*, **22** (No.26), 483-517.

Mitra, S.K., and H. Rakshit (1938). *Ind. J. Phys.*, **12**, 47-61.

Mitra, S.K., and A.K. Banerjee (1939). *Ind. J. Phys.*, **13**, 107-144.

Modali, S.B., J.C. Brandt, and S.O. Kastner (1972). *Astrophys. J.*, **175**, 265-274.

Moffett, R.J., and J.A. Murphy (1973). *Planet. Space Sci.*, **21**, 43-52.

Moos, H.W., and W.G. Fastie (1967). *J. Geophys. Res.*, **72**, 5165-5171.

Moos, H.W., and G.J. Rottman (1971). *Astrophys. J.*, **169**, L127-L130.

Moos, H.W., W.G. Fastie, and M. Bottema (1969). *Astrophys. J.*, **155**, 887-897.

Morton, D.C. (1962). *Planet. Space Sci.*, **9**, 459-460.

Morton, D.C., and J.D. Purcell (1962). *Planet. Space Sci.*, **9**, 455-458.

Nagy, A.F., and P.M. Banks (1972). *J. Geophys. Res.*, **77**, 4277-4279.

Nicolet, M. (1964). *Discussions Faraday Soc.*, **37**, 7-20.

Nicolet, M. (1965). *J. Geophys. Res.*, **70**, 691-701.
Nicolet, M., G. Kockarts, and P.M. Banks (1969). *Handbuch der Physik (Geophysik)*, **49**.
Nicolet, M. (1970). *Ann. Geophys.*, **26**, 531-546.
Nicolet, M. (1971). In G. Fiocco (Ed), *Mesospheric Models and Related Experiments*, Reidel Publ. Co. Holland.
Ogawa, T., and T. Tohmatsu (1966). *Rept. Ionos. Space Res. Japan*, **20**, 395-417.
Opik, E.J., and S.F. Singer (1959). *Phys. Fluids*, **2**, 653-655.
Opik, E.J., and S.F. Singer (1960). *Phys. Fluids*, **3**, 486-488.
Opik, E.J., and S.F. Singer (1961). *Phys. Fluids*, **4**, 221-233.
Paresce, S. Kumer, and S. Bowyer, *Planet. Space Sci.*, **20**, 297-299.
Park, C.G. (1970). *J. Geophys. Res.*, **75**, 4249-4260.
Patterson, T.N.L. (1966a). *Planet. Space Sci.*, **14**, 417-423.
Patterson, T.N.L. (1966b). *Planet. Space Sci.*, **14**, 425-431.
Patterson, T.N.L. (1970). *Rev. Geophys. and Space Phys.*, **8**, 461-467.
Patterson, T.N.L., F.S. Johnson, and W.B. Hanson (1963). *Planet. Space Sci.*, **11**, 767-778.
Prior, E.J. (1971). *Observed Effects of Earth Reflected Radiation and Hydrogen Drag on the Orbital Energies of Balloon Satellites*. Paper presented at Symposium on the use of artificial satellites for Geodesy, Washington, April 1971.
Prokudina, V.S. (1959). In *Spectral, Electrophotometrical and Radar Researches of Aurora and Airglow*, **1**, 43-44. (Results of IGY, USSR Acad. Sci., Moscow).
Prokudina, V.S. (1960). In *Review of Observational Results on the Airglow and Aurorae*, *Trans. IAU*, **10**, 327-328. (Conf. Moscow, 1958), Cambridge Univ. Press.
Purcell, J.D., and R. Tousey (1960). *J. Geophys. Res.*, **65**, 370-372, 1960; *Space Res. I*, 590-593, 1960.
Purcell, J.D., and R. Tousey (1961). *Mem. Soc. Roy. Sci. Liège*, **4**, 283-294.
Quessette, J.A. (1972). *J. Geophys. Res.*, **77**, 2997-3000.
Reber, C.A., J.E. Cooley, and D.N. Harpold (1968). *Space Res.*, **8**, 993-995.
Rossbach, A. (1969). *Zeit, Geophys.*, **35**, 557-564.
Schunk, R.W., and J.C.G. Walker (1972). *Planet. Space Sci.*, **20**, 581-589.
Serbu, G.P., and E.J.R. Maier (1970). *J. Geophys. Res.*, **75**, 6102-6113.
Shcheglov, P.V. (1963a). *Aurora and Airglow*, **10**, 40-43. (Results of IGY, USSR Acad. Sci. Moscow.)
Shcheglov, P.V. (1963b). *Nature*, **199**, 990.
Shcheglov, P.V. (1963c). *Astron. Circular USSR*, **273**, 1-4.
Shcheglov, P.V. (1964). *Astron. Zhurnal USSR*, **41**, 371-377. *Soviet Astronomy* (transl.), **8**, 289-294, 1964.
Shcheglov, P.V. (1967a). *Astron. Circular USSR*, **414**, 1-8.
Shcheglov, P.V. (1967b). *Astron. Circular USSR*, **427**, 5-7.
Shcheglov, P.V. (1968). *Astron. Circular USSR*, **489**, 3-4.
Sheffer, E.K. (1968). *Kosmicheskie Issledovaniya*, **6**, 735-737. *Cosmic Research* (transl.), **6**, 619-621, 1968.
Shefov, N.N. (1959). *Spectral, Electrophotometrical and Radar Researches of Aurora and Airglow*, **1**, 25-29. (Results of IGY, USSR Acad. Sci. Moscow.)
Shefov, N.N. (1970). *Space Res.*, **10**, 623-632.
Shen C.G. (1963). *J. Atmos. Sci.*, **20**, 69-72.
Shimizaki, T. (1971). *J. Atmos. Terr. Phys.*, **33**, 1383-1401.
Shimizaki, T., and A.R. Laird (1970). *J. Geophys. Res.*, **75**, 3221-3235. Erratum *J. Geophys. Res.*, **77**, 276-277, 1972.
Shimizaki, T., and A.R. Laird (1972). *Radio Science*, **1**, 23-43.
Shklovsky, I.S. (1959). *Planet. Space Sci.*, **1**, 63-65.
Shklovsky, I.S. (1960). In *Review of Observational Results on the Airglow and Aurorae*, *Trans. IAU*, **10**, 327-328. (Conf. Moscow, 1958), Cambridge Univ. Press.
Spitzer, L. (1949). In E.P. Kuiper (Ed), *The Atmospheres of the Earth and Planets*, 1st ed., Univ. Chicago Press.
Stebbings, R.F., and J.A. Rutherford (1968). *J. Geophys. Res.*, **73**, 1035-1038.
Stoney, G.J. (1868). *Proc. Roy. Soc. London*, **17**, 1-57.
Stoney, G.J. (1898). *Trans. Roy. Dublin Soc.*, **6**, 305-328. *Astrophys. J.*, **7**, 25-55, 1898.

Strobel, D.F. (1969). *J. Atmos. Sci.*, **26**, 906-911.
Strobel, D.F. (1972). *Radio Sci.*, **7**, 1-21.
Swider, W. (1965). *J. Geophys. Res.*, **70**, 4859-4873.
Swisher, R.L., and L.A. Frank (1968). *J. Geophys. Res.*, **73**, 5665-5672.
Taylor, H.A., L.H. Brace, H.C. Brinton, and C.R. Smith (1963). *J. Geophys. Res.*, **68**, 5339-5347.
Thomas, G.E. (1963a). *Radiation Transfer in the Hydrogen Geocorona*, Ph.D. Thesis, University of Pittsburgh, University Microfilms, Ann Arbor, Michigan.
Thomas, G.E. (1963b). *J. Geophys. Res.*, **68**, 2639-2660.
Thomas, G.E. (1970). *Space Research*, **10**, 602-607.
Thomas, G.E., and R.F. Krassa (1971). *Astron. Astrophys.*, **11**, 218-230.
Thomas, G.E., and R.C. Bohlin (1972). *J. Geophys. Res.*, **77**, 2752-2761.
Tinsley, B.A. (1967). *Planet. Space Sci.*, **15**, 1757-1776.
Tinsley, B.A. (1968). *J. Geophys. Res.*, **73**, 4139-4149.
Tinsley, B.A. (1969a). *Planet. Space Sci.*, **17**, 769-772.
Tinsley, B.A. (1969b). *Planet. Space Sci.*, **17**, 1320-1322.
Tinsley, B.A. (1969c). In B.M. McCormac and A. Omholt (Eds), *Atmospheric Emissions*, Van Nostrand Reinhold.
Tinsley, B.A. (1970). *Space Research*, **10**, 582-590.
Tinsley, B.A. (1971). *Rev. Geophys. and Space Phys.*, **9**, 89-102.
Tinsley, B.A. (1973). *Planet. Space Sci.*, **21**, 686-691.
Tinsley, B.A., and R.R. Meier (1971). *J. Geophys. Res.*, **76**, 1006-1016.
Tohmatsu, T. (1970). *Space Research*, **10**, 608-622.
Tousey, R. (1963). *Space Sci. Rev.*, **2**, 3-69.
Urey, H.C. (1959). *Handbuch der Physik*, **52**, 363-415.
Van Valen, L. (1971). *Science*, **171**, 439-443.
Venkateswaran, S.V. (1971). *Planet. Space Sci.*, **19**, 275.
Vidal-Madjar, A., and J.L. Bertaux (1972). *Planet. Space Sci.*, **20**, 1147-1162.
Vidal-Madjar, A., J.E. Blamont, and B. Phissamay (1973). *J. Geophys. Res.*, **78**, 1115-1144.
Wallace, L. (1969). *J. Geophys. Res.*, **74**, 115-131.
Wallace, L. (1971). *Planet. Space Sci.*, **19**, 377-398.
Wallace, L., C.A. Barth, J.B. Pearce, K.K. Kelly, D.E. Anderson, Jr., and W.G. Fastie (1970). *J. Geophys. Res.*, **75**, 3769-3777.
Wallace, L., F.E. Stewart, R.H. Nagel, and M.D. Larson (1971). *Astrophys. J.*, **168**, L29-L31.
Wallace, L., and D.F. Strobel (1972). *Planet. Space Sci.*, **20**, 521-531.
Waterson, J.J. (1846). *Phil. Trans. Roy. Soc. London*, **5**, 604. (Abstract only.) *Phil. Trans. Roy. Soc. London*, **183**, 1-79, 1892 (full 1846 paper with introduction by Lord Rayleigh).
Wegener, A. (1911a). *Phys. Zeitschr. Leipzig*, **12**, 170-178, 1911a; *Thermodynamik der Atmosphare, Leipzig*, p.46, 1911b.
Weller, C.S., R.R. Meier, and B.A. Tinsley (1971). *J. Geophys. Res.*, **76**, 7734-7744.
Winter, T.C., Jr., and T.A. Chubb (1967). *J. Geophys. Res.*, **72**, 4405-4414.
Wofsy, S.C., J.C. McConnell, and M.B. McElroy (1972). *J. Geophys. Res.*, **77**, 4477-4493.
Yarin, V.I. (1960). *Spectral, Electrophotometrical and Radar Researches of Aurora and Airglow*, 2-3, 72. (Results of IGY, USSR Acad. Sci., Moscow.)
Yčas, M. (1972). *Nature*, **238**, 163-164.
Young, J.M.G.R. Carruthers, J.C. Holmes, C.Y. Johnson, and N.P. Patterson (1968). *Science*, **160**, 990-991.

The Equatorial Electrojet[†]

DEREK M. CUNNOLD

Consultant to Lincoln Laboratory MIT[‡]

The equatorial electrojet is an intense electric current which flows at approximately 100 km altitude over the magnetic equator. It consists of a daytime flow of electrons in the westward direction. The theory of the electrojet is presented. It is driven by an eastward electric field which leads to the production of a vertical polarization field required in order to prevent the vertical flow of current. This polarization field is mathematically equivalent to the enhancement of the direct conductivity of the iono-sphere between 95 and 120 km and is believed to be directly responsible for the intense eastward current. Recent calculations which stress the importance of meridional currents in electrojet theory are reviewed. Since it is generally believed that the electrojet is a part of the mid-latitude S_q current system which is driven by tidal motions of the neutral atmosphere, the theory of the S_q current system is briefly described. A limited discussion is also given of the variability of the electrojet and of techniques used in its measurement. The electrojet is observed to contain irregularities of electron density of a few to several hundred meters in size which are strongly aligned with the magnetic field. Mechanisms which may lead to unstable conditions in the electrojet and the growth of irregularities are briefly described.

1. INTRODUCTION

The intensity of the horizontal component of the geomagnetic field measured at the surface of the earth is observed to undergo regular daily variations of intensity tens of gamma (1 gamma $= 10^{-5}$ gauss). These diurnal fluctu-ations are produced by currents flowing in the earth's upper atmosphere at a height of approximately 100 km. The currents moreover are thought to be produced by the tidal motions of the neutral atmosphere which force charged particles to move across geomagnetic field lines at speeds which are dependent on both particle mass and electronic charge and on the frequency of collisions between neutral and ionized particles. The concept that upper atmosphere currents are produced by tidal motions of the neutral gas is usually referred to as the dynamo theory. Of particular note is the obser-vation that within a narrow latitude belt around the geomagnetic equator

---

† This work was sponsored in part by the Department of the Army.
‡ Permanent address, Department of Meteorology, MIT, Cambridge, Massachusetts.

101

the amplitude of the magnetic field fluctuations is enhanced (by roughly a factor of two) with respect to that which would be predicted by interpolating the mid-latitude fluctuations to the equatorial region. This effect is illustrated in Figure 1 in which the diurnal range of the horizontal component of the geomagnetic field is plotted as a function of latitude.† The equatorial

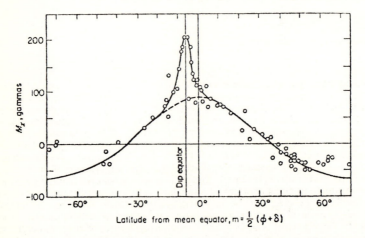

FIGURE 1   Variation of M_x, the daily range of the horizontal component of the magnetic field, with latitude from the mean equator (midway between the geographic and dip equators) on international quiet days during September and October, 1958. The equatorial enhancement is the survey of Forbush and Casaverde (1961) reduced to the same epoch with Huancayo observatory data. The position of the dip equator at about 76 °W is shown (Onwumechilli, 1967).

enhancement is the result of a strong (daytime) current flowing towards the east—but which is principally the result of electrons flowing towards the west. The current is confined to a region within a few degrees of the magnetic equator. It is referred to as the equatorial electrojet (Chapman, 1951) and its existence has been experimentally verified via several rocket flights (Singer et al., 1951; Cahill, 1959; Maynard et al., 1965).

---

† It may be noted that the abscissa used in Figure 1 is distance from the mean equator. Whereas the location of the equatorial electrojet is determined by structure of the geomagnetic field, the daily magnetic field variations over the entire globe are related not only to the geomagnetic field (which determines the motions of ionization produced by an applied force), but also to geographic location (which determines the solar insolation). Onwumechilli (1964) found that the least scatter of the global observations of the daily magnetic field variations was obtained by plotting the magnetic field fluctuations against latitude measured from the mean equator.

2. THEORY OF THE ELECTROJET

The conductivity of the ionosphere is not only height dependent but also anisotropic because of the (roughly dipolar) geomagnetic field. Typical night-time and daytime electron density profiles, produced as a result of the absorption of solar EUV radiation (see Figure 2), are shown in Figure 3. In order to derive expressions for the conductivity of the ionosphere, we start from the equations of motion for ions and electrons

$$n_i m_i \frac{\partial \mathbf{v}_i}{\partial t} + n_i m_i \mathbf{v}_i \cdot \nabla \mathbf{v}_i = -\nabla p_i + n_i e(\mathbf{E} + \mathbf{v}_i \times \mathbf{B}_0)$$

$$+ n_i m_i \mathbf{g} + n_i \frac{m_i m_n}{m_i + m_n} \nu_{in}(\mathbf{v}_n - \mathbf{v}_i) \tag{1}$$

$$n_e m_e \frac{\partial \mathbf{v}_e}{\partial t} + n_e m_e \mathbf{v}_e \cdot \nabla \mathbf{v}_e = -\nabla p_e - n_e e(\mathbf{E} + \mathbf{v}_e \times \mathbf{B}_0)$$

$$+ n_e m_e \mathbf{g} + n_e \frac{m_e m_n}{m_e + m_n} \nu_{en}(\mathbf{v}_n - \mathbf{v}_e) \tag{2}$$

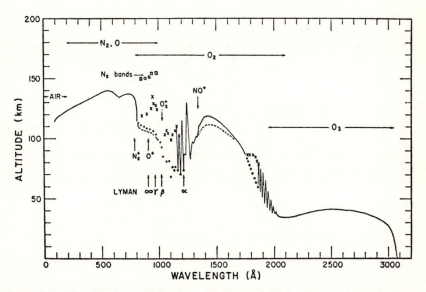

FIGURE 2   The altitude at which solar radiation vertically incident on the atmosphere falls to $1/e$ of its initial value. Also shown are the molecular species responsible for the absorption and the ionized species which result from that absorption (Watanabe, 1958).

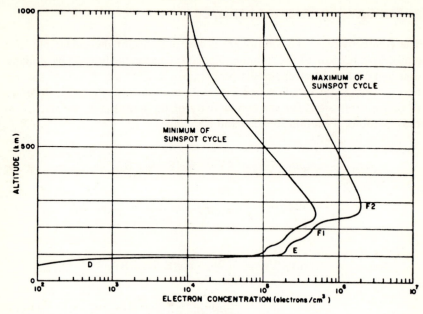

a. Daytime.

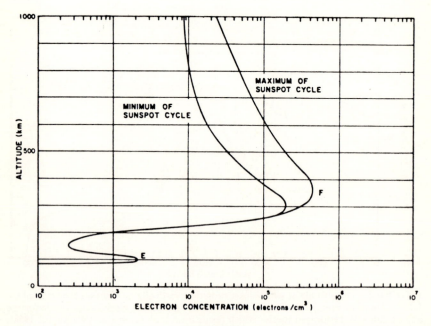

b. Nighttime.

FIGURE 3   Normal electron density distributions at extremes of the solar cycle (Johnson, 1965).

where

$$n_i, n_e = \text{ion, electron density,}$$
$$m_i, m_e, m_n = \text{ion, electron, neutral mass,}$$
$$\mathbf{v}_i, \mathbf{v}_e, \mathbf{v}_n = \text{ion, electron, neutral velocity,}$$
$$t = \text{time,}$$
$$p_i, p_e = \text{partial pressure of ions, electrons,}$$
$$e = \text{ionic charge } (= -\text{electronic charge}),$$
$$\mathbf{E} = \text{electric field,}$$
$$\mathbf{B}_0 = \text{geomagnetic field,}$$
$$\nu_{in} = \text{collision frequency between ions and neutral particles,}$$
$$\nu_{en} = \text{collision frequency between electrons and neutral particles,}$$
$$\mathbf{g} = \text{acceleration due to gravity.}$$

We shall be considering motions having a horizontal length scale of hundreds of kilometers and a temporal scale of hours. Because the lower ionosphere is only a weakly ionized plasma (there is only one charged particle for every 10^4–10^6 neutrals), we have neglected collisions between charged particles and have represented collisions between charged and neutral particles by simple relaxation terms. Collisions are also so frequent (see Figure 4) that the inertial and pressure terms in Eqs. (1) and (2) may be neglected. A

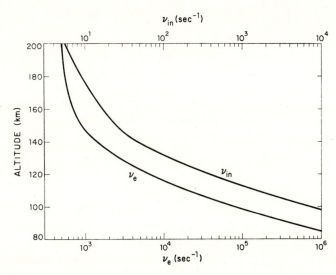

FIGURE 4.   The collision frequencies between charged and neutral particles as a function of altitude. ν_e differs slightly from ν_{en} of the text because ν_e includes collisions between electrons and ions which become important above 150 km (Based upon Johnson, 1965).

simple dimensional consideration also reveals that the gravitational force is negligible and that the perturbation in the magnetic field which is only $<10^{-3}$ of the primary geomagnetic field ($\mathbf{B_0}$) is virtually insignificant. If, in addition to neglecting these terms we also change the reference frame to one moving with the neutral gas, Eqs. (1) and (2) become

$$-e\mathbf{v}_i \times \mathbf{B_0} + \frac{m_i m_n}{m_i + m_n} \nu_{in} \mathbf{v}_i = e(\mathbf{E} + \mathbf{v}_n \times \mathbf{B_0}) \tag{3}$$

$$e\mathbf{v}_e \times \mathbf{B_0} + \frac{m_n m_e}{m_e + m_n} \nu_{en} \mathbf{v}_e = -e(\mathbf{E} + \mathbf{v}_n \times \mathbf{B_0}) \tag{4}$$

These two equations may be used to deduce the current density

$$\mathbf{j} = e(n_i \mathbf{v}_i - n_e \mathbf{v}_e) = \sigma \cdot (\mathbf{E} + \mathbf{v}_n \times \mathbf{B_0}) \equiv \sigma \cdot \mathbf{E}' \tag{5}$$

where σ is the conductivity tensor, $\mathbf{E}$ is the electric field in a frame of reference at rest whereas $\mathbf{E}'$ is the total electric field seen by an observer moving with the neutral wind. We can rewrite Eq. (5) as a sum of physically meaningful terms using Baker and Martyn's (1953) notation:

$$\mathbf{j} = \sigma \left(\frac{\mathbf{E}' \cdot \mathbf{B_0}}{|\mathbf{B_0}|^2} \right) \mathbf{B_0} + \sigma_1 \left[\mathbf{E}' - \frac{\mathbf{E}' \cdot \mathbf{B_0}}{|\mathbf{B_0}|^2} \mathbf{B_0} \right] + \sigma_2 \mathbf{B_0} \times \mathbf{E}' \tag{6}$$

It is to be noted that an extremely small charge separation is required in order to produce typical ionospheric electric fields (10^{-3} volts/m) and hence we set $n_i = n_e$. The conductivities σ_0, σ_1, and σ_2 may be expressed in terms of the ion gyrofrequency ($\omega_i = e|\mathbf{B_0}|/m_i$) and the electron gyrofrequency ($\omega_e = e|\mathbf{B_0}|/m_e$ and the collision frequencies. Now, however, the collision frequencies will be understood to mean the momentum transfer collision frequencies

$$\nu_{in} = \frac{m_n}{m_i + m_n} \nu_{in} \approx \tfrac{1}{2} \nu_{in}$$

which expresses the fact that if a positive ion and a neutral particle possess differing momenta prior to a collision between them, after the collision that momentum difference will be evenly distributed between two particles; also

$$\nu_{en} = \frac{m_n}{m_e + m_n} \nu_{en} \approx \nu_{en}.$$

Then

$$\sigma_0 = \frac{n_e e}{|\mathbf{B}_0|}\left(\frac{\omega_e}{\nu_{en}} + \frac{\omega_i}{\nu_{in}}\right). \tag{7}$$

σ_0 is the conductivity along the magnetic field lines and is several orders of magnitude larger in the ionosphere than σ_1 and σ_2. Thus ionospheric magnetic field lines tend to be equipotentials.

$$\sigma_1 = \frac{n_e e}{|\mathbf{B}_0|}\left(\frac{\nu_{en}\omega_e}{\omega_e^2 + \nu_{en}^2} + \frac{\nu_{in}\omega_i}{\omega_i^2 + \nu_{in}^2}\right) \tag{8}$$

is referred to as the Pedersen conductivity (1927) for motion in the direction of an electric field perpendicular to the magnetic field, and

$$\sigma_2 = \frac{n_e e}{|\mathbf{B}_0|}\left(\frac{\omega_e^2}{\omega_e^2 + \nu_{en}^2} - \frac{\omega_i^2}{\omega_i^2 + \nu_{in}^2}\right) \tag{9}$$

is the Hall conductivity for motion perpendicular to both the magnetic field and the electric field. Ionospheric values of σ_0, σ_1, and σ_2 are given in Figure 5. The large value of σ_2 between 95 and 120 km leads to the existence of strong currents in that region and thus defines the ionospheric dynamo

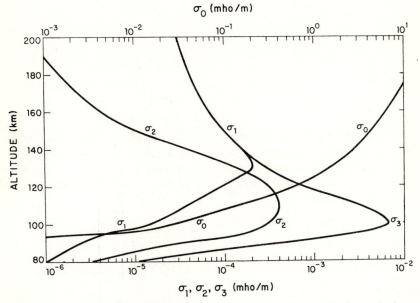

FIGURE 5 Ionospheric conductivities, σ_0, σ_1, σ_2, and σ_3, for average ionospheric conditions (Based upon Johnson, 1965).

region. We see from Eq. (9) that in the absence of collisions both electrons and ions drift at the same speed in the direction $\mathbf{E}' \times \mathbf{B}_0$ with the result that there is no current flow in that direction. That charged particles are capable of moving in the direction perpendicular to an applied electric field is the result of the spiralling motion of the particles around the geomagnetic field lines. For example, consider an individual particle circling about a magnetic field line. As a result of an applied electric field it will be accelerated over one half of its orbit and decelerated over the other half. Thus the particle will obtain a greater speed over one half of its orbit than over the other. Then since the radius of the particle orbit is proportional to the particle speed, the particle will not remain in a circular orbit but will acquire a net drift in the direction $\mathbf{E}' \times \mathbf{B}_0$.

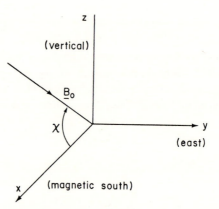

FIGURE 6 Coordinate system used in calculations. The x axis is chosen to be in the plane of the magnetic field, $\mathbf{B}_0$, and the vertical.

Consider the right-angled coordinate system shown in Figure 6 in which the z axis is directed vertically upwards, the x axis is directed towards magnetic south and the y axis points approximately east. In this coordinate system the geomagnetic field lies in the xz plane and makes an angle (the dip angle) with the x axis.

$$\sigma = \begin{bmatrix} \sigma_0 \cos^2\chi + \sigma_1 \sin^2\chi & \sigma_2 \sin\chi & (\sigma_0 - \sigma_1) \cos\chi \sin\chi \\ -\sigma_2 \sin\chi & \sigma_1 & \sigma_2 \cos\chi \\ (\sigma_0 - \sigma_1) \cos\chi \sin\chi & -\sigma_2 \cos\chi & \sigma_0 \sin^2\chi + \sigma_1 \cos^2\chi \end{bmatrix} \quad (10)$$

Equation 5 together with two of Maxwell's equations,

$$\nabla \cdot j = 0 \quad (11)$$

and

$$\text{curl } \mathbf{E} = 0 \tag{12}$$

may be solved to obtain the ionospheric current distribution implied by a given global distribution of neutral wind velocity $\mathbf{v}_n$. From a physical viewpoint, the fact that we have allowed an arbitrary electric field (subject to Eq. (12)) to exist is necessary in order that the current implied by Eq. (5) will satisfy Eq. (11). That is to say a polarization field will exist.

For our purposes, the ionosphere may be regarded as a thin conducting spherical shell since the horizontal length scale for tidal oscillations which are responsible for driving the currents is of the order of an earth radius. From Eq. (11) $j_z \approx 0$ and therefore let us set $j_z = 0$. Such an approximation follows the historical development of the theory and although not strictly correct leads to a simple mathematical explanation of the enhancement of the conductivity in the region of the magnetic equator. Then

$$
\begin{aligned}
j_x &= \sigma_{xx} E'_x + \sigma_{xy} E'_y \\
j_y &= -\sigma_{xy} E'_x + \sigma_{yy} E'_y
\end{aligned}
\tag{13}
$$

where

$$\sigma_{xx} = \sigma_0 \sigma_1 / (\sigma_0 \sin^2 \chi + \sigma_1 \cos^2 \chi)$$

$$\sigma_{xy} = \sigma_0 \sigma_2 \sin \chi / (\sigma_0 \sin^2 \chi + \sigma_1 \cos^2 \chi)$$

$$\sigma_{yy} = [\sigma_1 \sigma_0 \sin^2 \chi + (\sigma_1{}^2 + \sigma_2{}^2) \cos^2 \chi] / (\sigma_0 \sin^2 \chi + \sigma_1 \cos^2 \chi)$$

At the magnetic equator, $\chi = 0$, and

$$\sigma_{xx} = \sigma_0, \qquad \sigma_{xy} = 0, \qquad \sigma_{yy} = (\sigma_1{}^2 + \sigma_2{}^2)/\sigma_1 = \sigma_3.$$

σ_3 is plotted in Figure 5 from which it is clear that in the region between 100 and 130 km, σ_3 is much greater than both σ_1 and σ_2. The physics of this situation is as follows: suppose there exists an eastward electric field which drives the electrojet. Then, in addition to east–west motions, electrons and ions will drift vertically at differing speeds due to the conductivity σ_2. As a result of the non-uniform conductivity profile a vertical polarization field will be set up which is just sufficient to force j_z to be zero. This field will in turn lead to the drift of charged particles (particularly electrons) in the east–west direction due to the conductivity σ_2. In fact, in the 100–120 km region the east–west current thus induced will be considerably larger than that which is directly produced by the driving electric field. Mathematically at least the effect of the polarization field is equivalent to increasing the direct conductivity in the east–west direction from σ_1 to σ_3. Such an equatorial enhancement of the conductivity resulting from polarization fields was originally proposed by Martyn (1948) as the explanation of the equatorial electrojet. This theory has been further discussed by Baker and Martyn (1953), Hirono (1952), and Fejer (1953) among others. It is to be noted that this enhancement will only occur for angles

$$|\chi| \lesssim \tan^{-1} \sqrt{\frac{\sigma_1}{\sigma_0}} \approx 2°.$$

Although this explanation of the electrojet is generally believed to be essentially the correct one, it is mathematically inconsistent because of the inappropriate assumption that $j_z = 0$. It may be easily verified that the conductivities σ_{xx}, σ_{xy}, and σ_{yy} are strongly dependent on χ (and hence on x) in the region of the magnetic equator. Therefore the local horizontal length scale for variations in the x direction is not of the order of an earth radius but instead is comparable with the scale for variations in the vertical direction. Consequently the calculation must permit meridional currents as was pointed out by Price (1968). The scale for variations in the y direction remains an earth radius and thus a two dimensional solution (independent of y) may be sought. Then, from Eq. (12), it may be shown that E_y is a constant over the equatorial region. E_y may be regarded as the driving mechanism for the equatorial current system including the electrojet.

Consider the $\mathbf{u} \times \mathbf{B}_0$ contribution to the total electric field over the equator. Tidal velocities ($\mathbf{u}$) are almost entirely horizontal and $\mathbf{u} \times \mathbf{B}_0$ is directed nearly vertically. This dynamo component may be compared with the vertical polarization field E_z (due to E_y). At a magnetic dip angle of zero, j_z is small and therefore $E_z \sim (\sigma_2/\sigma_1)E_y$ which is much greater than E_y in the dynamo region around 100 km. Now E_y which is a constant over the equatorial region is the result of dynamo action at mid-latitudes and thus is of order $|\mathbf{B}_0| \{|\mathbf{u}|$ mid-latitude$\}$. It is generally believed that the dominant tidal mode which drives the mid-latitude (S_q) current system is associated with wind velocities which maximize at high latitudes (we shall have more to say on this point shortly), and therefore the $\mathbf{u} \times \mathbf{B}_0$ contribution to the total electric field over the equator is small. Thus tidal winds at equatorial latitudes make only a small contribution to driving the electrojet. However local winds at equatorial latitudes may influence the *structure* of the electrojet. For example transequatorial winds may make the electrojet asymmetric about the magnetic equator (Untiedt, 1967).

Untiedt (1967), Sugiura and Poros (1969) and Richmond (1973a) have solved Eqs. (5), (11) and (12) for the equatorial region subject to the assumption $\partial/\partial y = 0$. At the base of the ionosphere (80 km) j_z is set equal to zero and j_z is made to tend to zero for large z. The boundary conditions at the latitudinal boundaries dividing the equatorial region from the mid-latitude region are more difficult to define. At these boundaries, the currents and fields, specifically j_x, E_y, and E_z, must be continuous. That is to say that the equatorial current system must join on to the mid-latitude (S_q) current system. Moreover E_y is constant over the equatorial region and we thus see that the field which drives the electrojet is controlled by the mid-latitude current system—and by the mechanism responsible for it. Sugiura and Poros adopted latitude boundaries at $\pm 10°$ but neither they nor the other investigators attempted to obtain the current system at equatorial and mid-latitudes simultaneously. Thus they were forced to postulate conditions at the latitudinal boundaries. They assumed that the current system there was just that which would exist in a thin spherical shell. On this basis an expression for E_x at the boundaries was obtained. It is not possible to fully justify this condition but fortunately it has been found that the solutions obtained, except within the immediate vicinity of the boundaries, are apparently independent of the assumed boundary condition. A cross section of the meridional current system obtained by Sugiura and Poros (1969) is depicted in Figure 7. The existence of this predicted current system awaits experimental verification. The calculated form of the (zonal) electrojet current is given in Figure 8. In these calculations the value of E_y is an adjustable parameter for which Sugiura and Poros used a value of 2.4×10^{-3} volt/m. These currents may be a factor of 5 larger than what will be typically observed because this value of E_y may be too large. The peak current density shown is little different from that which would be predicted by the simpler model with $j_z = 0$ using the same value of E_y. However the more exact theory yields an electrojet which is more extensive in latitude, in agreement with observation (e.g.

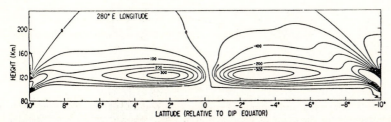

FIGURE 7   The flow pattern of the meridional current associated with the equatoria electrojet over Peru. The current flowing between successive contours in 50 amperes per km-thickness. The driving electric field is 2.4×10^{-3} volt/m. The current flows upward over the dip equator (Sugiura and Poros, 1969).

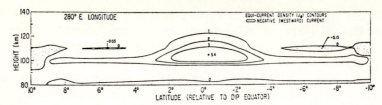

FIGURE 8   The eastward current density over Peru in units of 10^{-5} amp/m². The driving electric field is 2.4×10^{-3} volt/m. The shaded areas represent reversed, westward currents (Sugiura and Poros, 1969).

Yacob, 1966), and which thus contains a larger total current. In Figures 7 and 8, the solution between 8 and 10° latitude is dependent upon the choice of boundary conditions but the presence of small regions of reversed currents around 6° latitude in Figure 8 is apparently a real effect associated with a reversal of the (vertical) polarization field over a narrow latitude range in this model. This feature is however not found in Richmond's (1973a) model and would moreover tend to be masked by the substantial variations of current away from the equator associated with irregular wind systems. Richmond (1973b) has compared model predictions of the electrojet current with rocket observations by Shuman (1970), Maynard (1967), and Davis *et al.* (1967) and obtains approximate agreement for $E_y = 3.7 \times 10^{-4}$ volts/m

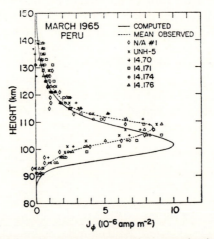

FIGURE 9   Observed and computed eastward current density profiles near noon at the dip equator off the coast of Peru in March 1965. The currents are normalized to correspond to a 100γ variation in H at Huancayo. The profile measured on rocket flight N/A #1 is from Shuman (1970); that on flight UNH-5 is from Maynard (1967); and those on flights 14.70, 14.171, 14.174 and 14.176 are from Davis *et al.* $E_y = 0.00037$ Vm^{-1} (Richmond, 1973b).

(see Figure 9). However theory produces the current center at approximately 102 km compared with the 108 km observed. This altitude discrepancy is currently unresolved.

3. TIDAL THEORY AND THE S_q CURRENT SYSTEM

As has previously been pointed out, the equatorial electrojet is closely related to the mid-latitude S_q current system and in order to understand the ultimate source of the electrojet, we must understand how the S_q system is driven. It is generally believed that the neutral wind velocity (**u**) responsible for these ionospheric currents results from atmospheric tidal oscillations. Now in the absence of viscosity and under the assumption that the tidal oscillations are perturbations about a horizontally uniform steady state temperature distribution, the tidal equations are of the variable separable form. The solutions in a coordinate system rotating with the earth may be expressed as a sum of modes of the form

$$G^{\sigma}(z,\theta,\phi,t) = \sum_{n,m} L_n{}^m(z)\,\Theta_n{}^m(\theta)\,\exp[i(\sigma t \pm m\phi)] \qquad (14)$$

where z, θ, and ϕ are the altitude, colatitude, and longitude respectively and t is the time. $2\pi/\sigma$ represents a solar (or lunar) day or fraction thereof, m is a positive integer, and n is an integer which is positive except when considering modes having periods in excess of 12 hours. The Θ functions are referred to as Hough functions and would in fact be Legendre polynomials (or spherical harmonics for the combined θ, ϕ variation) in the absence of the earth's rotation.

The vertical structure function, $L_n(z)$, for each tidal mode is given by an equation

$$H\frac{d^2L_n{}^m}{dz^2} + \left(\frac{dH}{dz}-1\right)\frac{dL_n{}^m}{dz} + \frac{1}{h_n{}^m}\left(\frac{dH}{dz}+k\right)L_n{}^m = \frac{k}{\gamma g H h_n{}^m}J_n{}^m \qquad (15)$$

where H = the atmospheric scale height which is a function of altitude,
$\quad k = (\gamma-1)/\gamma = 2/7$,
$\quad \gamma = c_p/c_v = 7/5$,
$\quad g$ = acceleration due to gravity,
$\quad h_n{}^m$ = a constant (called the equivalent depth) resulting from the separation of the equations in z and θ. The value of $h_n{}^m$ is determined by the condition that $\Theta_n{}^m(\theta) = \Theta_n{}^m(2\pi+\theta)$.
$\quad J_n{}^m$ = the thermotidal heating (or atmospheric heating resulting from the absorption of solar energy)/unit mass/unit time with period $2\pi/\sigma$, longitudinal period $2\pi/s$, and latitudinal structure $\Theta_n{}^m(\theta)$. The derivation of

Eq. (15) is lengthy and is not included because tidal theory is not the principal subject of this article. A derivation may be found in Chapman and Lindzen (1969). It is customary in tidal theory to write Eq. (15) in a simpler form by replacing z by $x = -\log|p_0(z)/p_0(0)|$, where p_0 is the pressure at altitude z and $p_0(0)$ is taken as 1000 mb, and replacing L_n^m by $y_n^m = \exp(x/2)L_n^m$. The equation for y_n^m is

$$\frac{d^2 y_n^m}{dx^2} - \frac{1}{4}\left\{1 - \frac{4}{h_n^m}\left(kH + \frac{dH}{dx}\right)\right\}y_n^m = \frac{kJ_n^m}{\gamma g h_n^m}\exp(-x/2) \tag{16}$$

The term on the right-hand side of Eq. (16) is the forcing function responsible for the particular tidal mode. Note that the contributions from each height must be summed (or integrated) to give the amplitude of the tidal mode at each height. The form of Eq. (16) is such that, depending upon whether the expression in curly brackets is positive or negative, the solution is evanescent (trapped) or propagating. It should further be noted that this bracketed expression is a function of height and thus that some tidal modes may propagate in certain regions of the atmosphere but be trapped (as a result of the temperature gradient) in others.

The strongest tides of the upper atmosphere are generated as a result of the absorption of solar energy rather than via the gravitational attraction of the sun or moon. The absorption of solar energy in the atmosphere has been depicted in Figure 2. Two parts of the solar spectrum are particularly important for producing the upper atmosphere tides. The portion of the solar spectrum between approximately 2000 and 3000 Å is largely absorbed by atmospheric ozone at roughly 50 km. This absorption results in the generation of strong diurnal and semi-diurnal oscillations (and of smaller amplitude oscillations of shorter and longer periods). The tidal energy propagates upwards and because kinetic energy per unit volume ($\frac{1}{2}\rho u^2$) tends to remain constant during the propagation (i.e. $|L_n^m(z)| \sim \exp[x/2]$), substantial velocities (~ 100 m/sec) may be produced at ionospheric heights.

Above 100 km in the atmosphere, non-linear interactions between tidal modes may become important and atmospheric viscosity is large enough to affect the propagation of tidal energy. Thus each tidal mode ceases to be characterized by a velocity which increases with the increasing height in the ionosphere. The altitude at which these effects become significant depends upon the particular tidal mode. The energy propagated upward into the ionosphere competes with the in situ absorption of solar EUV (100–1750 Å) energy in the ionosphere to produce the net diurnal motions of the neutral gas in the ionosphere at middle and low latitudes. At the present time insufficient data exist to be able to determine in any detail the distribution of energy among tidal modes or in what proportion EUV and UV radiation contribute to the motions of the neutral gas (say) between 90 and 150 km. An attempt to answer some of these problems was made by Lindzen (1970), Lindzen and Blake (1971), and Lindzen (1971).

The global distribution of incident solar flux, ozone and other atmospheric species which absorb solar energy will be principally represented by the lowest order spherical harmonics and thus the lowest order tidal modes should possess the largest amplitudes. This conclusion is reinforced by the fact that the region over which significant tidal generation occurs possesses a vertical extent of tens of kilometers and since modes having large vertical wavelengths also possess large horizontal wavelengths, only the lowest order modes will not undergo extensive destructive interference as a result of the generation of tidal energy over a range of heights.

A knowledge of the atmospheric tidal motions permits Eqs. (5), (11), and (12) to be used to obtain the ionospheric current system generated as a result of these neutral gas motions and based upon the conclusions of the previous paragraph only the lowest order tidal modes need be considered. Such calculations have been made by Matsushita (1969) and one of his students (Tarpley, 1970a, b), and by Richmond et al. (1976). In such theoretical treatments a separate calculation is usually made for each tidal mode. The variation of the amplitude of the mode with height is postulated based upon a few available observations and upon some approximate theoretical considerations from tidal theory. To a large extent the ability of a tidal mode to generate ionospheric currents is dependent upon its vertical wavelength—the latter is obtained from tidal theory with viscous dissipation and non-linear effects neglected (i.e. from the bracketed expression in Eq. (16)). A mode of short vertical wavelength (e.g. 1–10 km) will tend to produce ionospheric currents which reverse every few kilometers. The vertically integrated horizontal currents will therefore be small and will result in an insignificant perturbation of the geomagnetic field at ground level. Thus only tidal oscillations which remain roughly in phase over the ionospheric height range of large conductivity, σ_2 (95–120 km), will produce significant ionospheric currents. Therefore the tidal modes which are important for the generation of ionospheric currents are the first few diurnal and semi-diurnal modes possessing only a few zeros as a function of latitude. In fact, even the first (propagating) diurnal mode has a vertical wavelength of only about 25 km while the first two semi-diurnal modes possess wavelengths of roughly 100 and 40 km respectively. Thus if these modes possess wind velocities of similar amplitudes, the first semi-diurnal mode would dominate the production of ionospheric currents. Such a component of the current resulting from the gravitational attraction of the moon has been observed (e.g. in the electrojet by Bartels and Johnson (1940) and by Onwumechilli (1964)) and has received a detailed examination by Tarpley and Balsley (1972).

Tidal theory also predicts the existence of an additional class of diurnal (trapped) modes. These modes are evanescent being characterized by wave numbers which are purely imaginary (for these modes $h_n{}^m$ is negative and the bracketed expression in Eq. (16) positive at all heights). Consequently, they tend to be significant only in the region in which they are produced as a result of the in situ absorption of solar energy. Therefore trapped modes resulting from the absorption of UV radiation will not be a significant source of ionospheric tidal motions or currents. On the other hand, trapped modes resulting from EUV absorption will be directly produced in the ionospheric dynamo region and having a very large vertical wavelength may be an important source of ionospheric currents. Richmond *et al.* (1976), Tarpley (1970b) and others have shown that the first diurnal trapped mode produces an S_q current system (see Figure 10) in general agreement with that predicted from observations. The diurnal trapped modes possess largest wind amplitudes at high latitudes (see for example Figure 11) and rocket observations (e.g. Ainsworth *et al.*, 1961) do tend to contain evidence of diurnal wind amplitudes in the E region which are larger at high altitudes. Thus it is believed that the first diurnal trapped mode is responsible for the S_q current system and the equatorial electrojet. It should be noted that while this mode will oscillate in an approximately sinusoidal fashion over a 24 hour period, the electrojet velocity will not vary in a similar way because of the modulation resulting from the variation of the ionospheric electron density from day to night. However the electrojet should undergo a reversal in direction at night. Despite the weakness of the nighttime electrojet (because of the small nighttime conductivity) this feature has been observed (Balsley, 1969a, e.g. see Figure 12).

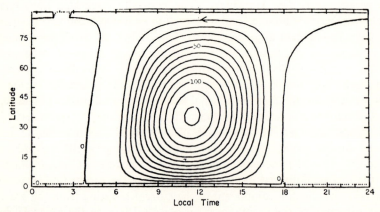

FIGURE 10   The electric current generated by the first diurnal trapped tidal mode for a wind amplitude of 130 m/sec in the dynamo region over the pole. The contour interval is 10,000 amps. The current flows in a counterclockwise direction (Tarpley, 1970b).

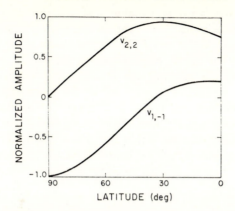

FIGURE 11   Latitude dependence of the eastward wind velocity, normalized to a peak amplitude of unity, for two tidal modes. $v_{1, -1}$ refers to the first diurnal trapped mode, while $v_{2,2}$ is the first semidiurnal mode (Based on Chapman and Lindzen, 1969).

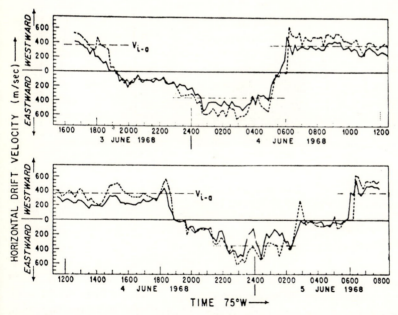

FIGURE 12   Electrojet drift velocities obtained at Jicarmarca Observatory in Peru during the period June 3–5, 1968. The solid curve shows data obtained by directing the radar antenna to the east, at the zenith angle of 60°; the dashed curve shows corresponding data obtained to the west. It may be noted that the westerly velocities are consistently greater than or equal to those obtained to the east. v_{i-a} is the acoustic velocity of ions in the electrojet; it is relevant to the discussion of instabilities in the electrojet (Balsley, 1970).

The day to day and seasonal variability of the electrojet is probably associated more with the variability of the driving electric field E_y than with conductivity variations. While it is probably true that a single tidal mode is primarily responsible for the S_q current system, other tidal modes almost certainly influence it. For example the first few semi-diurnal modes are observed to possess significant amplitudes in the lower ionosphere (e.g. Bernard, 1974). Richmond *et al.* (1976) and Tarpley (1970b) have calculated the current systems which might result from these modes (see for example Figure 13). The present estimate is that the largest amplitude semi-diurnal mode makes only a roughly 30 percent contribution to the S_q current system despite its dominance over the diurnal mode for producing ionospheric winds at 120 km.

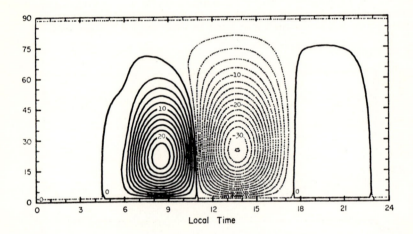

FIGURE 13   Electric current generated by the first solar semi-diurnal mode assuming a maximum wind amplitude of 20 m/sec in the dynamo region. The contour interval is 2000 amps. Solid and broken curves show counterclockwise and clockwise currents respectively (Tarpley, 1970b).

There may also exist variable wind systems which are not symmetric about the magnetic equator. The resulting currents would not form closed loops in the manner of the symmetric modes. Instead because of the large conductivity along magnetic field lines, charge will tend to flow from one hemisphere to the other along magnetic field lines which extend far into the magnetosphere. For a wind system which is antisymmetric about the magnetic equator, the current system would consist of two components (van Sabben, 1966) as depicted in Figure 14. Each component would form a loop which is closed through the magnetosphere. An antisymmetric wind system could also contain transequatorial winds and these would produce a latitudinal asymmetry of the electrojet (Untiedt, 1967) by driving ionization across the equator. Asymmetrics of the ionospheric currents which may be the result of this process have been deduced from observations for both the S_q current system (see for example Hutton, 1967a, b) and the electrojet (e.g. Shuman, 1970).

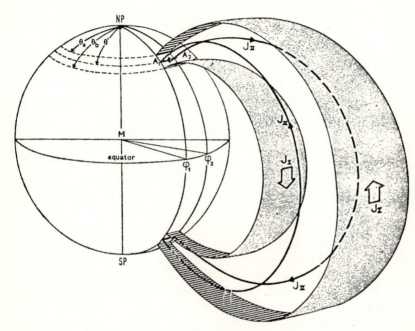

FIGURE 14   Meridional current system J_I and zonal current system J_{II} belonging to an elementary region in the ionosphere and passing through the magnetosphere along magnetic lines of force (van Sabben, 1966).

There exists another ionospheric current system which can occasionally affect the equatorial electrojet. This current system is generated at high latitudes in the vicinity of the auroral zone as a result of motions in the magnetosphere. During magnetically disturbed conditions this current system may undergo considerable enhancement and may extend to equatorial latitudes. Variations of the equatorial electrojet resulting from this process have been noted by many authors including Chapman and Akosofu (1963), Mayaud (1963), and Cain and Sweeney (1973).

4. OBSERVATIONS OF THE ELECTROJET AND ITS VARIABILITY

The first indication of the existence of the electrojet was obtained from ground-based magnetometer records of the horizontal component of the magnetic field (Johnston and McNish, 1932). The analysis of this type of data remains the most important technique of obtaining global statistical information about the electrojet because of its low cost. When interpreting such records, it should be noted that variations of the magnetic field are not only produced directly by the ionospheric current flow but also arise from currents induced in the earth as a result of that current flow.

The transient fields associated with these currents may be separated into an external (i.e. ionospheric) contribution and an internal (i.e. within the earth) contribution (e.g. Chapman and Bartels, 1940). The ratio of these contributions may be measured for each of several spherical harmonic components of the magnetic field variations and give a mean value of approximately 2.5 (Rikitake, 1966). The variation of conductivity within the earth implied by this result may be evaluated using potential theory. In the simplest treatment a spherically symmetric conductivity distribution is assumed and the earth is supposed to consist of a uniformly conducting interior surrounded by a non-conducting shell. Chapman's (1919) analysis implies a conductivity of 3.6×10^{-13} e.m.u. at depths greater than 250 km while more recent work by Rikitake (1966) leads to a conductivity of 5×10^{-12} e.m.u. for depths greater than 400 km. Attempts have been made (Lahiri and Price, 1939) to include the high conductivity of the oceans in the model by adding an outer spherically symmetric conducting surface layer of approximately 1 km in depth. This model led to significantly different results and implies a very large conductivity ($\geqslant 10^{-11}$ e.m.u.) for the earth's interior for depths in excess of approximately 600 km.

The induced current system will in fact be spatially non-uniform—e.g. as a result of conductivity contrasts between land and sea. This effect may be responsible for observed spatial variations of the magnetic field under the electrojet (Forbush and Casaverde, 1961; Gassman and Wagner, 1966). It has even been suggested that observed variations in the local longitudinal structure of the electrojet may result from modulation of the electrojet itself imposed by the current system induced within the earth (Balsley, 1970; see Figure 12).

It was originally believed that the magnetic field perturbation at ground level arising from the *electrojet*'s effect on the current system induced within the earth was small compared to that resulting from the external electrojet because of the great depth (~ 1000 km) at which the induced electrojet was produced (Onwumechilli and Ogbuehi, 1967). On the other hand because of their much greater horizontal extent induced currents resulting from the S_q *current system* produced magnetic field perturbations which were estimated to be about 40 percent of those due to the ionospheric S_q currents (Untiedt, 1967; Davis *et al.*, 1967). Thus at the equator induced currents were responsible for a 20–30 percent enhancement of the magnetic field perturbations at the ground. However Oni (1973) has pointed out that conductivity measurements of the earth indicate that in fact the electrojet induces a current in a relatively shallow layer near the earth's surface. Thus additional investigations are needed to better define the current system induced within the earth and its effect on observed equatorial magnetic field variations.

The return current for the equatorial electrojet has not been observed but model studies suggest that it must flow at latitudes greater than 10° since little return current is calculated at equatorial latitudes (Sugiura and Poros, 1969; Richmond, 1973a). Some confusion about the nature of the electrojet has arisen from the observation (Ogbuehi *et al.*, 1967) that the magnetic field perturbations under the electrojet are poorly correlated with those at other latitudes. Reasonable explanations of this feature however exist. Richmond (1973b) has pointed out that variable winds will modify the current a few degrees away from the equator without producing a significant effect on the electrojet directly over the equator. Moreover latitudinal variations of the magnetic field perturbations will result from variations in the locations of the current loops forming the S_q system (Price and Wilkins, 1963; Price and Stone, 1964; Hutton, 1967a, b; Tarpley, 1973). Thus the S_q current system will undergo meridional displacements with the location of the electrojet remaining fixed. In particular Hutton (1967a, b) concluded that the electrojet may at times be part of the northern hemisphere current loop and at other times be associated with the southern hemisphere current loop.

A seasonal variation in the strength of the electrojet with maxima in the equinoxes and of magnitude approximately 30 percent has been observed (Yacob, 1966; Tarpley, 1973; see Figure 15). Tarpley has interpreted this seasonal variation as being the result of the North–South movement of the S_q current loops due to seasonal variations in the diurnal and semi-diurnal winds responsible for the S_q current system. Day to day electrojet fluctuations of a similar magnitude have also been observed (Hutton, 1967b; Cain and Sweeney, 1973) and these variations also are probably related to variations in the neutral winds at ionospheric heights. The electrojet also possesses a solar cycle variation of approximately a factor of two (Onwumechilli, 1967).

The strength of the electrojet is observed to be a function of longitude (Onwumechilli, 1967; Cain and Sweeney, 1973). Calculations by Sugiura and Cain (1966) and by Sugiura and Poros (1969) yield a longitudinal variation

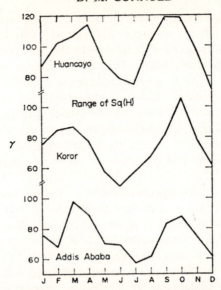

FIGURE 15   The range of averaged quiet-day $S_q(H)$ for three electrojet stations for each month.   The quiet days are from 1965 and include only days for which no value of K_p exceeds 2.   The stations are widely separated in longitude (Tarpley, 1973).

of 30 percent with maximum amplitude over Peru and minimum amplitude over India.   This effect results from longitudinal non-uniformity of the geomagnetic field, the curvature of magnetic field lines being smaller over Peru than over India.   Although the observations roughly conform to this predicted variation in electrojet amplitude there are indications of a small secondary maximum over India (Cain and Sweeney, 1973).   It is possible that this is associated with longitudinal variations in the conductivity within the earth about which there is insufficient data at the present time.

Rocket observations have suggested that at some equatorial latitudes currents flow at higher altitudes as well as at the electrojet level (Cahill, 1959; Maynard, 1967).   Such current layers have received some theoretical attention (Whitehead, 1966; Singh and Misra, 1967) but further measurements and calculations are required to establish such currents as significant features of the undisturbed equatorial ionosphere.

5. ELECTROJET IRREGULARITIES

A major feature of the electrojet which we have yet to discuss is that it is turbulent and contains a spectrum of irregularity sizes.   This feature is important not only because it allows plasma instability mechanisms to be studied under various conditions in an unlimited environment but also because the strength of the electrojet should be limited by the occurrence of turbulence (Rogister, 1971). *In situ* observations of electrojet irregularities have been made by Prakash *et al.* (1970). They observed structure at virtually all scales to which their instrument was sensitive between a few km and 1 m. Between 30 and 300 m the irregularities had a typical magnitude of 10 percent of the background electron density. However these irregularities tended to occur in layers which at night were mostly located in regions in which the electron density (on the larger scale) decreased with height. The theory of these irregularities—many of which had been previously identified with radar techniques—is based upon studies of two plasma instability mechanisms both of which contribute to the production of electrojet irregularities. Many observational studies indicate however that the actual atmospheric situation is complex and that these two instability mechanisms do not operate independently of one another.

The first of these instabilities was identified by Farley (1963) and Buneman (1963) as a result of the radar studies of the electrojet over Peru by Bowles *et al.* (1963). They observed 3 m irregularities aligned with the magnetic field and having a phase velocity of approximately 360 m/sec. Moreover the Doppler shift introduced by these irregularities was the same for radar elevation angles of 30 and 60° indicating the presence of waves travelling at 360 m/sec at angles up to 60° from the horizontal. Farley noted that for a fully-ionized plasma it was possible to transfer drift energy of the electrons to a longitudinal plasma wave propagating at the acoustic velocity of the ions provided that the drift speed exceeded the thermal velocity of the electrons. *A priori* one might suppose that drift energy could be transferred to wave energy provided that the drift speed exceeded the wave phase velocity; the electrical fields between the charged particles however produce the more stringent condition. In the ionosphere on the other hand neutral particles exert a strong control over charged particle motions as does the geomagnetic field so that for instability the electron drift velocity need only slightly exceed the ion acoustic velocity. These oscillations are found to possess wavelengths of a few meters and because of the high mobility of electrons along magnetic field lines, the irregularities of plasma density are strongly field aligned by a factor of order 100.

Additional radar studies of the electrojet over Peru revealed the presence of other irregularities which were present even when the previously described irregularities were not. Balsley (1969b) labelled these additional irregularities type II and showed that they travelled at the electron drift velocity. The explanation suggested for these irregularities was that they were produced by the motion of electrons perpendicular to a (vertical) gradient of electron density (Maeda *et al.*, 1963; Knox, 1964; Rogister and D'Angelo, 1969;

Whitehead, 1971). The observations of Prakash *et al.* (1970) provide strong evidence that this explanation is correct. To understand this instability mechanism consider the situation depicted in Figure 16 and suppose the ambient electron density increases in the vertical direction. Assume the geomagnetic field is in the direction of the y axis. Now suppose that electrons are drifting towards the west and let us impose a sinusoidal perturbation of electron density in the y direction. A charge separation will be produced as a result of the electron drift leading to an electric field E_y as depicted in the figure. As a result of this field electrons and ions will acquire a Hall drift in the vertical direction. If this Hall drift is such that a downward drift occurs

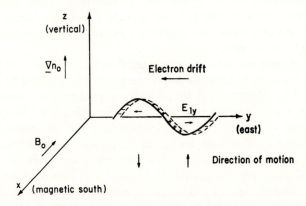

FIGURE 16   Model for drift-gradient instability. The maximum of an electron density perturbation moves downward into a region of lower density while the minimum moves upward into a region of higher density (Cunnold, 1970).

at the maximum of the density perturbation while an upward drift occurs at the minimum, relative to the ambient electron concentration the maximum is increasing while the minimum is decreasing. Instability thus results. Irregularities produced in this way will also be strongly field aligned and have wavelengths perpendicular to the magnetic field lines from tens to hundreds of meters.

The instability condition for the gradient drift instability is such that if electrons are flowing towards the west (as they are in the electrojet in the daytime) then the electron density must be increasing upward. At night when the electrojet reverses the irregularities would tend to form on the upper side of the electrojet where the electron density decreases with height. However the in situ observations of Prakash *et al.* (1970) do show that there are smaller scale vertical gradients which exceed the average vertical gradient and it is those gradients which should be used in evaluating conditions for instability. The diffusion of electrons tends to prevent the formation of the short wavelength (3 m) irregularities observed by Balsley (1969b) and thus strong smaller scale gradients, which may arise from down-scale transfer in a process analogous to that which occurs in fully-developed turbulence, are probably responsible for the generation of irregularity wavelengths less than roughly 10 m.

The observation which has led to considerable refinement of the original theory of electrojet irregularities was the radar evidence that type I irregularities can propagate at all angles to the horizontal even including vertically. This observation produced much work on the nonlinear development of the irregularities and eventually led Sudan *et al.* (1973) to propose that the gradient drift instability can produce irregularities of sufficient amplitude as to locally lead to secondary production of both type I and type II irregularities. These secondary irregularities could propagate at any angle to the horizontal. Additional details of this process as well as of the remainder of electrojet irregularity theory may be found for example in Balsley and Farley (1973) and Sato (1973).

6. SUMMARY

The equatorial electrojet consists of an intense daytime westward flow of electrons at about 105 km altitude over the equator which is driven by an eastward electric field. The electrojet is observed to reverse at night. The direct ionnospheric conductivity is an order of magnitude too small to account for the observed electrojet current and the ionospheric conductivity must therefore be enhanced considerably. It is generally accepted that this enhancement is the result of the existence of a strong vertical polarization field required in order to prevent the flow of vertical current over the equator. It is this polarization field which leads to the strong westward flow of electrons. The existence of this field is mathematically equivalent to the enhancement of the direct conductivity. The source of the *eastward* electric field is believed to be tidal motions at mid-latitudes which also produce the S_q current system. The electrojet in fact appears to be just an enhancement of the S_q current occurring directly over the magnetic equator.

Studies of the electrojet using ground based magnetometers, rockets and radars have provided much useful information on the variability of the electrojet as well as on the ionospheric current systems and the tidal motions of the neutral atmosphere. Observations of irregularities in the electrojet suggest that the amplitude of the electrojet is limited by turbulence. Moreover studies of these irregularities have led to the identification of plasma instability mechanisms functioning in the ionosphere. It appears that the irregularities may be understood in terms of a complex interaction between irregularities produced by the gradient-drift and modified two stream instabilities.

References

Ainsworth, J. E., D. E. Fox and H. E. Lagow (1961) NASA-TN-D-670.
Akasofu, S. I. and S. Chapman (1963) *J. Geophys. Res.* **68**, 2441.
Baker, W. G. and D. F. Martyn (1953) *Phil. Trans. Roy. Soc. London* **A246**, 281.
Balsley, B. B. (1969a) *J. Atmos. Terrest. Phys.* **31**, 475.
Balsley, B. B. (1969b) *J. Geophys. Res.* **74**, 2333.
Balsley, B. B. (1970) *J. Geophys. Res.* **75**, 4291.
Balsley, B. B. and D. T. Farley (1973) *J. Geophys. Res.* **78**, 7471.
Bartels J. and H. F. Johnston (1940) *J. Geophys. Res.* **45**, 485.
Bernard, R. (1974) *J. Atmos. Terrest. Phys.* **36**, 1105.
Bowles, K. L., B. B. Balsley and R. Cohen (1963) *J. Geophys. Res.* **68**, 2485.
Buneman, O. (1963) *Phys. Rev. Lett.* **10**, 285.
Burrows, K. (1970) *J. Geophys. Res.* **75**, 1319.
Cahill, L. J. (1959) *J. Geophys. Res.* **64**, 489.
Cain, J. C. and R. E. Sweeney (1973) *J. Atmos. Terrest. Phys.* **35**, 1231.
Chapman, S. (1951) *Proc. Phys. Soc. London* **B64**, 833.
Chapman, S. and R. S. Lindzen (1970) *Atmospheric Tides*, D. Reidel Publ. Co., Dordrecht, Holland, also published in *Space Sci. Rev.* **10**, 1, 1969.
Chapman, S. and J. Bartels (1940) *Geomagnetism*, Oxford University Press, chap. 22.
Cohen, R. and K. L. Bowles (1963) *J. Geophys. Res.* **68**, 2503.
Cunnold, D. M. (1970) *J. Geophys. Res.* **74**, 5709.
Davis, T. N., K. Burrows and J. D. Stolarik (1967) *J. Geophys. Res.* **72**, 1845.
Farley, D. T. (1963) *J. Geophys. Res.* **68**, 6083.
Fejer, J. A. (1953) *J. Atmos. Terrest. Phys.* **4**, 184.
Forbush, S. E. and M. Casaverde (1961) Carnegie Inst. Wash. Publ. 620.
Gassman, G. J. and R. A. Wagner (1966) *J. Geophys. Res.* **71**, 1879.
Hirono, M. J. (1952) *J. Geomag. Geoelec.* **4**, 7.
Hutton, R. J. (1967a) *J. Atmos. Terrest. Phys.* **29**, 1411.
Hutton, R. J. (1967b) *J. Atmos. Terrest. Phys.* **29**, 1429.
Johnson, F. S. (1961) *Satellite Environment Handbook*, Stanford Univ. Press, Stanford, Calif.
Johnston, J. F. and A. G. McNish (1932) *C. R. Congr. Inst. Electricité Paris* **12**, 41.
Knox, F. B. (1964) *J. Atmos. Terrest. Phys.* **26**, 239.
Lahiri, B. N. and A. T. Price (1939) *Phil. Trans. Roy. Soc. London* **A237**, 509.
Lindzen, R. S. (1970) *Geophys. Fluid Dyn.* **1**, 303.
Lindzen, R. S. (1971) *Geophys. Fluid Dyn.* **2**, 89.
Lindzen, R. S. and D. Blake (1971) *Geophys. Fluid Dyn.* **2**, 31.
Maeda, K., T. Tonda and H. Maeda (1963) *Rep. Ionos. Res. Space Res. Jap.* **17**, 3.
Martyn (1948) *Nature* **162**, 142.
Matsushita, S. (1969) *Radio Science* **4**, 771.
Mayaud, P. N. (1963) *Ann. Geophys.* **19**, 164.
Maynard, N. C. (1967) *J. Geophys. Res.* **72**, 1863.
Maynard, N. C., L. J. Cahill Jr. and T. S. G. Sastry (1965) *J. Geophys. Res.* **70**, 1241.
Ogbuehi, P. O., A. Onwumechilli and S. O. Ifeldi (1967) *J. Atmos. Terrest. Phys.* **29**, 149.
Oni, E. (1973) *J. Atmos. Terrest. Phys.* **35**, 1267.
Onwumechilli, A. (1964) *J. Atmos. Terrest. Phys.* **26**, 729.

Onwumechilli, A. (1967) In *Physics of Geomagnetic Phenomena*, *I*, ed. S. Matsushita and W. H. Campbell, Academic Press, New York, 425.

Onwumechilli, A. and P. O. Ogbuehi (1957) *J. Geomag. Geoelec., Japan* **19**, 15.

Pedersen, P. O. (1927) Propagation of radio waves, etc., Danmarks Natur. Samf., Copenhagen.

Prakash, Satya, S. P. Gupta and B. H. Subbaraya (1970) *Planet. Space Sci.* **18**, 1307.

Price, A. T. and G. A. Wilkins (1963) *Phil. Trans. Roy. Soc. London* **A256**, 31.

Price, A. T. and D. J. Stone (1964) *Ann. I. G. Y.* **35**, 62.

Price, A. T. (1968) *Geophys. J. Roy. Astron. Soc.* **15**, 93.

Richmond, A. D., S. Matsushita and J. D. Tarpley (1976) *J. Geophys. Res.* **81**, 547.

Richmond, A. D. (1973a) *J. Atmos. Terrest. Phys.* **35**, 1083.

Richmond, A. D. (1973b) *J. Atmos. Terrest. Phys.* **35**, 1105.

Rikitake, T. (1966) *Electromagnetism and the Earth's Interior*, Elsevier Pub. Co., Amsterdam.

Rogister, A. (1971) *J. Geophys. Res.* **76**, 7754.

Rogister, A. and N. d'Angelo (1970) *J. Geophys. Res.* **75**, 3879.

Sato, T. (1973) *J. Geophys. Res.* **78**, 2232.

Shuman, B. M. (1970) *J. Geophys. Res.* **75**, 3889.

Singer, S. F., E. Maple and W. A. Bowen (1951) *J. Geophys. Res.* **56**, 265.

Singh, R. N. and K. D. Misra (1967) *Nature* **214**, 375.

Sudan, R. N., J. Akinrimisi and D. T. Farley (1973) *J. Geophys. Res.* **78**, 240.

Sugiura, M. and J. C. Cain (1966) *J. Geophys. Res.* **71**, 1869.

Sugiura, M. and D. J. Poros (1969) *J. Geophys. Res.* **74**, 4025.

Tarpley, J. D. (1973) *J. Atmos. Terrest. Phys.* **35**, 1063.

Tarpley, J. D. and B. B. Balsley (1972) *J. Geophys. Res.* **77**, 1951.

Tarpley, J. D. (1970a) *Planet. Space Sci.* **18**, 1075.

Tarpley, J. D. (1970b) *Planet. Space Sci.* **18**, 1091.

Untiedt, J. (1967) *J. Geophys. Res.* **72**, 5799.

van Sabben, D. (1966) *J. Atmos. Terrest. Phys.* **28**, 965.

Watanabe, K. (1958) *Adv. in Geophys.* **5**, 153.

Whitehead, J. D. (1966) *Planet. Space Sci.* **14**, 519.

Whitehead, J. D. (1971) *J. Geophys. Res.* **13**, 3116.

Yacob, A. (1966) *J. Atmos. Terrest. Phys.* **28**, 581.

Electron Plasma Resonances in the Topside Ionosphere

J. R. MCAFEE

National Oceanic and Atmospheric Administration, Boulder, Colorado 80302, U.S.A.

A very large variety of responses can follow rf pulses in the topside ionosphere. Most of those which can not be identified as normal electromagnetic echoes show a resonant-like behavior, or a ringing for times long after the exciting pulse. The general characteristics of the major ionospheric resonances, at the plasma, upper hybrid, and cyclotron harmonic frequencies, can be explained by treating the phenomena as one of electrostatic wave propagation and reflection. The well-developed theory for the resonances at the plasma and upper hybrid frequencies, presented in detail, may also point to similar explanations at the cyclotron harmonics and also at the maximum frequencies of the Bernstein modes. The theory not only provides a check on the dispersion relations for electrostatic waves, but raises the possibility of using the observations to measure electron temperatures. Some of the other resonant-like behavior, observed more sporadically, can be explained in principle by considering possible nonlinear effects. These include the low-frequency subsidiary resonances and the diffuse resonances, as well as the resonance at twice the upper hybrid frequency. Others, like the resonance at the cyclotron frequency and proton echoes are unexplained.

1 INTRODUCTION

Bottomside ionosondes have been a standard tool for ionospheric measurement and monitoring for many years. However, since radio waves return to the ground only from the portion of the ionosphere below the level of maximum electron density (bottomside), ionosondes provide no information about the ionosphere above this level (topside). With the advent of useful research satellites, it became obvious to put a sounder in orbit above this level to perform a function in the topside ionosphere entirely analogous to the bottomside sounders. Topside sounder satellites were then launched under the ISIS program (Jackson, 1965; Chapman, 1965; Calvert, 1966; Chapman and Warren, 1968) to measure electron density profiles by radio sounding (further information on topside sounding in general may be found in the *Proceedings of the IEEE*, special issue on topside sounding and the ionosphere, Vol.57, No.6, June, 1969).

The situation turned out to be less analogous than expected because of one substantial difference: a bottomside sounder is located basically outside of the ionospheric plasma while the topside sounder is located in the plasma itself. Because of this a number of additional phenomena were observed, the most notable of which is the apparent resonant behavior of the surrounding plasma at certain frequencies. Normally the sounder transmits a short rf pulse and then listens over a certain length of time for one or more echoes from the more dense ionosphere below. In the case of the resonant-like behavior, the same, short rf pulse is followed by a ringing in the receiver which may be very strong (often saturating the receiver) and persist for a long time (> 30 msec for a 100 μsec pulse).

This stimulation and observation of ionospheric plasma resonance phenomena is interesting from a number of points of view. It is first a phenomena which, by its mere presence in a scientific experiment, demands an explanation if there is to be complete confidence in the experimental results. Additionally, since the explanations involve a good deal of plasma physics theory, the resonance observations become an experiment in themselves, relating the theory to the observation. In fact, the topside ionosphere makes an excellent laboratory plasma, with infrequent collisions, uniformity on a scale of kilometers, and comparable plasma and electron cyclotron frequencies. And finally, since various parameters of the plasma are involved in the explanation, the observations could become a diagnostic tool for measurement in the ionosphere. With this in mind, the observations of the major (or consistently seen) resonances will be discussed, their explanation in terms of plasma wave propagation presented, and then their possible application to ionospheric measurement explored. Finally, observations of the other nonelectromagnetic-echo type responses and their possible explanations will be presented.

2 TOPSIDE SOUNDER OBSERVATIONS

Topside sounders consist of an rf transmitter, a receiver tuned to the same frequency, and an appropriate antenna. A short pulse is transmitted and the subsequent response detected by the receiver. In the case of a swept-frequency sounder, the frequency is then changed slightly and the process repeated. If a sufficient frequency range is covered, echos will be received from all parts of the topside ionosphere below the sounder. The echo time delay or range as a function of frequency can be interpreted as an electron density profile, the normal aim of a sounder.

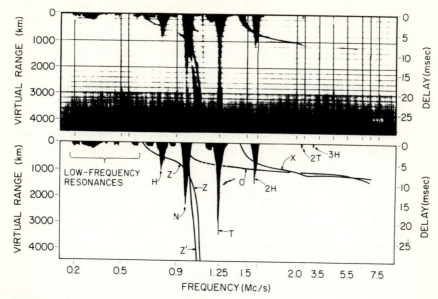

FIGURE 1   A swept-frequency ionogram showing the electromagnetic echo traces, O, X, Z and Z' and the resonances H, N, T, $2H$ and $3H$.

For purposes of analysis, the detected receiver output is put on film in the form of a raster called an ionogram. An example of a swept-frequency ionogram is shown in Figure 1. The pulse occurs at the top of the record and the response is recorded below with the darkness related to the signal strength. As the pulse frequency changes the responses are placed side by side. Hence the abscissa is the frequency and the ordinate (reading down) the delay time. Since the frequency steps are small, the individual echoes merge to form a trace. These are labeled O, X, and Z. Analysis of these traces gives the electron density profile. An additional trace, labeled Z', is caused by the oblique propagation and reflection of an electromagnetic wave.

The resonant behavior, appearing as a 'spike', is very obvious at certain frequencies. These resonances are labeled N, T, H, $2H$, and $3H$ in Figure 1. Some additional responses are also present.

Resonant-like behavior was first observed in rocket sounder experiments (Knecht et al., 1961), but it was not until the first satellite sounder, Alouette I, that extensive observations became available. By comparing the frequencies at which the 'spikes' occurred with the natural frequencies of the local plasma, a correspondence was found at the electron plasma frequency (f_N), upper hybrid frequency (f_T), and electron cyclotron frequency (f_H) and its harmonics (nf_H) (Lockwood, 1963; Calvert and

Goe, 1963). The N, T, H, $2H$ and $3H$ labels in Figure 1 refer to these
frequencies. Lockwood used the descriptive term 'spike' for the observations.
Calvert and Goe chose the alternative 'resonance' because of their proposed
explanation: the excitation of electrostatic plasma oscillations at f_N and
f_T, and cyclotron resonances at f_H and nf_H.

The wave propagation near these frequencies is characterized by a vanishing
group velocity either along or across the field. This occurs at the electron
plasma, cyclotron, and upper hybrid frequencies in the cold plasma theory.
The introduction of a finite temperature in the form of a Maxwell-Boltzmann
distribution gives this type of behavior additionally at the harmonics of the
electron cyclotron frequency (Bernstein modes). Since such slow waves exist
at the proper frequencies, a very simple explanation consisted of the excitation
of standing waves or waves moving with the sounder which interact continuously to provide the ringing (Fejer and Calvert, 1964; Sturrock, 1965;
Nuttall, 1965; Dougherty and Monaghan, 1965; Deering and Fejer, 1965;
Shkarofsky, 1968, 1970; Dougherty, 1969).

There is also more to be considered in discussing the resonance observations
than the persistent ringing. Further details are available from the results of
the fixed frequency sounders. With the frequency held constant the changes
in the response are due to the spatial changes in the ionosphere itself as the
sounder moves along its orbit. Since the ionospheric plasma normally changes
slowly compared with the frequency sweep rate of a swept-frequency sounder,
the effect is to stretch out any particular response over a much longer period
of time. Instead of passing through a resonance over a few pulses, then, the
resonance may be observed over many consecutive pulses (Calvert and Van-
Zandt, 1966). The significance of the term, spike, is also lost, and the term,
resonance, becomes more suitable.

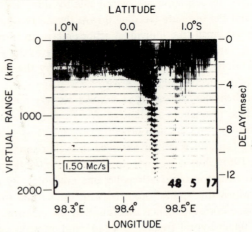

FIGURE 2   A fixed-frequency ionogram of the resonance at the plasma frequency.

An example of a resonance observation at the plasma frequency by the Explorer XX topside sounder is shown in Figure 2. The format is the same as for the swept-frequency ionogram except that the abscissa is now real time and position rather than frequency. The individual pulses and responses can be resolved as vertical lines. The resonance actually extends over a much longer time period than the 45 seconds shown. The change in the resonance strength and duration is actually due to the rotation of the dipole antenna. Of particular interest is the modulation which is typical of the resonance at the plasma frequency. The resonance at the upper hybrid frequency is quite similar, as may be seen in Figure 3, which shows both the plasma (top) and upper hybrid resonances over a period of about 4 minutes. The resonances at harmonics of the electron cyclotron frequency show a much more complicated modulation pattern, however, as is evident in Figure 4, illustrating the resonances at 3 (top) and $4f_H$. The periodicity is again due to the antenna rotation. Two distinct patterns seem evident.

Much of the material concerning the observations of the major resonances and their early explanations may be found in reviews (Calvert and McAfee, 1969; Thomas and Landmark, 1969; Thomas and Landmark, 1970). These theories were unable to account for much of the behavior in a resonance. The strength and duration of the resonances seemed far too large and no explanation of the modulation was put forward with any enthusiasm. Because of the empirical relationship between the observed frequencies and the natural plasma resonant frequencies, it is still obvious, however, to look for better interpretations in terms of the behavior of the plasma waves near these frequencies, and this requires a knowledge of the dispersion relation.

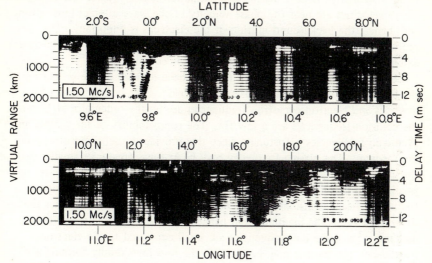

FIGURE 3   Fixed-frequency ionograms of the resonances at the plasma (top) and upper hybrid frequencies.

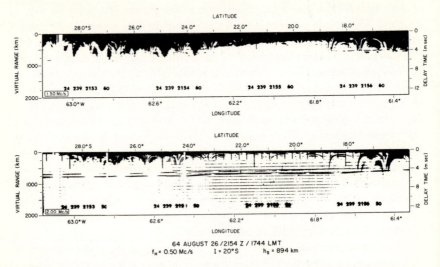

FIGURE 4   Fixed-frequency ionograms of the resonances at three (top) and four times
the electron cyclotron frequency.

3 THE DISPERSION RELATION FOR ELECTROSTATIC WAVES

Since the explanation for the major topside sounder resonance observations will
involve waves with both group and phase velocities small compared to the
velocity of light, the cold plasma approximation is inadequate. Those particles
whose velocities are near the phase velocity of the wave interact most strongly
and, hence, the background motion of the electrons is important. The
problem of waves in the plasma must be treated by combining the Boltzmann
equation with Maxwell's equations (the Vlasov equations).

Several simplifying assumptions may be made because of the properties of
the ionospheric plasma in the observation region. The typical plasma frequencies *and* electron cyclotron frequencies are on the order of a MHz, while
the collision frequencies are a few Hz. The comparability of the plasma and
cyclotron frequencies is fortunate since this provides a useful anisotropy of
the medium. The plasma frequency may be at times greater or smaller than
the cyclotron frequency and the upper hybrid frequency may be greater or
smaller than harmonics of the cyclotron frequency. The small collision
frequency means collisions may be neglected. This not only simplifies the
theory but keeps the medium from being very dissipative due to collisional
damping. Since the ion frequencies are small compared to the electron
frequencies, ion motion may be ignored, the ions then simply forming a
neutralizing background. Another simplification involves using the electrostatic approximation, which has proved sufficient for the plasma resonance.

Maxwell's equations for plane waves

$$\nabla \times \mathbf{B} = \frac{4\pi \mathbf{J}}{c} + \frac{1}{c} \frac{\partial \mathbf{E}}{\partial t} \tag{1}$$

$$\nabla \times \mathbf{E} = -\frac{1}{c} \frac{\partial \mathbf{B}}{\partial t} \tag{2}$$

may be Fourier analyzed in both time and space and combined to give

$$\frac{c^2}{\omega^2} \mathbf{k} \times (\mathbf{k} \times \mathbf{E}) + \frac{4\pi i}{\omega} \mathbf{J} + \mathbf{E} = 0 \tag{3}$$

where plane wave solutions varying as $\exp i\,(\mathbf{k} \cdot \mathbf{r} - \omega t)$ have been assumed so that $\mathbf{k}$ is the propagation vector and ω the angular frequency. The form of (3) is usually simplified to

$$\mathbf{n} \times (\mathbf{n} \times \mathbf{E}) + \mathbf{K} \cdot \mathbf{E} = 0 \tag{4}$$

where $\mathbf{n} \equiv kc/\omega$ has the magnitude of the refractive index and the direction of the propagation vector. The tensor $\mathbf{K}$ is an equivalent dielectric tensor for the medium which relates the electric field with the electron currents in the form

$$\mathbf{K} \cdot \mathbf{E} = \mathbf{E} + \frac{4\pi i}{\omega} \mathbf{J} . \tag{5}$$

In the cold plasma theory, $\mathbf{K}$ is a function of the wave frequency and the plasma and cyclotron frequencies. By including the background motion it becomes not only a function of the additional plasma property, temperature, but also of the propagation vector itself. This last quality makes the dispersion relation difficult to analyze unless the electrostatic approximation can be employed.

Electrostatic waves are characterized by their electric field being derivable from a potential

$$\mathbf{E} = -\nabla \varphi = -i\mathbf{k}\,\varphi$$

so that $\mathbf{E}$ is parallel to $\mathbf{k}$. The electrostatic approximation is obtained by taking the scalar product of $\mathbf{n}$ into Eq. (4) giving

$$\mathbf{n} \cdot \mathbf{K} \cdot \mathbf{E} = 0 . \tag{6}$$

If the electric field is split into components parallel and perpendicular to $n(k)$, i.e. $E = E_{\parallel} + E_{\perp}$, and when $|E_{\parallel}| \gg |E_{\perp}|$, then (6) becomes

$$n \cdot K \cdot n = 0 \tag{7}$$

which is the electrostatic approximation. The validity of (7) may be seen by rewriting (4) in terms of $E_{\parallel}$ and $E_{\perp}$

$$(n^2 - K \cdot) E_{\perp} = K \cdot E_{\parallel} \ .$$

Then as long as $|n^2| \gg |K_{ij}|$ for all i and j, $|E_{\perp}| \ll |E_{\parallel}|$ and Eq. (7) is valid.

The electrostatic approximation of (7) may be written in terms of its components. If it is assumed that the propagation vector lies in the $x-z$ plane of a Cartesian coordinate system xyz with the background magnetic field in the z-direction, then (7) becomes

$$k_x^2 \, K_{xx} + k_x k_z \, (K_{xz} + K_{zx}) + k_z^2 \, K_{zz} = 0 \ .$$

This further simplifies, since by the Onsager relations $K_{xz} = K_{zx}$, so that

$$k_x^2 \, K_{xx} + 2 k_x k_z \, K_{xz} + k_z^2 \, K_{zz} = 0 \tag{8}$$

is the dispersion relation in the electrostatic approximation.

It is now necessary to find the current, J, due to the electric field, E, in order to determine K from (5). Writing J in terms of the electron distribution function, f,

$$J = -e \int f(v) \, v \, d^3 v \tag{9}$$

where $f(v)$ is found from the collisionless Boltzmann equation

$$\frac{\partial f}{\partial t} + v \cdot \frac{\partial f}{\partial r} - e \left(E + \frac{v}{c} \times B \right) \cdot \frac{\partial f}{\partial v} = 0 \ . \tag{10}$$

Handling of the Boltzmann equation is difficult and involved even for a first order solution. To illustrate the procedure the first order perturbation to the velocity distribution may be calculated in the Lagrange system of coordinates, i.e., coordinates following the zero-order electron trajectories.

For the trajectory, $r = r(t)$, the rate of change of the distribution function along the trajectory is

$$\frac{df}{dt} = \frac{\partial f}{\partial t} + \frac{\partial f}{\partial \mathbf{r}} \cdot \frac{d\mathbf{r}}{dt} + \frac{\partial f}{\partial \mathbf{v}} \cdot \frac{\partial \mathbf{v}}{\partial t} .$$

The zero-order trajectory, calculated in a static magnetic field, $\mathbf{B}_0$, gives

$$\left(\frac{df_0}{dt}\right)_0 = \frac{\partial f}{\partial t} + \mathbf{v} \cdot \frac{\partial f}{\partial \mathbf{r}} - \frac{e \mathbf{v}}{m c} \times \mathbf{B}_0 \cdot \frac{\partial f}{\partial \mathbf{v}}$$

which is the collisionless Boltzmann equation, i.e. $(df_0/dt)_0 = 0$, and has a general solution independent of $\mathbf{r}$ and t, $f_0 = f_0 (v_{\parallel}, v_{\perp}^2/2)$ where $\parallel$ and $\perp$ refer to directions with respect to $\mathbf{B}_0$.

The first order equation along the trajectory is then

$$\left(\frac{df_1}{dt}\right)_0 + \left(\frac{df_0}{dt}\right)_0 = \frac{e}{m} \left(\mathbf{E}_1 + \frac{\mathbf{v}}{c} \times \mathbf{B}_1\right) \cdot \frac{\partial f_0}{\partial \mathbf{v}} = 0$$

and since the second term is identically zero

$$\left(\frac{df_1}{dt}\right)_0 = \frac{e}{m} \left[\mathbf{E}_1 + \frac{\mathbf{v}}{c} \times \mathbf{B}_1\right] \cdot \frac{\partial f_0}{\partial \mathbf{v}}$$

which when integrated along the trajectory from $t' = -\infty$ to $t' = t$ gives

$$f_1(\mathbf{r},\mathbf{v},t) = \frac{e}{m} \int_{-\infty}^{t} dt' \left[\mathbf{E}_1(\mathbf{r}',t') + \frac{\mathbf{v}}{c} \times \mathbf{B}_1(\mathbf{r}',t')\right] \cdot \frac{\partial f_0(\mathbf{v}')}{\partial \mathbf{v}'} \qquad (11)$$

This choice of limits is valid for growing solutions, $f_1(t) \to 0$ in the $t \to -\infty$ limit. Other solutions are obtained by the analytic continuation of (11). A further discussion of the justification of this method is given in Stix (1962), along with the details of calculating f_1 and its moments to find $\mathbf{J}$ and hence $\mathbf{K}$. In doing so the only assumptions made are that the background distribution function is a Maxwellian across the field and a shifted Maxwellian along the field

$$f_0 = N\left(\frac{m}{2\pi k}\right)^{3/2} \left(\frac{1}{T_{\perp}}\right)\left(\frac{1}{T_{\parallel}}\right)^{1/2} \exp\left[-\frac{m(v_x^2 + v_y^2)}{k T_{\perp}} - \frac{m(v_z - V_z)^2}{k T_{\parallel}}\right]$$

when N is the electron density, the magnetic field is in the z direction, and $\parallel$ and $\perp$ again refer to along and across the field, respectively. The possibility of streaming along the background field is included in a streaming velocity, V_z.

The elements of the dielectric tensor needed in (8) are

$$K_{xx} = 1 + \frac{\omega_p^2}{\omega^2} e^{-\lambda} \sum_{l=-\infty}^{\infty} \frac{l^2 I_l(\lambda)}{\lambda} \left[\frac{T_\perp}{T_\parallel} - 1 + i \left[\alpha_l \frac{T_\perp}{T_\parallel} - \frac{l\Omega}{k_z} \left(\frac{m}{2 k T_\parallel} \right)^{1/2} \right] f_0 \right.$$

$$K_{xz} = \frac{\omega_p^2}{\omega^2} e^{-\lambda} \frac{k_x}{k_z} \sum_{e=-\infty}^{\infty} \frac{l^2 I_l(\lambda)}{\lambda} \left\{ \left(1 - \frac{T_\perp}{T_\parallel} \right) - i \frac{(\omega + l\Omega)}{l\Omega} \right.$$

$$\left. \left[\alpha_l \frac{T_\perp}{T_\parallel} - \frac{(l\Omega)}{k_z} \left(\frac{m}{2 k T_\parallel} \right)^{1/2} \right] F_0(\alpha_l) \right\}$$

$$K_{zz} = 1 + \frac{\omega_p^2 e^{-\lambda}}{\omega^2 k_z^2} \frac{m}{k T_\perp} \sum_{l=-\infty}^{\infty} I_l(\lambda) \left\{ \omega^2 + l^2 \Omega^2 \left(\frac{T_\perp}{T_\parallel} \right) \right.$$

$$\left. + i(\omega + l\Omega)^2 \left[\alpha_l \frac{T_\perp}{T_\parallel} - \frac{l\Omega}{k_z} \left(\frac{m}{2 k T_\parallel} \right)^{1/2} \right] F_0(\alpha_l) \right\} \tag{12}$$

where $\omega_p^2 = \dfrac{4\pi N e^2}{m}$ $\qquad \Omega = \dfrac{e B_0}{mc}$

$$\lambda = \frac{k_x^2 k T_\perp}{m\Omega^2} \qquad \alpha_l \equiv \frac{\omega - k_z V_z + l\Omega}{k_z} \left(\frac{m}{2 k T_\parallel} \right)^{1/2}$$

I_l = Bessel function of imaginary argument

and $\quad F_0(x) = \left[\pi^{1/2} \dfrac{k_z}{|k_z|} + 2i \displaystyle\int_0^x dt \exp(t^2) \right] \exp(-x^2)$.

Substituting (12) in (8) gives

$$k_x^2 + k_z^2 + \frac{\omega_p^2 m \, e^{-\lambda}}{k T_\parallel} \sum_{l=-\infty}^{\infty} I_l(\lambda) \left\{ 1 + i \left[\alpha_l - \frac{l\Omega}{k_z} \left(\frac{m}{2 k T_\parallel} \right)^{1/2} \frac{T_\parallel}{T_\perp} \right] \right.$$

$$\left. F_0(\alpha_l) \right\} = 0 \tag{13}$$

Eq. (13) is the complete electrostatic dispersion relation, but although it is much simpler in form than might be expected, it must be reduced even further by the proper approximations to become tractable.

The integral function, $F_0(\alpha_l)$, for $|\alpha_l| \gg 1$ is given by

$$F_0(\alpha_l) = \pi^{1/2} \frac{k_z}{|k_z|} \exp(-\alpha_l^2) + \frac{i}{\alpha_l} \left(1 + \frac{1}{2\alpha_l^2} + \frac{3}{4\alpha_l^4} + \cdots \right)$$

The first term contains the damping and may conveniently be neglected for large $|\alpha_l|$. For propagation across the field, $k_z = 0$, and $\alpha_l \to \infty$, so that it may be neglected rigorously. In this case, using the fact that $I_{-l}(\lambda) = I_l(\lambda)$, (13) becomes

$$1 - 2\omega_p^2 e^{-\lambda} \sum_{l=1}^{\infty} \frac{I_l(\lambda)}{\lambda} \frac{l^2}{(\omega^2 - l^2\Omega^2)} = 0 \tag{14}$$

Eq. (14) describes the familiar Bernstein modes (Bernstein, 1958; Crawford, 1965). For $\lambda \ll 1$ the Bessel functions may be approximated by

$$I_l(\lambda) \approx \frac{1}{l!} \left(\frac{\lambda}{2}\right)^l + \text{higher order terms}$$

and (14) reduces to

$$1 - \frac{\omega_p^2}{\omega^2 - \Omega^2} - 2\omega_p^2 \sum_{l=2}^{\infty} \frac{\lambda^{l-1} l^2}{2^l l!(\omega^2 - l^2\Omega^2)} = 0 \tag{15}$$

One solution for $\lambda \to 0$ ($k_x \to 0$) is $\omega = (\omega_p^2 + \Omega^2)^{1/2}$, which is the upper hybrid frequency. Other solutions occur for ω near $l\Omega$, but note that $l > 1$ which makes the fundamental different from the harmonics of the cyclotron frequency. Letting $\omega = L\Omega(1 + \Delta)$ where $L > 1$ is a particular integer and $|\Delta| \ll 1$ so that ω is near a cyclotron harmonic, and dropping all terms in the series except the Lth, (15) may be solved for Δ to give

$$\Delta \approx \frac{\omega_p^2}{2\Omega^2 L!} \frac{(\omega^2 - \Omega^2)}{(\omega^2 - \omega_T^2)} \left(\frac{\lambda}{2}\right)^{L-1}$$

where ω_T is the upper hybrid frequency ($\omega_T^2 = \omega_p^2 + \Omega^2$). Since $\omega > \Omega$, for $\lambda > 0$, Δ is positive for $\omega > \omega_T$ and negative for $\omega < \omega_T$. Hence there are solutions for frequencies just below or above each cyclotron harmonic depending upon whether the harmonic is greater than or less than the upper hybrid frequency, respectively. This type of analysis may also be carried out for large λ giving solutions just above each cyclotron harmonic (including the fundamental).

Propagation entirely along the field, $k_x = 0$, may also be considered, in which case (13) reduces to

$$k_z^2 + \omega_p^2 \frac{m}{\ell T_\parallel} \left[1 + i\alpha_0 \, F_0(\alpha_0) \right] = 0 \tag{16}$$

For $|\alpha_0| \gg 1$ $(k_z \to 0)$

$$F_0(\alpha_0) \simeq \frac{1}{\alpha_0} \left(1 + \frac{1}{2\alpha_0^2} + \frac{3}{4\alpha_0^4} \right)$$

and (16) becomes

$$1 - \frac{\omega_p^2}{(\omega - k_z V_z)^2} - \frac{3\omega_p^2 \, k_z^2 \, \ell T_\parallel}{(\omega - k_z V_z)^4 \, m} \approx 0 \, .$$

As $k_z \to 0$, $\omega \to \omega_p$, which is the plasma resonance frequency.

The parallel and perpendicular propagation described by (16) and (14) gives an indication of the resonant behavior at the plasma, upper hybrid and cyclotron harmonic frequencies. In general however, propagation will not be conveniently along or across the field, so it is necessary to consider arbitrary directions. Restricting the analysis to the plasma and upper hybrid resonances so that the wave frequency is not near a cyclotron harmonic (cases where there is coincidence between these are ignored), Eq. (13) may be simplified by assuming both k_x and k_z are reasonably small, so that $\lambda \ll 1$ and $|\alpha_l| \gg 1$ for all l, and keeping only the first-order correction terms both along and across the field.

With the following approximations

$$e^{-\lambda} \approx 1 - \lambda + \ldots$$

$$I_l(x) \approx \frac{1}{l!} \left(\frac{\lambda}{2} \right)^l + \ldots$$

$$F_0(\alpha_l) \approx \frac{i}{\alpha_l} \left(1 + \frac{1}{2\alpha_l^2} + \frac{3}{4\alpha_l^4} + \ldots \right)$$

and keeping terms linear in $\ell T_\parallel/m$ or $\ell T_\perp/m$ Eq. (13) becomes

$$Pk_z^2 + Sk_x^2 - \omega_p^2 \left\{ \frac{\ell T_\parallel}{m} k_z^2 \left[\frac{3k_z^2}{\omega^4} + \frac{(3\omega^2 + \Omega^2) k_x^2}{(\omega^2 - \Omega^2)^3} \right] \right.$$

$$\left. + \frac{\ell T_\perp}{m} k_x^2 \left[\frac{3k_x^2}{(\omega^2 - \Omega^2)(\omega^2 - 4\Omega^2)} + \frac{k_z^2(3\omega^2 - \Omega^2)}{\omega^2(\omega^2 - \Omega^2)^2} \right] \right\} = 0 \qquad (17)$$

where $P \equiv 1 - \dfrac{\omega_p^2}{\omega^2}$, $\quad S \equiv 1 - \dfrac{\omega_p^2}{\omega^2 - \Omega^2}$

and a coordinate transformation has been made so that $V_z = 0$. The terms P and S are zero at the plasma and upper hybrid frequencies, respectively. This allows k_z or k_x to become large and the terms involving temperature to become significant. Assuming that both k_z and k_x do not become large simultaneously, the two cross terms involving $k_x^2 k_z^2$ may be ignored and (17) becomes

$$Pk_z^2 + Sk_x^2 - 3\omega_p^2 \left[\frac{\ell T_\parallel k_z^4}{m\omega^4} + \frac{\ell T_\perp k_x^4}{m(\omega^2 - \Omega^2)(\omega^2 - 4\Omega^2)} \right] = 0. \qquad (18)$$

The second temperature term may be dropped for the plasma resonance and the the first for the upper hybrid resonance. Eq. (18) may be rewritten in terms of the total refractive index, $n^2 = n_x^2 + n_z^2$, the angle between the propagation vector and the magnetic field (with axial symmetry), $\theta \equiv \tan^{-1} k_x/k_z$, and the velocities normalized to the speed of light, $\gamma_\parallel^2 \equiv \ell T_\parallel/mc^2$ and $\gamma_\perp^2 \equiv \ell T_\perp/mc^2$, to become

$$n^2 = \frac{\omega^2 \sec^2\theta \ [P + S \tan^2\theta]}{3\omega_p^2 \left[\gamma_\parallel^2 + \dfrac{\omega^4 \tan^4\theta \ \gamma_\perp^2}{(\omega^2 - \Omega^2)(\omega^2 - 4\Omega^2)} \right]} \qquad (19)$$

The numerator in (19) is the cold plasma dispersion relation:

$$\tan^2\theta = -P/S.$$

An example of the appearance of the refractive index surface, n vs. θ, is shown in Figure 5 for the plasma resonance with $\omega \gtrsim \omega_p > \Omega$ and for the upper hybrid resonance with $\omega \lesssim \omega_T < 2\Omega$. The electromagnetic part has been included in the sketches (the hub on the left and the inner surface on the right). The electrostatic mode fills in the cold plasma resonance cone (dotted line) at some large but finite refractive index. Of particular interest is the strong dependence of refractive index on the angle with the magnetic field.

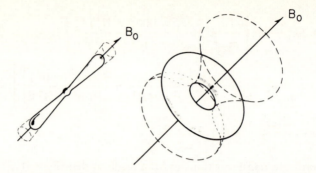

FIGURE 5   The refractive index surface when just above the plasma frequency (left) and just below the upper hybrid frequency.

The behavior of the refractive index as a function of frequency as a para-meter is shown in Figure 6 for the plasma resonance in terms of the normal-ized frequency difference $(\omega - \omega_p)/\omega = \Delta$. The sensitivity of n to small changes in frequency is obvious, both in the large value of n along the field direction and in the resonance cone angle, while the electromagnetic region (dotted line) is relatively insensitive. The upper hybrid resonance is similarly sensitive, but near $\theta = 90°$ rather than $\theta = 0$ and for ω below the resonant frequency.

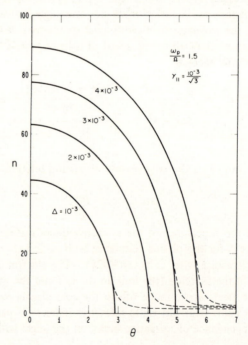

FIGURE 6   Refractive index, n, vs. propagation angle from the magnetic field, θ.

The importance of the large values of refractive index and its sensitivity to the direction of the propagation vector and to its frequency when near the resonant frequency become clearer when the resulting group velocities are considered. The group velocity $\mathbf{v}_g$ is derivable from the dispersion relation (18) or (19) through the standard relation (Stix, 1962)

$$\mathbf{v}_g = \frac{\partial \omega}{\partial \mathbf{k}} = \hat{x}\frac{\partial \omega}{\partial k_x} + \hat{y}\frac{\partial \omega}{\partial k_y} + \hat{z}\frac{\partial \omega}{\partial k_z}$$

so that, for the plasma resonance only, the dispersion relation and the three components of group velocity are

$$Pn^2 \cos^2\theta + Sn^2 \sin^2\theta - Qn^4 \cos^4\theta = 0$$

$$u_x = -\frac{2n_x Sc}{D}$$

$$u_y = -\frac{2n_y Sc}{D} \tag{20}$$

$$u_z = 2n_z \frac{(Qn^2 \cos^2\theta + S \tan^2\theta)c}{D}$$

where

$$D \equiv n^2\omega \left(\cos^2\theta \frac{\partial P}{\partial \omega} + \sin^2\theta \frac{\partial S}{\partial \omega}\right) + 4Qn^4 \cos^4\theta$$

$$Q \equiv 3\left(\gamma_\parallel \frac{\omega_p}{\omega}\right)^2 .$$

The y component has been included for completeness by using the axial symmetry and letting

$$k_x^2 \to k_x^2 + k_y^2 .$$

Examples of the group velocity magnitude and direction are given in Figures 7 and 8 for the same circumstances as in Figure 6. The normalized magnitude, v_g/c, is very small along the field direction, even compared to the normalized thermal velocity, $\gamma_\parallel = 10^{-3}$. The direction of the group velocity goes from parallel to the propagation vector ($\varphi = \theta$) when $\theta \approx 0$ to perpendicular to the propagation vector ($\varphi = \theta + 90°$) when $\theta > 0$. The small velocity magnitude and the large variation in its direction for small changes in the propagation vector account for the properties observed by the topside sounders. To relate these to the received signal requires the determination of the energy flow in an inhomogeneous medium through ray tracing.

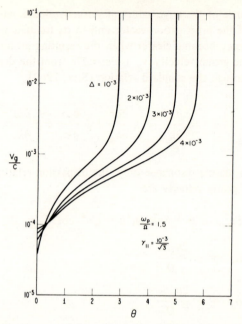

FIGURE 7　Normalized group velocity magnitude, V_g/c , vs. propagation angle from the magnetic field, θ .

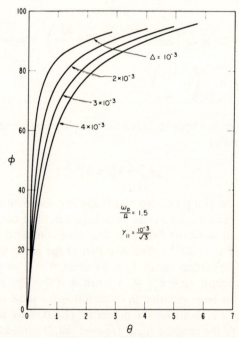

FIGURE 8　Group velocity angle from the magnetic field, ϕ , vs. propagation angle from the magnetic field, θ .

4 RAY TRACING NEAR THE PLASMA AND UPPER HYBRID FREQUENCIES

Ray tracing is the process of determining the trajectory of a particular wave packet. In a homogeneous media this energy travels in a straight line. In an inhomogeneous media, such as the ionosphere, there may be a great deal of refraction and reflection. The problem is simplified whenever geometrical optics may be used, i.e. when the wavelength is short compared to the scale of changes in refractive index (Budden, 1961). This is not always possible since in some cases there are abrupt changes in the plasma and in others the refractive index changes rapidly because of a particular sensitivity to some plasma parameter.

When the use of geometrical optics is proper, a ray may be traced by using the local group velocity, v_g . The ray trajectory is then defined by

$$\frac{d\mathbf{r}}{dt} = \mathbf{v}_g(\mathbf{r}) . \tag{21}$$

The group velocity may be determined from the dispersion relation, as in (20), for the particular plasma parameters of the local medium. Since there are a continuum of solutions to the dispersion relation, a particular ray corresponds to a particular choice of initial conditions, for instance, an initial propagation vector. At other positions in the plasma the refractive index will change, but is related by Snell's law, which states that the refractive index components along the directions of non-changing plasma are constant. Eq. (21) may be integrated to give ray trajectories. First, however, much insight into the types of trajectories may be gained graphically.

H. Poeverlein (1948) introduced a graphical method which makes use of the fact that the direction of the group velocity is always perpendicular to the refractive index surface $n(\theta)$. By graphing the refractive index surface for various points in the plasma, and by making use of Snell's law, a qualitative idea of the trajectories may be obtained. An example is given in Figure 9 for the plasma resonance. Only a small part of the refractive index surfaces are shown; the origin is far to the left. The assumptions made are that the horizontal magnetic field is constant, the electron density varies vertically, and propagation is in the magnetic meridian (the plane defined by ∇N and $\mathbf{B}_0$) to make the problem two-dimensional. The three curves then correspond to three levels in the ionosphere. The inner one corresponds to the highest density (smallest $\omega_p - \omega$ difference). In the topside ionosphere this corresponds to a low altitude level. The outer curve is then a higher level and the middle curve an intermediate level. In considering a particular ray trajectory, an initial propagation vector is chosen. This then defines the

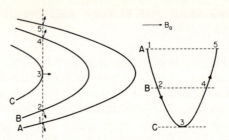

FIGURE 9   Poeverlein construction and possible ray path for the resonance at the plasma frequency.

propagation vector at any other level through Snell's law, represented by the vertical dotted line conserving the horizontal component. Starting at point 1, the group velocity is initially downward and to the right. Since the wave moves downward, the plasma frequency increases and the wave packet reaches the level defined by the middle curve, point 2, where the group velocity is still downward but more toward the right. Eventually, the wave packet reaches a level where the group velocity is horizontal, point 3, which is the reflection. The wave then travels upward through point 4 to 5, or the original level. The simple trajectory chosen here is sketched in at the right. Other choices of the initial wave can lead to more complex trajectories, however.

A second example is shown in Figure 10, again with the origin to the left but with a larger initial angle between the propagation vector and the magnetic field. In this case the group velocity is initially downward and to the left, point 1. By the time the wave reaches point 2 it is vertical. At reflection, point 3, it is horizontal and to the right, as in the first example. It then returns to the initial level through point 4 to point 5. Since there is travel both to the left and right during the trajectory, the point at which the wave returns to the initial level may either be to the left or the right of the point at which it began depending upon the magnitude of the group velocity along the trajectory. Two possible trajectories are illustrated.

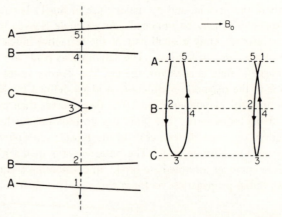

FIGURE 10   Poeverlein construction and two possible ray paths for the resonance at the plasma frequency.

The two graphical examples point out the types of possible trajectories in this simple case. In the first example the wave ends up on the right. In the second it ends up on the left (or a larger initial propagation angle may be chosen to achieve this). For some intermediate case, then, the wave can return to its point of origin. Also implicit is the fact that the refractive index changes very little during the trajectories shown. Hence, the wavelength is insensitive and the use of geometrical optics is quite reasonable. To substantiate these possibilities it is necessary to calculate actual ray trajectories (McAfee, 1968; McAfee, 1969a; McAfee, 1969b; McAfee, 1970a; Aubry *et al.*, 1970; Bitoun *et al.*, 1970; Fang and Andrews, 1971).

In general the magnetic field and electron density gradient will not be perpendicular. It is then more convenient in ray tracing to use an orthogonal coordinate system (x_1, x_2, x_3) based upon the electron density gradient rather than the (x, y, z) system based upon the field. If x_3 is taken along the gradient and the $x_1 - x_3$ plane to contain the magnetic field, since x and y are arbitrary, x_2 may be chosen along y. If α is the angle between x_1 and the field, the two coordinate systems are related by a simple rotation through α about y.

The refractive index components are then related by

$$n_1 = n_x \sin \alpha + n_z \cos \alpha$$

$$n_2 = n_y \tag{22}$$

$$n_3 = n_x \cos \alpha - n_z \sin \alpha$$

and similarly the group velocity by

$$u_1 = u_x \sin \alpha + u_z \cos \alpha$$

$$u_2 = u_y \tag{23}$$

$$u_3 = u_x \cos \alpha - u_z \sin \alpha$$

when the field direction has a negative 3 component. Since the plasma does not vary horizontally, n_1 and n_2 are constants along a trajectory. Using x_3 as a variable of integration, the trajectory may be written in a parametric form from (21) as

$$x_{1,2}(x_3) = \int_0^{x_3} dx_3' \frac{u_{1,2}(x_3')}{u_3(x_3')} \tag{24}$$

$$t(x_3) = \int_0^{x_3} dx_3' \frac{1}{u_3(x_3')}$$

(24)

giving the other two coordinates and the time of travel.

In order to do the integrations of (24) conveniently, a good deal of simplification may be made. Restricting the analysis to the resonance at the plasma frequency, (18) is

$$Pn_z^2 + S(n_x^2 + n_y^2) - Qn_z^4 \approx 0$$

(25)

where $Q \equiv 3(\omega_p/\omega)^2 \gamma_\parallel^2$, $S \equiv 1 - \omega_p^2/(\omega^2 - \Omega^2)$, and $P \equiv 1 - \omega_p^2/\omega^2$.

If $\omega \approx \omega_p$ both Q and S are insensitive to changes in both ω_p and $\Omega(\omega_p \neq \Omega)$, at least compared to the sensitivity of P . (In the upper hybrid resonance, the sensitivity is in S with respect to both ω_p and Ω , however.) The variation of the dispersion relation is all due to the variation in $P(\omega_p^2)$. Since $\omega_p^2 \propto N$, the electron density, P may be written

$$P(N) = 1 - \frac{\omega_{po}^2}{\omega^2} \left(\frac{N}{N_0}\right)$$

(26)

where the zero subscript refers to any convenient point. The assumption is now made that the density varies linearly with x_3 so that $N = N_0(1 + x_3/H)$ where H is a scale factor. In an exponentially varying ionosphere, H would be the scale height and the linear variation would be appropriate as long as $|x_3| \ll H$. Now (26) becomes

$$P(x_3) = 1 - \frac{\omega_{po}^2}{\omega^2} - \frac{x_3}{H} = P_0 - \frac{\omega_{po}^2}{\omega^2} \frac{x_3}{H}$$

(27)

where $P_0 \equiv P(x_3 = 0)$ and x_3 is positive in the direction of increasing N .

The ray trajectory may now be written with (27) in (24) as

$$x_{1,2}(\beta) = -\frac{H\omega^2}{\omega_{po}^2} \int_{\beta_0}^{\beta} d\beta' \frac{u_{1,2}(\beta')}{u_3(\beta')} \frac{\partial P(\beta')}{\partial \beta'}$$

(28)

$$x_3(\beta) = \frac{H\omega^2}{\omega_{po}^2} [P(\beta_0) - P(\beta)]$$

$$t(\beta) \quad = \quad -\frac{H\omega^2}{\omega_{po}^2} \int_{\beta_0}^{\beta} d\beta' \; \frac{1}{u_3(\beta')} \; \frac{\partial P(\beta')}{\partial \beta'} \tag{28}$$

and the integration may be over any convenient parameter, β.

Using the quantities

$$\beta \equiv n_x/n_z$$

$$a \equiv n_1/n_z = \cos\alpha + \beta \sin\alpha \tag{29}$$

$$\epsilon \equiv n_2/n_1$$

(25) may be written

$$P(\beta) = \frac{Qn_1^2}{a^2} - S\epsilon^2 a^2 - S\beta^2 \; . \tag{30}$$

This choice of β makes the ray trajectory single-valued. The other quantities, a and ϵ, appear naturally. Substituting (20) and (29) into (23) gives the group velocity components

$$u_1(\beta) = \frac{2cn_1}{Da^3} \left[Qn_1^2 \cos\alpha + Sa^2 \beta(\beta \cos\alpha - \sin\alpha) + S\epsilon^2 a^4 \cos\alpha \right]$$

$$u_2(\beta) = -\frac{2c \, S\epsilon n_1}{D} \tag{31}$$

$$u_3(\beta) = -\frac{2cn_1}{Da^3} \left[Qn_1^2 \sin\alpha + Sa^3 \beta + S\epsilon^2 a^4 \sin\alpha \right]$$

where

$$D = \frac{n_1^2}{a^4} \left[4Qn_1^2 + \omega a^2 \frac{\partial P}{\partial \omega} + \omega a^2 (\beta^2 + a^2 \epsilon^2) \frac{\partial S}{\partial \omega} \right]$$

and differentiating (30) gives

$$\frac{\partial P}{\partial \beta} = -\frac{2}{a^3} \left[Qn_1^2 \sin\alpha + Sa^3 \beta + S\epsilon^2 a^4 \sin\alpha \right] \tag{32}$$

By substituting (3) and (32) into (28) and integrating, the parametric equations for the ray path become

$$x_1(\beta) = H \frac{\omega^2}{\omega_{po}^2} \cot \alpha \left\{ \left[\frac{Qn_1^2}{a_0^2 a^2} + S\epsilon^2 + \frac{S}{\sin^2 \alpha} \right] \left[a_0^2 - a^2 \right] \right. $$

$$\left. - \frac{2S(1 + \cos^2 \alpha)}{\sin \alpha \cos \alpha} (\beta_0 - \beta) - \frac{2S}{\sin^2 \alpha} \ln\left(\frac{a}{a_0}\right) \right\}$$

(33)

$$x_2(\beta) = -2H \frac{\omega^2}{\omega_{po}^2} \epsilon S (\beta_0 - \beta)$$

$$x_3(\beta) = H \frac{\omega^2}{\omega_{po}^2} \left[P(\beta_0) - P(\beta) \right]$$

$$t(\beta) = \frac{H}{c} \frac{\omega^2}{\omega_{po}^2} n_1 \left\{ \frac{4Qn_1^2}{3 \sin \alpha} \left(\frac{1}{a^3} - \frac{1}{a_0^3} \right) + \frac{\omega \partial P}{\partial \omega} (\beta_0 - \beta) \right.$$

$$+ \omega \frac{\partial S}{\partial \omega} \left[(\epsilon^2 + \csc^2 \alpha)(\beta_0 - \beta) + \frac{\cos^2 \alpha}{\sin^3 \alpha} \left(\frac{1}{a} - \frac{1}{a_0} \right) \right.$$

(33)

$$\left. \left. + \frac{2 \cos \alpha}{\sin^3 \alpha} \ln\left(\frac{a}{a_0}\right) \right] \right\}$$

$$P(\beta) = Qn_1^2 - S\epsilon^2 - S\beta^2 \ .$$

The set (33) is the general solution for the ray path for a frequency ω given the plasma properties, ω_p , Ω , $T_\parallel$, α , and H , along with a choice of initial conditions β_0 , n_1 , and ϵ satisfying (30). The scale factor H appears only as a normalization. All frequencies appear as ratios, so they may also be normalized. If it is assumed that the wave was generated by a sounder pulse at $t = 0$, $t(\beta)$ corresponds to a delay time. The problem now is to find the proper ray trajectories so that energy returns to the sounder. This could be found by setting $x_1(\beta) = x_2(\beta) = x_3(\beta) = 0$ in (33) and finding the proper initial wave conditions. The case where $\alpha = 0$ will be illustrated because it leads to simple solutions and the results illustrate the general behavior. The set (33) then becomes

$$x_1(\beta) = +\frac{2H\omega^2}{\omega_{po}^2}\left[(Qn_1^2 + S\epsilon^2)(\beta_0 - \beta) + \frac{S}{3}(\beta_0^3 - \beta^3)\right]$$

$$x_2(\beta) = -\frac{2H\omega^2}{\omega_{po}^2}\,\epsilon S(\beta_0 - \beta) \qquad\qquad (34)$$

$$x_3(\beta) = -\frac{H\omega^2}{\omega_{po}^2}\,S(\beta_0^2 - \beta^2)$$

$$t(\beta) = \frac{H\omega^2}{\omega_{po}^2}\frac{n_1}{c}\left[\left(4Qn_1^2 + \frac{\omega\partial P}{\partial\omega} + \frac{\omega\partial S}{\partial\omega}\,\epsilon^2\right)(\beta_0 - \beta) + \frac{\omega}{3}\frac{\partial S}{\partial\omega}(\beta_0^3 - \beta^3)\right].$$

To return to the starting point $(x_1 = x_2 = x_3 = 0)$ requires $\epsilon = 0 \rightarrow x_2 = 0$, $\beta = -\beta_0 \rightarrow x_3 = 0$ and then

$$x_1(-\beta_0) = \frac{4H\omega^2\beta_0}{\omega_{po}^2}\left[Qn_1^2 + \frac{S}{3}\beta_0^2\right]$$

giving (for $x_1 = 0$)

$$n_1^2 = -\frac{S\beta_0^2}{3Q}.$$

Substituting into the dispersion relation

$$P(-\beta_0) = Qn_1^2 - S\beta_0^2 = -\frac{4}{3}S\beta_0^2$$

the time of propagation is then

$$t(-\beta_0) = \frac{2H\omega^2}{\omega_{po}^2}\frac{n_1}{c}\beta_0\left[4Qn_1^2 + \frac{\omega\partial P}{\partial\omega} + \frac{\omega}{3}\frac{\partial S}{\partial\omega}\beta_0^2\right]$$

giving

$$t(\omega) = \frac{H}{2c}\frac{\omega^2}{\omega_{po}^2}\left(\frac{3}{-SQ}\right)\left[\frac{\omega\partial P}{\partial\omega} + \left(1 - \frac{\omega}{4S}\frac{\partial S}{\partial\omega}\right)\left(1 - \frac{\omega_{po}^2}{\omega^2}\right)\right]\left(1 - \frac{\omega_{po}^2}{\omega^2}\right) \qquad (35)$$

Hence, a frequency-delay time relationship has been established for this case. The basic mechanism for the strength and persistence of the resonance observation may be seen when this relationship is investigated.

An example of $t(\omega)$ for $\omega_p/\Omega = 1.5$, $\gamma_{\parallel} = 3^{-1/2} \times 10^{-3}$ and $H = 500$ km is shown in Figure 11 (solid line). The sensitivity is apparent in the wide range of delay times for a relatively small frequency band. The resonance observation may now be looked at not in terms of resonant interaction between sounder and plasma at a particular frequency, but as a wave propagation phenomena which occurs over a range of frequencies. Since a sounder pulse lasts a finite time (normally 100 μsec giving a power spectrum bandwidth of 20 kHz) energy is available over a range of frequencies which may give rise to wave propagation and return over all listening times. If $f = 1$ MHz, for example, the frequency bandwidth for $0 < t < 20$ msec is about 3 kHz, a band easily excited by a single pulse. On the other hand the 3 kHz band is sufficiently wide to allow a large part of the pulse spectrum to contribute to the returning energy. This increases the amount of energy available in the observation. The plasma, in effect, sorts out frequencies in the power spectrum of the transmitted pulse to keep energy returning constantly. Since in practice the returning signals are detected (only the envelope is seen), the result is what appears to be a ringing in the receiver. In essence, the resonances are completely analogous to the electromagnetic traces (O, X, and Z in Figure 1), but with a slope that appears to be infinite.

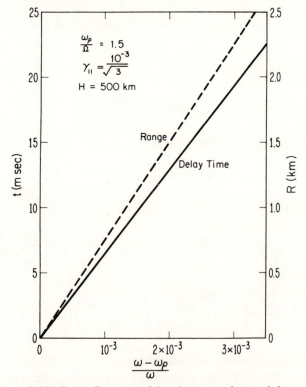

FIGURE 11   Frequency-delay time-range characteristics.

The distance down to the level of reflection, the range, is also shown in Figure 11 as the dashed curve. Even though the delay times are large, the waves do not travel far because of the low group velocity magnitude. This is an important consideration since a linearly varying electron density has been assumed. This assumption is excellent since even the 2.5 km is ½% of the 500 km scale height.

The low group velocity brings up another consideration, however, which is not normally of any significance: namely, the motion of the sounder. Usually, the fact that the sounder has moved between the emitted pulse and the received echo is unimportant because the sounder velocity is much smaller than the wave group velocity. With electrostatic waves this is no longer true, and, besides making the choice of the correct ray trajectories a little more difficult, it leads to an interesting result.

As an example, assume that $\alpha = 0$ so that (34) is valid, and let the satellite velocity be along the field line. Then for a wave packet to intercept the satellite it must return to the level of emission and in the original magnetic meridian but be displaced along the field direction by the distance the sounder has moved while the ray trajectory is being traversed. The first two considerations give $x_2(\beta) = x_3(\beta) = 0$ to again yield $\epsilon = 0$, $\beta = -\beta_0$. Instead of $x_1(\beta) = 0$ being the third condition, however, it is now $x_1(\beta)/t(\beta) = V_s$, the sounder velocity.

Up to this point only half of the refractive index surface $(-90° < \theta < 90°)$ has been considered. In the first example, there are solutions corresponding to $\theta' = 180° - \theta$ in which the ray paths are traversed in the opposite directions, all other things being equal. This direct correspondence disappears in the present example since the sounder velocity introduces an asymmetry into the system. If there are also solutions for $x_1(\beta)/t(\beta) = -V_s$, these will contribute from the half of the refractive index surface opposed to the direction of the sounder. In Figure 10, for example, a ray path is shown which ends up to the left of the origin. The same ray path may be traversed in the opposite direction by using the other half of the refractive index surface so that reflection occurs with the group velocity to the left while the ray path ends up on the right. This type of trajectory is referred to as the backward wave and the more normal type as the forward wave.

The motion of the sounder has one other effect. Even though its velocity is small, the refractive index may be large enough to make the Doppler shift of the received signal of significance. Because the propagation vectors are nearly opposite for the two types of ray paths, forward and backward, the Doppler shifts are opposite and the importance is magnified.

These effects may be seen in the example of Figure 12 for the same conditions as Figure 11 but with the sounder motion considered. The two sets of curves refer to the solutions for the forward and backward waves with the dashed curves being the solutions in the plasma rest frame and the solid curves being the solutions in the sounder frame with the Doppler shifting of the received signal. The sounder motion splits the solutions from the two halves of the refractive index surface by an amount which is small compared to the wave frequency but is on the same order as the frequency variations with time. The Doppler shifts enhance the splitting.

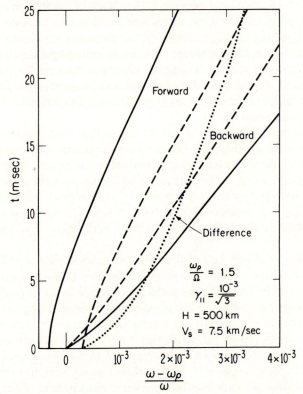

FIGURE 12   Frequency-delay time characteristics with a receiver velocity included. The dashed line is not Doppler shift corrected.

All of the previous discussion about the reason for the strength and persistence of the resonance observation pertains here also. Energy may be returned to the sounder during the entire listening period with contributions from a bandwidth which would cover a reasonable portion of the transmitted power spectrum. In addition, however, a new effect is possible. Since at any given delay time, t , there are two returning waves of different frequencies, it might be expected that the two waves would add to give a 'beating'. The detection of this signal would give rise to a modulation whose frequency is the difference frequency of the two waves. The normalized difference frequency is also plotted in Figure 12 as the dotted curve and is seen to increase with delay time.

The beating of the two waves accounts for the fringe pattern observed in the resonance at the plasma frequency. The plasma resonance typically has a modulation with a period which decreases with delay time, in agreement with the typically increasing beat frequency in Figure 12.

An additional feature of the resonance at the plasma frequency is the difference in behavior depending upon whether the plasma frequency is greater or less than the electron cyclotron frequency. The previous examples are general for $\omega_p > \Omega$. When $\omega_p < \Omega$ the ray paths are entirely different because of the different sign of the quantity S at $\omega = \omega_p$:

$$S(\omega = \omega_p) = -\Omega^2/(\omega_p^2 - \Omega^2)$$

which is negative for $\omega_p > \Omega$, positive for $\omega_p < \Omega$. Looking at (20) for example, as θ increases away from zero (along the field), n either increases or decreases depending upon the sign of S . Both u_x and u_y are directly proportional to S , and $u_z = 0$ is a possibility for $\theta \neq 90°$ only when S is negative. In (35) there are no solutions for positive S , and with the sounder velocity included solutions are possible only for delay times up to some small value (McAfee, 1970b). The apparent weakness of the plasma resonance for $\omega_p < \Omega$ has also been supported observationally (Benson, 1971). The change in behavior is, of course, not accidental since the Whistler mode is present at the plasma frequency only when $\omega_p < \Omega$.

The results of the analysis for the simple examples presented here carry over into the more complicated general case of arbitrary gradient-field geometry and arbitrary sounder motion, although care must be taken as the dip angle becomes large. When the sounder motion is out of the magnetic meridian, the two curves for the forward and backward waves tend to merge until, when the motion is perpendicular to the field lines, the solutions are identical due to the restoration of symmetry.

This type of analysis may be carried out for the upper hybrid resonance, as well. Since the equations are of the same form as in the plasma resonance, similar ray paths are possible. There are some important differences however. The dependence upon *both* ω_p and Ω as expressed in S has already been noted.   Since the upper hybrid resonance is basically across the field (the propagation vector is nearly perpendicular) the ray paths are rotated by 90° from the plasma resonance. Another difference which is not so obvious is the fact that the upper hybrid resonance is caused by reflection from the *less* dense and/or *lower* magnetic field region of the plasma. This is because electrostatic waves propagate below the upper hybrid resonant frequency, rather than above. The inner curve in Figure 9 for the upper hybrid resonance is still the closest to the resonant frequency, but this means the resonant frequency is smaller since $\omega < \omega_T = (\omega_p^2 + \Omega^2)^{1/2}$. The upper hybrid exhibits a change in character for $\omega_T >$ or $< 2\Omega$ similar to the plasma resonance for $\omega_p >$ or $< \Omega$. This is caused by the change in the sign of the temperature term in (18) at $\omega = 2\Omega$ (P is always positive at the upper hybrid resonance). The ray paths are possible for $\omega_T < 2\Omega$. The difference in behavior is a result of the different properties of the Bernstein mode above and below the second cyclotron harmonic.

Although the electrostatic approximation proves adequate at the plasma frequency, an electromagnetic wave with vanishing group velocity is present near the upper hybrid resonance. This affects the refractive index surface sufficiently (see Figure 5) to necessitate its inclusion in the dispersion relation. This is accomplished by adding the term

$$1 - \frac{\omega_p^2(2\omega^2 - \omega_p^2)}{\omega^2(\omega^2 - \Omega^2)} \approx P(\omega = \omega_T)$$

to (18). Other terms from the full dispersion relation do not significantly contribute. Integrating (24) in the case of the hybrid resonance gives then the parametric ray paths

$$
\begin{aligned}
x_1(\beta) = \frac{2H\,\omega_p^2\,n_0^2}{\sin\alpha\,\omega_T^2} &\left\{ -Q(\beta - \beta_0)\left[\frac{n_1}{n_0} - \frac{\cos\alpha}{2}\,(\beta + \beta_0)\right] \right. \\
&+ \frac{P\sin\alpha\tan\alpha}{2\,n_0^2}\left(\frac{1}{n_\perp^2} - \frac{1}{n_{\perp 0}^2}\right)\left(1 - n_0^2\tan^2\alpha\right) \\
&+ P\tan^2\alpha\,\frac{n_1}{n_0}\left(\frac{\beta}{n_\perp^2} - \frac{\beta_0}{n_{\perp 0}^2}\right) \\
&- P\tan\alpha\sec^2\alpha\,\frac{n_1}{n_2 n_0^2}\left[\arctan\left(\frac{\beta n_0 - n_1\sec\alpha}{n_2\tan\alpha}\right)\right. \\
&\left. - \arctan\left(\frac{\beta_0 n_0 - n_1\sec\alpha}{n_2\tan\alpha}\right)\right] - \frac{P\sec\alpha\tan^2\alpha}{2\,n_0^2}\ln\left(\frac{n_\perp^2}{n_{\perp 0}^2}\right)\Bigg\}
\end{aligned}
\tag{36}
$$

$$
\begin{aligned}
x_2(\beta) = \frac{2H\,\omega_p^2\,n_0 n_2}{\sin\alpha\,\omega_T^2} &\left\{ -Q(\beta - \beta_0) - \frac{P\sec\alpha\,n_1}{n_0 n_2^2}\left(\frac{1}{n_\perp^2} - \frac{1}{n_{\perp 0}^2}\right)(1 + n_0^2\tan^2\alpha) \right. \\
&+ \frac{P}{n_2^2}\left(\frac{\beta}{n_\perp^2} - \frac{\beta_0}{n_{\perp 0}^2}\right)(1 - n_0^2\tan^2\alpha + 2n_1^2\sec^2\alpha)
\end{aligned}
$$

$$+ \frac{P \tan \alpha}{2n_0 n_2^3} \ (1 + n_0^2 \tan^2 \alpha) \left[\arctan \left(\frac{\beta n_0 - n_1 \sec \alpha}{n_2 \tan \alpha} \right) \right.$$

$$\left. - \arctan \left(\frac{\beta_0 n_0 - n_1 \sec \alpha}{n_2 \tan \alpha} \right) \right] \Bigg\}$$

$$x_3(\beta) = \frac{H \omega_p^2 \ n_0^2 (\beta - \beta_0)}{\omega_{T_0}^2} \left\{ \frac{2 \cot \alpha}{\sin \alpha} \left(\frac{P}{n_1^2 n_{10}^2} - Q \right) \left[\frac{n_1}{n_0} - \frac{\cos \alpha}{2} (\beta + \beta_0) \right] \right.$$

$$\left. + \frac{P n_0^2}{n_1^2 n_{10}^2} \left[\left(\beta + \beta_0 - \frac{2 n_1}{n_0} \ \beta \beta_0 \csc \alpha \cot \alpha \right) \right] \right\}$$

$$t(\beta) \quad = \quad \frac{2 H n_0}{c \sin \alpha} \ (\beta - \beta_0)$$

$$S(\beta) \quad = \quad - \frac{P(1 + n_0^2 \beta^2)}{n_1^2} - Q n_1^2$$

where $\quad \beta \quad \equiv \quad n_z / n_0$

$$n_0^2 \quad \equiv \quad n_1^2 \csc^2 \alpha + n_2^2$$

$$n_1^2 \quad \equiv \quad n_0^2 - 2 n_1 n_0 \beta \csc \alpha \cot \alpha + n_0^2 \beta^2 \cot^2 \alpha$$

$$Q \quad \equiv \quad \left(\frac{\omega_T^2}{4 \Omega^2 - \omega_T^2} \right) \left(\frac{3 k \ T_\perp}{m c^2} \right)$$

$$\omega_T^2 \equiv \quad \omega_{T_0}^2 \ (1 + x_3 / H) \ .$$

Both the plasma and upper hybrid resonances can be interpreted in terms of the propagation and reflection of electrostatic waves in the plasma. The basic assumption made is that electrostatic waves are excited by the pulse. The excitation of the wave by the sounder and the applicability of the wave packet treatment may be seen in more detail by the calculation of the response of a nonhomogeneous plasma to a short pulse.

5 CALCULATION OF THE ELECTRIC FIELD

Calculations of the response of a plasma to a small, pulsed dipole formed a part of the early explanations of resonance observations (Nuttall, 1965; Sturrock, 1965; Deering and Fejer, 1965). In all cases, however, the medium was assumed to be homogeneous. The simple inclusion of a slowly varying medium in the calculation (Fejer and Yu, 1970) confirmed the ray tracing results and yielded additional information.

The field arising from an infinitesimal impulsive source at the origin is given in terms of a Fourier integral by

$$E(\mathbf{r},t) \propto \int \frac{\exp\,[-i(\omega t - \mathbf{k}\cdot r)]}{D(\omega,\mathbf{k})} \, d\omega \, dk_1 \, dk_2 \, dk_3 \qquad (37)$$

where $D = 0$ is the dispersion equation in a homogeneous medium. Integration of (37) over k_3 would give

$$E(\mathbf{r},t) \propto \sum_j \int \frac{\exp\,[-iF(\omega, k_1, k_2, x_3)]\; d\omega \, dk_1 \, dk_2}{\left[\dfrac{\partial}{\partial k_3} D(\omega, k_1, k_2, k_3)\right]_{k_3 = k_{3j}}} \qquad (38)$$

where $F \equiv \omega t - k_1 x_1 - k_2 x_2 - k_{3j} x_3$ and the k_{3j} are the roots of D at x_3. If the medium is allowed to vary slowly in the x_3 direction, the plane wave solutions may be replaced by WKB solutions and (38) then becomes

$$E(\mathbf{r},t) \propto \sum_j \int \frac{M(\omega, k_1, k_2, x_3) \exp\,[-iF(\omega, k_1, k_2, x_3)]\; d\omega \, dk_1 \, dk_2}{\left[\dfrac{\partial}{\partial k_3} D(\omega, k_1, k_2, k_3)\right]_{k_3 = k_{3j}}}$$

$$(39)$$

where $\quad f = \omega t - k_1 x_1 - k_2 x_2 - \displaystyle\int_0^{x_3} k_{3j}(\omega, k_1, k_2, x)\, dx \qquad (40)$

and M is a slowly varying amplitude factor which is unity for $x_3 = 0$.

The evaluation of (39) may be carried out by a method such as that of stationary phase (Budden, 1961). This is done by finding the values ω_s, k_{1s}, k_{2s} which satisfy

$$\frac{\partial F}{\partial \omega} = \frac{\partial F}{\partial k_1} = \frac{\partial F}{\partial k_2} = 0 \qquad (41)$$

and represent the saddle points of F in the complex ω, k_1, k_2 planes. The integration in (39) may then be carried out by expanding F about a saddle point. The connection with ray tracing is through ω_s, k_{1s} and k_{2s} which represent a wave packet whose group velocity trajectory takes it from the origin to $\mathbf{r}$ in a time t.

Since the dispersion equations are given in Section 3, the roots of D may be found and used to evaluate (40), (41) and then (39). The simplest example is the case of the resonance at the plasma frequency with a magnetic field in the x_1 direction (horizontal) and no receiver motion $(x_1 = x_2 = x_3 = 0)$.

The dispersion relation is from (25)

$$D = \frac{k_1^2 c^2}{\omega^2} P(x_3) + \frac{(k_2^2 + k_3^2)}{\omega^2} c^2 S - \frac{k_1^4 c^4}{\omega^4} Q = 0 \qquad (42)$$

and since it is only sensitive to ω in the term $P(x_3)$, (42) may be simplified, with $\omega \approx \omega_{po}$, to

$$D = \frac{c^2 S}{\omega_{po}^2} \left\{ k_3^2 - \left[\frac{k_1^4 c^2 Q}{\omega_{po}^2 S} - \frac{k_1^2 P(x_3)}{S} - k_2^2 \right] \right\} \qquad (43)$$

where $\quad P \approx 1 - \dfrac{\omega_{po}^2}{\omega^2} - \dfrac{\omega_{po}^2}{\omega^2} \dfrac{x_3}{H}$

$$S \approx \frac{-\Omega^2}{\omega_{po}^2 - \Omega^2}$$

$$Q \approx 3 \gamma_{\parallel}^2 .$$

The roots of (43) are then

$$k_3 = \pm \left[\frac{k_1^4 c^2 Q}{\omega_{po}^2 S} - \frac{k_1^2 P(x_3)}{S} - k_2^2 \right]^{1/2} . \qquad (44)$$

For the integral in (40) to have meaning when $x_3 = 0$, both roots must be used. The crossover occurs for $k_3 = 0$ which corresponds to a reflection level

$$x_r = \frac{H}{\omega_{po}^2} \left[\omega^2 - \omega_{po}^2 - k_1^2 c^2 Q + S\, \omega_{po}^2 \frac{k_2^2}{k_1^2} \right]. \tag{45}$$

Because of symmetry, the integral in (40) is twice its value between 0 and x_r and with $x_1 = x_2 = 0$ (40) then becomes

$$F = \omega t - \frac{4H\,|k_1|}{3\omega_{po}^3(-S)^{1/2}} \left[\omega^2 - \omega_{po} - k_1^2 c^2 Q + S\, \omega_{po}^2 \frac{k_2^2}{k_1^2} \right]^{3/2}. \tag{46}$$

The application of (41) to (46) then gives the saddle point values

$$\omega_s = \omega_{po} \left[1 + \frac{ct}{2H} \left(\frac{-SQ}{3} \right) \right]^{1/2}$$

$$k_{1s} = \frac{\omega_{po}}{2} \left(\frac{t}{Hc} \right)^{1/2} \left(\frac{-S}{3Q} \right)^{1/4} \tag{47}$$

$$k_{2s} = 0$$

where the last term restricts the propagation to the magnetic meridian.

The function F may now be expanded about this point defined by (47)

$$F = [F]_s + \frac{1}{2} \left[\frac{\partial^2 F}{\partial \omega^2} \right]_s (\omega - \omega_s)^2 + \frac{1}{2} \left[\frac{\partial^2 F}{\partial k_1^2} \right]_s (k_1 - k_{1s})^2$$

$$+ \frac{1}{2} \left[\frac{\partial^2 F}{\partial k_2^2} \right]_s (k_2 - k_{2s})^2 + \left[\frac{\partial^2 F}{\partial \omega\, \partial k_2} \right]_s (\omega - \omega_s)(k_1 - k_{1s})$$

$$+ \left[\frac{\partial^2 F}{\partial \omega\, \partial k_2} \right]_s (\omega - \omega_s)(k_2 - k_{2s}) + \left[\frac{\partial^2 F}{\partial k_1\, \partial k_2} \right] (k_1 - k_{1s})(k_2 - k_{2s})$$

$$+ \cdots \tag{48}$$

where the subscript s refers to evaluation at the saddle point described in (47).

With the use of (46) and (47) in (48) and approximating the other terms in (39) by their values at the saddle point (39) may be integrated to give

$$E \propto t^{-1/2} \exp i \, [F]_s$$

(49)

$$[F]_s = \omega_{po} t + \frac{ct}{4H} \left(\frac{-SQ}{3} \right)^{1/2} \omega_{po} t$$

Thus, the decay of the field is proportion to $t^{-1/2}$ while the wave frequency increases linearly with time. This decay, in contrast to $t^{-5/2}$ for a homogeneous plasma (Deering and Fejer, 1965), is more consistent with observation. The actual wave frequency is given by

$$\omega = \frac{\partial [F]_s}{\partial t} = \omega_{po} \left[1 + \frac{ct}{2H} \left(\frac{-SQ}{3} \right)^{1/2} \right] = \omega_s$$

(50)

and is the same as (35) with $\omega \partial P/\partial \omega \approx 2$ and neglecting the second term.

Several checks are available concerning the range of validity for the assumptions involved in this derivation (some apply to the ray tracing method as well). The assumption that $\omega \approx \omega_{po}$ requires from (50) that

$$t \ll \frac{2H}{c} \left(\frac{3}{-SQ} \right)^{1/2} .$$

(51)

The use of the WKB approximation, slow spatial variation, follows from $x_r(k_{3j})_s \gg 1$ with the use of (44), (45), and (47) to give

$$t^2 \gg \frac{16H}{\omega_{po} c(-3SQ)^{1/2}} .$$

(52)

The requirement in (52) is also necessary for the self-consistency of the stationary phase approximation.

The extension of this type of analysis to a general case for the resonances at both the plasma and upper hybrid frequencies is similar in principle. An arbitrary magnetic field direction can be included in the dielectric tensor (as in Section 4) and the motion included by letting $\mathbf{r} = \mathbf{V}_s t$, where $\mathbf{V}_s$ is the receiver (satellite) velocity. The results may be summarized for the two resonances as follows (Graff, 1970; Graff, 1971a; Graff, 1971b; Feldstein and Graff, 1972):

Near the plasma frequency

$$\omega(t) = \omega_p \left\{ 1 + \frac{1}{2} |\cos \alpha| \left(\frac{-SQ_p}{3} \right)^{1/2} \frac{ct}{H_p} + \frac{(1 + \sec^2 \alpha) V_3 t}{4H_p} \right.$$

$$\left. + \frac{1}{4} (1 + 2 \cos^2 \alpha)(V_1 - V_3 \tan \alpha) \left(\frac{-S}{3Q_p} \right)^{1/4} \left(\frac{t}{cH_p |\cos \alpha|} \right)^{1/2} \right\} \tag{53}$$

$$t \gg \frac{2}{(-3SQ_p |\cos \alpha|)^{1/4}} \left(\frac{H_p}{\pi \omega_p c} \right)^{1/2} \tag{54}$$

$$t \gg \left(\frac{3}{-SQ_p} \right)^{1/2} \frac{(V_1 - V_3 \tan \alpha)^2 H_p}{8\pi Q_p c^3 |\cos^3 \alpha|} \tag{55}$$

$$4\pi \left(\frac{-SQ_p}{3} \right)^{1/2} |\cos^3 \alpha| \frac{c}{|V_3|} \gg 1$$

Near the upper hybrid frequency

$$\omega(t) = \omega_T \left[1 - b(PQ_T)^{1/2} \frac{(1 + 2q^2)}{(1 + q^2)^{1/2}} + \frac{b^2 V_3 t}{4H_T} \right.$$

$$\left. \mp \left(\frac{V_2 b^{1/2}}{2c \tanh \eta} \right) \left(\frac{P}{Q_T} \right)^{1/4} \frac{q^2}{(1 + q^2)^{3/4}} \right] \tag{56}$$

with

$$q \equiv \frac{b^2 \sin \alpha \, ct}{4.3^{1/2} H_T}$$

and

$$\eta \equiv \sinh^{-1} \left[\frac{V_2}{V_1 \sin \alpha + V_3 \cos \alpha} \right]$$

$$\tag{5$$

$$t \gg \frac{4H_T}{cb^{3/2}} \left(\frac{P}{Q_T} \right)^{1/4} \left(\frac{c}{\omega_T H P \sin^2 \alpha} \right)^{1/3} \left[b^{1/2} \left(\frac{P}{Q_T} \right)^{1/4} (1 + q^2)^{1/4} \mp \frac{V_2}{4c Q_T \tanh \eta} \right]^{-1/}$$

$$t \ll \frac{2H(PQ_T)^{1/2}}{|V_3|b^2}$$

(58)

$$Q_T^{3/4} \left| \frac{c \tan \eta}{V_2} \right| \gg 1$$

where all values are local and the parameters are defined as follows: ω = received frequency; t = time from pulse; Ω, ω_p, ω_T = electron cyclotron, plasma, upper hybrid frequencies; c = speed of light; m = electron mass; k = Boltzmann's constant; $T_\parallel$, $T_\perp$ = electron temperature along and across the field (Maxwell-Boltzmann distribution); subscripts 1, 2, 3 refer to an orthogonal coordinate system with the gradient of the resonant frequency (ω_p or ω_T) in the 3 direction and the magnetic field direction in the 1-3 plane; α = angle between the magnetic field direction and the 1 direction, positive away from the 3 direction; V_1, V_2, V_3 = 3 components of receiver velocity; H_p = scale height of plasma frequency; H_T = scale height of upper hybrid frequency (involves both the gradients in electron density and magnetic field strength);

$$Q_p = \frac{3kT_\parallel}{mc^2} = 5.05 \times 10^{-10} \ T_\parallel(°K) \ ;$$

$$Q_T = \frac{3\omega_p^2 \ kT_\perp}{(4\Omega^2 - \omega_T^2) \ mc^2} = \frac{5.05 \times 10^{-10} \ \omega_p^2}{(4\Omega^2 - \omega_T^2)} \ T_\perp(°K) \ ;$$

$$S = 1 - \frac{\omega_p^2}{\omega^2 - \Omega^2} \to \frac{-\Omega^2}{\omega_p^2 - \Omega^2} \ (\omega \approx \omega_p) \ ;$$

$$P = 1 - \frac{\omega_p^2}{\omega^2} \to \frac{\Omega^2}{\omega_T^2} \ (\omega \approx \omega_T) \ ;$$

$$b = \frac{\omega_p}{\omega_T} \ ; \ \text{and}$$

Equations (53) and (56) are doubled valued, the upper sign referring to a forward type wave, the lower to a backward type wave. Equations (54) and (57) are the conditions necessary for the WKB approximations (and ray tracing) while (55) and (58) concern linearization approximations made in obtaining closed form solutions. When (55) or (58) do not hold it is necessary to solve the equations by numerical techniques.

6 OTHER RESONANT-LIKE BEHAVIOR

Most of the other resonances have not been investigated in as much detail as those at the plasma and upper-hybrid frequencies. Some can probably be explained in a similar manner, by electrostatic wave propagation, although the analysis is more difficult. Many of them are not consistently observed, requiring much more specific conditions. Some of these have been attributed to non linear effects and others have no explanation as yet.

The upper hybrid resonance is a special case of the general Bernstein mode resonances. The resonances at the harmonics of the electron cyclotron frequency should also involve the Bernstein mode. Although general ray tracing calculations have not been performed near a cyclotron harmonic, it has been shown that refractive index surfaces for $f \geqslant nf_H > f_T$ have the same characteristics as those for f_N and f_T (Andrews and Fang, 1971). This, at least, suggests that the same type of mechanism is responsible for observations of the cyclotron harmonics, also.

The reason for the lack of detailed calculations regarding the cyclotron harmonics is that the refractive index (and hence group velocity) is much more complicated. Near the plasma and upper hybrid frequencies, the damping (Landau) can be neglected as long as the phase velocity stays well above the electron thermal velocity. This only necessitates staying close to the resonant frequency, which is normally sufficient in explaining observations. The damping (cyclotron) near the cyclotron harmonics is, however, a function of the direction of propagation and is present at all frequencies. Perpendicular propagation is always undamped, but propagation slightly away from perpendicular soon becomes heavily damped. This not only complicates the refractive index surfaces, and hence the calculation of the ray paths, but introduces losses into waves which might be more important otherwise. The cyclotron harmonic resonances do decay more quickly and the damping may be the reason.

The Bernstein modes have additional frequencies at which the group velocity is zero. These occur at a maximum frequency, f_{Qn}, between each pair of cyclotron harmonics (above f_T) for a finite and large propagation vector. Resonance observations were actually predicted at these frequencies because of this vanishing group velocity (Dougherty and Monaghan, 1965) before they were identified on topside ionograms (Warren and Hagg, 1968). The frequencies of the Bernstein maxima can be found from Eq. (14). Since these occur for $\lambda \sim 1$, however, the solutions are not easily found. An approximate formula, $f_{Qn} = f_H(n + 0.464 \, f_N^2/n^2 f_H^2)$, is valid whenever the second term is much less than unity, or $n > 2$ for $f_N \sim f_H$. The existence of refractive index surfaces near these frequencies which are of the form to produce reflections has been demonstrated (McAfee, 1969b; Andrews and

Fang, 1971). Ray paths calculated by Muldrew (1972) do not show much refraction, however, and a simple group velocity — satellite velocity matching may be sufficient. Both a forward and backward wave solution can be found so that beating is also possible in these resonances.

It appears then that both the resonances at the cyclotron harmonics, nf_H , and at the Bernstein maxima, f_{Qn} , are a wave propagation phenomena. This, however, may not be true for the resonance at the cyclotron frequency itself. As was mentioned previously, the fundamental is quite different from its harmonics in terms of the dispersion relation. Attempts to find modes of propagation near f_H which could account for the resonance observations have not been successful. It may be that the resonance is simply due to the gyration of electrons in phase due to the pulse. These electrons are free to interact with the antenna until collisions or field inhomogeneities destroy the phase relationship (Tataronis and Crawford, 1970). Still, the resonance at the fundamental is so much in character with those at the harmonics, that it seems more likely that a wave mechanism is responsible. The fundamental, in fact, shows the same type of fringe pattern and spin modulation as appears in Figure 4 of the harmonics.

There are several types of resonance observations that seem to have their origins in non linear processes in the plasma. This should not be surprising considering the large amplitude ($\sim$ 1 kVolt) pulses which are applied to sounder antennas. An obvious case is the resonance at $2f_T$, twice the upper hybrid frequency, an example of which appears in Figure 1. There is no special significance in the cold or even warm plasma dispersion relation at this frequency. Since these are linearized equations, it is possible that the effect is due to a non linear plasma response. This resonance is observed fairly consistently, but is always rather weak. A counterpart at twice the plasma frequency has been observed only occasionally.

Another example, also labeled in Figure 1, involves the resonances at low frequencies, f_s , called subsidiary resonances. These are observed sporadically at sets of frequencies below the resonances at f_N , f_H , and f_T . Their first explanation (Barry et al., 1967) made use of the fact that a relationship seemed to exist between the resonant frequency, f_s , and the cyclotron frequency, f_H , taking the form $f_s = (n/m) f_H$, where n and m are probably small integers. The resonances were assumed natural and due to stimulated magnetic dipole emission from ionospheric constituents near the sounder, the fractions being the Landè g-values.

Further studies of the resonant frequencies (Barrington and Hartz, 1968; Barrington and Hartz, 1969; Hartz and Barrington, 1969) have, however, shown that they are related to f_N and f_T , as well as f_H , in a similar manner as $f_s = (n/m) (f_H, f_N, f_T)$. Usually two of these can be satisfied simultaneously, often only one can be satisfied, and occasionally none are satisfied. The proposed explanation attributes the resonances to images of the strong resonances caused by non linearities. Two of the strong resonances would have to be excited by different harmonics of the sounding frequency so that their difference, or beat, frequency would equal that of the receiver. The generation of harmonics by the transmitter would be quite normal, even without the plasma environment. To observe the beat frequency involves a non linear process either in the plasma or the receiver itself. The other cases are assumed to be a result of stimulation of one resonance through a transmitted harmonic and an energy transfer to the others, or even the stimulation of the strong resonances from an arbitrary transmitted frequency. Evidence for this type of stimulation has been provided by the fixed transmitter frequency-swept receiver frequency mode on the ISIS-II topside sounder (Muldrew, 1970a).

One of the subsidiary resonances, occuring at $f_s = 1/2\, f_T \approx f_N - f_H$, has the interesting characteristic that there is little or no response for the first millisecond or two. The descriptive term is that the resonance 'floats' and it was called the floating spike (Hagg and Muldrew, 1970). A possible explanation was that it took time for energy to degenerate from the upper hybrid resonance, stimulated by twice the transmitted frequency, to the plasma and cyclotron resonances where their beating could produce the response. The Q resonances also have a tendency to float. An alternative explanation for this case (Muldrew, 1970b) involves the perturbation of the sheath by the pulse. Since the Q resonances involve wavelengths short compared to free space wavelengths (large propagation vectors), they can be on the order of the sheath dimensions, and smaller if the pulse enlarges the sheath. Wave propagation through the sheath to the antenna may then have to await the collapse of the enlarged sheath.

An additional set of resonances at frequencies, f_{Dn} , which are sporadically observed are called diffuse resonances due to their appearance. They occur in frequency bands in a manner similar to the Q resonances. The first observation (Nelms and Lockwood, 1967) concerned the lowest band, and subsequent observations (Oya, 1970) were made in other bands. The similarity with the Q resonances does not extend to an explanation, however, since there is no plasma wave of low group velocity at the observed resonant frequencies. A complicated (but well-developed) theory has been proposed (Oya, 1971) which involves three wave resonant interaction, the waves being near $2f_H$, f_{Qn+2} , and f_{Dn} and f_H , f_{Qn+1} , and f_{Dn} . The strong pulse is assumed to produce plasma turbulence and a temperature anisotropy $(T_\perp > T_{\parallel})$ leading to electron cyclotron instability. The instability maintains the turbulence, which in turn produces the proper conditions for the wave-wave interaction.

One other type of observation exhibits a resonant-like behavior in the form of a spike, but this spike begins at the extraordinary trace rather than immediately after the pulse (Hagg, 1966). The response is actually in the form of a steep trace with largest delay just above $2f_H$ and merging into extraordinary trace at a slightly higher frequency. Because of this behavior it is called a remote-resonance trace and is attributed to a $2f_H$ resonance remote from the sounder. The explanation is based upon resonant oscillations being excited by the down-going extraordinary wave (Muldrew, 1967). The modulation in electron motions caused by the passing wave slowly changes, because of the gradient in f_H, into one corresponding to an upward-going wave. At this time a pulse is generated which can propagate back to the sounder. The derived delays are consistent with the observations. There is no explanation as to why these are observed only for the second harmonic of the electron cyclotron frequency.

Some of the other observations are not resonances in the sense of a 'ringing' behavior, but they are interesting nevertheless. One of these is analagous to the cyclotron echoes which have been investigated in laboratory plasmas (Herrmann *et al.*, 1967). In essence, the plasma remembers the period of time between two exciting pulses and returns an echo after that period. Normally, if the two exciting pulses are sounder pulses, the echo is obscured by the subsequent pulse, since the sounder pulse repetition rate is constant. Sometimes, however, a ducted echo is sufficiently strong to act as an exciting pulse along with a regular sounder pulse, and in these cases echoes are seen (Muldrew and Hagg, 1970).

A similar observation seems to involve the ions. These appear as additional signals at a delay equal to the proton cyclotron period, first seen as a spur on the low-frequency side of the plasma resonance (Graff, 1967; King and Preece, 1967) and later as echo traces between f_H and f_N (Nishizaki and Matuura, 1969). In the latter case there are often additional traces at multiples of the proton period. These observations are, as yet, unexplained.

7 APPLICATIONS

The measurements that can be obtained from an analysis of the resonances are several. First, the frequencies at which the major resonances occur are an accurate measure of two of the plasma properties, electron density and magnetic field strength. The frequency of the cyclotron resonance yields the magnetic field strength. The frequency of the plasma resonance yields the electron density directly, while the upper hybrid resonant frequency may be used in conjunction with the plasma frequency to give the magnetic field strength, or the electron density if the field is known. The accuracy of such measurements is limited only by the frequency accuracy of the instrument (receiver bandwidth, transmitted pulse power spectrum, and the transmitter receiver center frequencies) plus the fact that the received frequency in the resonance is not exactly the simple resonant frequency (and in fact is changing). Although the theory for the observation of the resonances at multiples of the electron cyclotron frequency is not as well formulated, the resonant frequencies agree between harmonics as well as with other magnetic field strength measurements and models. The plasma and upper hybrid resonances show a similar consistency.

Second, and more important, however, the explanation for the plasma and upper hybrid resonances involves waves which depend upon electron temperature for their particular behavior. This leads to the obvious possibility of using the resonances to measure temperature. Unfortunately, the normal topside sounders detect the received signal before telemetry to the ground, hence eliminating any frequency information which could be applied to the derived frequency-delay time characteristics (53) and (56). Even so, temperature has been measured by using the behavior of the beat pattern which survives detection (Warnock *et al.*, 1970; Feldstein and Graff, 1972).

The beat or interference comes about because of the arrival of two waves of slightly different frequencies at the same time. The two frequencies are given for the plasma resonance in (53) and the upper hybrid in (56). The beat frequency, ω_B, is the difference frequency between the two, in both cases simply twice the last term, so that for the plasma resonance

$$\omega_B = \frac{\omega_p}{2} (1 + 2 \cos^2 \alpha)(V_1 + V_3 \tan \alpha) \left(\frac{-S}{3Q_p}\right)^{1/4} \left(\frac{t}{cH_p |\cos \alpha|}\right)^{1/2} \quad (59)$$

and for the upper hybrid resonance

$$\omega_B = \left(\frac{V_2 b^{1/2}}{c \tanh \eta}\right) \left(\frac{P}{Q_T}\right)^{1/4} \frac{q^2}{(1 + q^2)^{3/4}} \quad (60)$$

with the terms as defined in the previous section and under the same restrictions (54), (55), and (57), (58).

The resonance near the plasma frequency shown in Figure 13 (Warnock *et al.*, 1970) contains an excellent example of the beat pattern. The left quarter of the record is a fixed-frequency ionogram at 1.0 mHz and the rest a swept-frequency ionogram also showing the resonance at 1.0 mHz. This example is from the ISIS-I satellite which operates in both fixed and swept modes. The mixed mode shown here has the advantage of providing good beat pattern measurements in the fixed mode while measurement of other parameters may be obtained from the swept mode.

The pertinent parameters are

$$
\left.
\begin{array}{rcl}
f_N & = & 1.00 \text{ mHz} \\
f_H & = & 0.448 \text{ mHz} \\
H_p & = & 992 \text{ km} \\
\multicolumn{3}{l}{\text{negligible horizontal gradient}}
\end{array}
\right\} \text{ from swept-frequency ionogram}
$$

$$
\left.
\begin{array}{rcl}
V_1 & = & 7.36 \text{ km/sec} \\
V_2 & = & 0.55 \text{ km/sec} \\
V_3 & = & 1.18 \text{ km/sec} \\
\alpha & = & -29.0^\circ
\end{array}
\right\} \text{ from satellite orbital data}
$$

and the only unknown parameter is then the temperature. The fact that the horizontal gradient is negligible is actually determined by looking at several successive ionograms. Before applying this to (59), a check on (54) and (55) gives the following three conditions

$$
\left.
\begin{array}{c}
t \gg 6 \, T^{-1/4} \text{ msec} \\[2mm]
t \gg 3 \times 10^4 \, T^{-3/2} \text{ msec} \\[2mm]
14 \, T^{1/2} \gg 1
\end{array}
\right\} \tag{61}
$$

and with $T = 1000^\circ$K as an example, these are $t \gg 1$ msec , $t \gg 1$ msec , and $440 \gg 1$. As long as one uses $T > 1000^\circ$K then, the approximations will all be valid for delay times greater than a few milliseconds, and this improves for larger temperatures.

Applying these parameters to (59) then gives

$$
f_B = \frac{1.89 \, t^{1/2} \text{(msec)}}{T^{1/4} \text{(}^\circ\text{K)}} \quad \text{kHz} \tag{62}
$$

or a beat period

$$
P(t) = \frac{0.529 \, T^{1/4} \text{(}^\circ\text{K)}}{t^{1/2} \text{(msec)}} . \tag{63}
$$

Eq. (63) defines a parametric set of curves of $P(t)$ for T . These may be compared to the measured period as a function of delay time to find the best value of temperature. The results shown in Figure 14 (Warnock *et al.*, 1970), point out the consistency between the theory and observation, and the least squares fit of $T = 4500 \pm 300°K$ is a reasonable temperature for the ionospheric conditions and also agrees with an electron probe measurement on the same satellite which yielded a temperature or $4050 \pm 200°K$ at the same time (Brace, private communication).

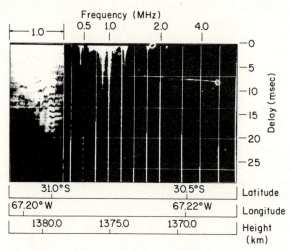

FIGURE 13   A mixed-mode ionogram of the resonance at the plasma frequency (1 MHz).

Even though this example yields a fairly accurate and reasonable value of electron temperature, the use of the beating as a standard method with present sounders is not very satisfactory because of the lack of good examples. As may be seen from (63) the beat period is not very sensitive to the temperature and hence the observations must be scaled fairly accurately for any comparisons. The scatter of the observation points in Figure 14 gives an indication of the difficulty in obtaining values from the records. The presence of the resonance over many consecutive pulses in this example gives a sufficient number of data points to allow a degree of confidence in the result. If the observation were limited to the swept frequency part of Figure 13, only a few points would be available, which indicates the desirability of having the resonance on a fixed frequency mode. This happens only occasionally by chance, so that good examples are rare.

Much better use of the theory could be made if the actual frequency-delay time characteristics described by (53) and (56) could be measured. First, this would be a direct check on the applicability and accuracy of the theory. Second, a much better temperature measurement would result. Although the last term in (53) or (56) is entirely responsible for the beat

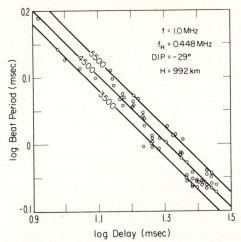

FIGURE 14   Calculated (line) and measured (circles) behavior of the beat period as a
function of delay time.

frequency, it is actually the second term in each case which usually dominates. This is to say that the variation in frequency with delay time of
either solution becomes larger than their difference. This second term is
proportional to $T^{1/2}$ and is therefore more sensitive. Eq. (53) would be
written as

$$f(t) - f_N = 0.86 \; T^{1/2} t - 0.7 \; t \mp 950 \; t^{1/2} \; T^{-1/4}$$

with f in H_z and t in msec and using the parameters in the previous
example. Because of the linear dependence on delay time for the first
term, it tends to dominate after a few milliseconds.

The intrinsic beating in the plasma and upper hybrid resonances is not
the only form of beating observed, nor the only form which has been used
to make a measurement. When the electron density is sufficiently low so
that $f_N \ll f_H$, the upper hybrid approaches the cyclotron frequency and
both can be excited and received after a single pulse. The beating then
between the two might then be interpretable in terms of the otherwise
unmeasurable electron density (Hagg, 1967). If the assumption is made that
the resonances occur at the exact upper hybrid and cyclotron frequencies,
then $f_N^2 \approx 2f_H (f_T - f_H) \approx 2f_H f_B$ where f_B is the beat frequency.
Unfortunately, not only has no basis been laid for such an assumption, but
previous experience with the upper hybrid would suggest that this is not the
case. The small deviation of the resonances from the exact frequencies may
cause a large error in the difference frequency and a corresponding large
error in the f_N measurement.

There are, however, several areas where the observations could still be used for comparison with theory as well as to make plasma measurements. One attempt has been made to derive an electron temperature from the diffuse resonances (Oya and Benson, 1972). A 'splitting' in frequency has been interpreted as a Doppler shift due to the satellite motion. The inferred propagation vectors can then be related to the temperature through the dispersion relation. Methods such as this, and that used to analyze beat patterns at f_N and f_T, may also be applied to the nf_H and Q resonances when their theoretical properties are more clearly defined.

The advantages of using wave propagation methods for determining plasma parameters are very important. First, an instrument such as a sounder has the capability of making measurements of several parameters electron density and temperature, magnetic field strength) near the sounder and even remotely. There is even redundancy in many cases. Second, the measurable properties of the ambient plasma are not affected by perturbations due to the instrument's presence (wakes, sheaths, photoelectrons, vehicle potentials) even when local measurements are made. For instance, the determination of temperature from the behavior of resonances involves waves which travel away from the sounder and spend most of their time in the ambient plasma. A third advantage lies in the fact that the temperature measurements refer to components. The plasma resonance, for instance, depends only on the temperature (velocity distribution) along the field direction, the Bernstein mode resonances on the temperature across the field. Measurement of both can give the temperature anisotropy.

Another important feature of radio sounder measurements is the lack of dependence on amplitude in the technique. The strength of an effect (at least normally) has little importance compared to the frequency at which it occurs. However, this is not to say that one should not be concerned about amplitudes. Despite the success of the electrostatic wave propagation model in explaining the observed resonances near the plasma and upper hybrid frequencies, there are some features which will require additional consideration. The plasma resonance is, for example, longest and strongest when the dipole antenna is *nearly* perpendicular to the magnetic field (see Figures 2 and 3). At first glance this seems contradictory to exciting waves with their electric fields aligned with the magnetic field. However, the dipoles are free-space half wavelength and hence become long when considering electrostatic waves of much shorter wavelength. The projection of the antenna along the field line does, in fact, match the wavelengths only when the antenna is nearly perpendicular, so that this characteristic may be due to the efficiency of excitation. Little is known about the antenna plasma interaction, particularly near resonant frequencies.

Another factor which may be important involves the finite antenna size compared to the spatial variation in the frequency of the returning waves. A spectrum of frequencies can be present along the antenna at any one time.

This property was first considered to be responsible for the lack of fringes (beating) in the plasma resonance for short delay times (McAfee, 1969a), but more than likely, the explanation in this case is more adequate from the standpoint of the response to the pulse (Fejer and Yu, 1970) as in Section 5.

The spin modulation is actually most pronounced in the cyclotron resonances (see Figure 4). The two distinct patterns seem to suggest that there could be two sets of waves which are contributing, one strong when the antenna is parallel, one when it is perpendicular. If the strength is a function of antenna projection-wave length matching, it would suggest solutions with both short and long wavelengths across or along the field or moderate wavelengths both along and across the field. The first possibility is not unreasonable since solutions exist to the Bernstein modes for both small and large wave numbers.

The question of amplitude is then not an academic one even though it may usually be ignored without losing the basic measuring capabilities of a sounder. The strong resonances often saturate the receivers and it is apparent that a good deal of energy must go into the electrostatic modes. Receiver saturation may actually be responsible for eliminating some data. For instance, the amount of modulation due to beating depends upon the relative amplitudes of the two waves involved. If the two are not quite equal, the modulation may not even be visible when the signal is strong and the receiver is saturated. It should be emphasized, however, that topside sounders are designed to transmit and receive electromagnetic pulses and are simply not as well suited for electrostatic waves. The fact that electrostatic waves contribute so much to the observations in spite of this is quite remarkable.

8 SUMMARY

The basis for the observations of at least the strongest resonant-like behavior by topside sounders is that the energy is being returned to the receiver by the reflection of electrostatic waves. The theory for the resonances at the plasma and upper hybrid frequencies has already been developed. The electrostatic wave explanation should also hold for the resonances at the harmonics of the cyclotron frequency and the frequency maxima in the Bernstein modes, although the complete theory here has not been carried out. Most of the other observed resonant-like behavior have been explained heuristically in terms of non linear processes.

Although there has been a good deal of work on the theory near the plasma and upper hybrid resonances, this is only a small part of our possible understanding of sounder observations. The first attempts at retrieving the entire frequency spectrum from the received signal during a resonance are currently being carried out. They should provide a much more definitive answer to the question of whether the electrostatic wave theory is correct and, if so, how it can be used as a measurement tool. There is a great deal of work to be done, both theoretically and experimentally, with the other resonances. The fact that they appear much more sporadically than the major resonances makes their interpretation difficult, even empirically. Many of these will not be on a firm basis until observational techniques are changed to emphasize the resonance aspects of sounders.

References

Andrews, M.K. and M.T.C. Fang, (1971). *J. Plasma Physics*, 6, 579.
Aubry, M.P., J. Bitoun and P. Graff, (1970). *Radio Sci.*, 5, 635.
Barrington, R.E. and T.R. Hartz, (1968). *Science*, 160, 181.
Barrington, R.E. and T.R. Hartz, (1969). In *Plasma Waves in Space and Laboratory*,
     Vol.1, J.O. Thomas and B.J. Landmark (Eds.), Edinburgh University Press,
     Edinburgh, pp.55–82.
Barry, J.D., P.J. Coleman, W.F. Libby and L.M. Libby, (1967). *Science*, 156, 1730.
Benson, R.F., (1971). *J. Geophys. Res.*, 76, 1083.
Bernstein, I.B., (1958). *Phys. Rev.*, 109, 10.
Bitoun, J., P. Graff and M. Aubry, (1970). *Radio Sci.*, 5, 1341.
Budden, K.G., (1961). *Radio Waves in the Ionosphere*, University Press, Cambridge.
Calvert, W., (1966). *Science*, 154, 228.
Calvert, W. and G.B. Goe, (1963). *J. Geophys. Res.*, 68, 6113.
Calvert, W. and J.R. McAfee, (1969). *Proc. IEEE*, 57, 1089.
Calvert, W. and T.E. VanZandt, (1966). *J. Geophys. Res.*, 71, 1799.
Chapman, J.H., (1965). In *Progress in Radio Science*, Vol.3, Elsevier, Amsterdam,
     pp.46–64.
Chapman, J.H. and E.S. Warren, (1968). *Space Sci. Rev.*, 8, 842.
Crawford, F.W., (1965). *Nuclear Fusion*, 5, 73.
Deering, W.D. and J.A. Fejer, (1965). *Phys. Fluids.*, 8, 2066.
Dougherty, J.P., (1969). In *Plasma Waves in Space and Laboratory*, Vol.1,
     J.O. Thomas and B.J. Landmark (Eds.), Edinburgh University Press,
     Edinburgh, pp.83-95.
Dougherty, J.P. and J.J. Monaghan, (1965). *Proc. Roy. Soc.*, A, 289, 214.
Fang, M.T.C. and M.K. Andrews, (1971). *J. Plasma Physics*, 6, 567.
Fejer, J.A. and W. Calvert, (1964). *J. Geophys. Res.*, 69, 5049.
Fejer, J.A. and W.M. Yu, (1970). *J. Geophys. Res.*, 75, 1919.
Feldstein, R. and P. Graff, (1972). *J. Geophys. Res.*, 77, 1896.
Graff, P., (1967). *C.R. Acad. Sci. Paris*, 265, 618.
Graff, P., (1970). *J. Geophys. Res.*, 75, 7193.
Graff, P., (1971a). *J. Plasma Physics*, 5, 427.
Graff, P., (1971b). *J. Geophys. Res.*, 76, 1060.
Hagg, E.L., (1966). *Nature*, 210, 927.
Hagg, E.L., (1967). *Can. J. Phys.*, 45, 27.
Hagg, E.L. and D.B. Muldrew, (1970). In *Plasma Waves in Space and Laboratory*,
     Vol.2, J.O. Thomas and B.J. Landmark (Eds.), Edinburgh University Press,
     Edinburgh, pp.69–75.
Hartz, T.R. and R.E. Barrington, (1969). *Proc. IEEE*, 57, 1108.
Herrmann, G.F., R.M. Hill and D.F. Kaplan, (1967). *Phys. Rev.*, 156, 118.
Jackson, J.E., (1965). In *Electron Density Distribution in the Ionosphere*,
     E. Thrane (Ed.), North-Holland Publishing Co., Amsterdam, pp.325–347.
King, J.W. and D.M. Preece, (1967). *J. Atmos. Terrest. Phys.*, 29, 1387.
Knecht, R.W., T.E. VanZandt and S. Russell, (1961). *J. Geophys. Res.*, 66, 3078.

Lockwood, G.E.K., (1963). *Can. J. Phys.*, **41**, 190.

McAfee, J.R., (1968). *J. Geophys. Res.*, **73**, 5577.

McAfee, J.R., (1969a). *J. Geophys. Res.*, **74**, 802.

McAfee, J.R., (1969b). *J. Geophys. Res.*, **74**, 6403.

McAfee, J.R., (1970a). In *Plasma Waves in Space and Laboratory*, Vol.2, J.O. Thomas and B.J. Landmark (Eds.), Edinburgh University Press, Edinburgh, pp.123–128.

McAfee, J.R., (1970b). *J. Geophys. Res.*, **75**, 4287.

Muldrew, D.B., (1967). *J. Geophys. Res.*, **72**, 3777.

Muldrew, D.B., (1970a). In *Space Research X*, T.M. Donahue, P.A. Smith and L. Thomas (Eds.), North-Holland Publishing Co., Amsterdam, pp.786–794.

Muldrew, D.B., (1970b). In *Progress in Radio Science 1966–1969*, Vol.1, URSI, Brussels,

Muldrew, D.B., (1972). *J. Geophys. Res.*, **77**, 1794.

Muldrew, D.B. and E.L. Hagg, (1970). In *Plasma Waves in Space and Laboratory*, Vol.2, J.O. Thomas and B.J. Landmark (Eds.), Edinburgh University Press, Edinburgh, pp.55–68.

Nelms, G.L. and G.E.K. Lockwood, (1967). In *Space Research VII*, R.L. Smith-Rose (Ed.), North-Holland Publishing Co., Amsterdam, pp.604–627.

Nishizaki, R. and N. Matuura, (1969). *J. Geophys. Res.*, **74**, 5169.

Nuttall, J., (1965). *J. Geophys. Res.*, **70**, 1119.

Oya, H., (1970). *J. Geophys. Res.*, **75**, 4279.

Oya, H., (1971). *Phys. of Fluids*, **14**, 2487.

Oya, H. and R.F. Benson, (1972). *Transactions*, American Geophysical Union, **53**, 472.

Poeverlein, H., (1948). *Akad. Wiss. Lit., Mainz. Abh. Math. Naturwiss, Kl.*, **1**, 175.

Shkarofsky, I.P., (1968). *J. Geophys. Res.*, **73**, 4859, and in *Plasma Waves in Space and Laboratory*, Vol.2, J.O. Thomas and B.J. Landmark (Eds.), Edinburgh University Press, Edinburgh, pp.111–121, (1970).

Stix, T.H., (1962). In *The Theory of Plasma Waves*, McGraw-Hill, New York.

Sturrock, P.A., (1965). *Phys. Fluids*, **8**, 88.

Tataronis, J.A. and F.W. Crawford, (1970). In *Plasma Waves in Space and Laboratory*, Vol.2, J.O. Thomas and B.J. Landmark (Eds.), Edinburgh University Press, Edinburgh, pp.91–110.

Thomas, J.O. and B.J. Landmark (Eds.), (1969). *Plasma Waves in Space and Laboratory*, Vol.1, Edinburgh University Press, Edinburgh.

Thomas, J.O. and B.J. Landmark (Eds.), (1970). *Plasma Waves in Space and Laboratory*, Vol.2, Edinburgh University Press, Edinburgh.

Warnock, J.M., J.R. McAfee and T.L. Thompson, (1970). *J. Geophys. Res.*, **75**, 7272.

Warren, E.S. and E.L. Hagg, (1968). *Nature*, **220**, 466.

Auroral Particle Precipitation-Observations

B. A. WHALEN and I. B. McDIARMID
Division of Physics, National Research Council of Canada, Ottawa K1A 0R6

I. INTRODUCTION

In the geomagnetic field and interplanetary space, nature has provided scientists with an opportunity to study properties of a high temperature, low density magnetized plasma. Of interest are such properties as confinement, particle energization, electro-magnetic wave propagation, wave-particle interactions and plasma instabilities. The interaction of this plasma with the earth's neutral atmosphere, which produces the aurora and other effects, is of interest to spectroscopists and meteorologists. There is also reason to believe that plasma processes which occur in regions influenced by the geomagnetic field are similar to various solar and other astrophysical acceleration processes. Thus the study of the near earth environment promises to improve our understanding of physical processes of a fundamental and diverse nature.

The aurora, which is one of the most visually spectacular of the naturally occurring phenomena, has been studied for centuries. Some photographic examples of the aurora appear in Figures 1 and 2. It was only with the advent of space technology and the ability to make direct measurements at high altitutdes that the complexity of the phenomena became evident. It is now clear that the aurora is only one manifestation of an array of phenomena which occur in the near earth environment as a result of the interaction between plasma from the sun (the solar wind) and the geomagnetic field.

The aurora is usually defined as sporadic electromagnetic radiation emitted from the upper atmosphere at middle to high latitudes caused by the incidence of energetic ionizing particles (mostly electrons). The processes by which the incident particle energy is converted into electromagnetic radiation is still a matter of some disagreement. Other high altitutde quasi-static electromagnetic emissions which generally have different energy sources are referred to as airglow and will not be considered in this report.

The majority of visible auroras are observed to have a maximum luminosity in the 100 km to 130 km altitude range and are excited by electrons which deposite most of their kinetic energy in this region, that is electrons with energies of the order of a few keV. The frequency of occurrence of visual auroras has a maximum at high latitudes, however they have been observed at latitudes as low as 10° from the equator. The frequencies of occurrence of auroras in the zenith when plotted against geomagnetic coordinates and local time form oval shaped patterns about the geomagnetic pole (see Figure 3, from Feldstein, 1966). Similar patterns are observed in both hemispheres. The region bounded by the greater than 70% frequency of occurrence contours is usually referred to as the auroral oval.

As can be seen from this figure the location of the auroral oval depends on local time it being between 60° to 70° geomagnetic latitude on the midnight meridian and between 70° and 80° on the noon meridian. This oval is considered to be fixed with respect to the earth-sun line with the earth rotating

FIGURE 1 Auroral arcs photographed near Ft. Churchill, Manitoba (courtesy E. E. Budzinski).

FIGURE 2 Unusual type A aurora photographed at Baker Lake, N. W. T., February 10, 1958 (courtesy E. E. Budzinski).

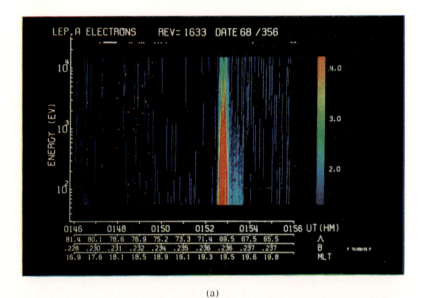

(a)

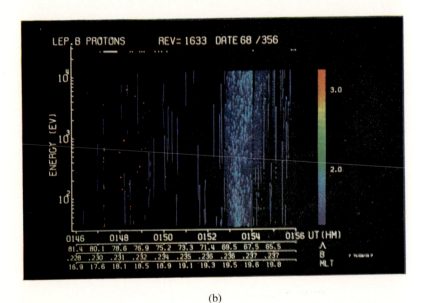

(b)

FIGURE 12 Colour-coded energy-time spectrograms for precipitating electrons (a) and protons (b). The detector responses are coded as indicated on the right hand side with the blue end of the spectrum representing low intensities and the red end high intensities (Ackerson and Frank, 1972).

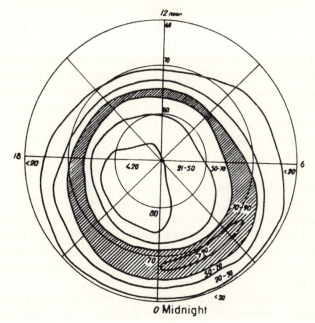

FIGURE 3   Frequency of appearance of aurora in the zenity as a function of local time and latitude (Feldstein, 1966).

beneath the pattern. The size and shape of the oval is not constant but expands (moves to lower latitude) and contracts with increasing and decreasing geomagnetic activity.

Although virtually all visible light from auroras results from electron bombardment, protons also precipitate into the upper atmosphere and produce detectable emissions. As the incident protons strike the atmosphere they emit atomic hydrogen lines which are Doppler shifted due to the incident proton velocity. Doppler shifted Balmer hydrogen lines (H_α and H_β) have been detected in auroral emissions and led earlier workers to speculate that the aurora resulted from proton bombardment of the upper atmosphere. The region of auroral hydrogen emissions is more or less coincident with the electron excited auroral oval.

An apparently distinct group of auroras occur between the auroral ovals and geomagnetic poles and are referred to as polar cap auroras. These auroras occur at higher altitudes and are produced by lower energy electrons than in the nightside auroral oval. Other characteristics which separate these auroras from those at lower latitudes are discussed in Section III.

A third class of auroras occur equatorward of the oval at mid-latitudes and are referred to as Stable Auroral Red Arcs (SAR arcs). These arcs may be visible or subvisible and usually occur in conjunction with major geo-magnetic storms. A detailed theoretical treatment of SAR arc formation has been given by Cornwall et al. (1971) but there has been no direct experi-mental determination of the source mechanism for these emissions and consequently they will not be discussed in any detail in the following.

Visual auroral emissions are also observed in the polar caps during periods of high solar activity. Energetic protons and electrons ejected from the sun gain access to the polar cap and strike the polar atmosphere creating the low altitude ionization responsible for radio "blackouts" and the diffuse aurora referred to as the "polar glow". During these periods radio receivers (riometers) in the 30 MHz frequency range which monitor cosmic radio noise show a decrease in intensity due to ionospheric absorption, thus the name normally applied to these events is Polar Cap Absorption (PCA). Another common name for these periods when energetic solar protons appear in the polar cap is Solar Proton Events or Solar Electron Events when energetic solar electrons are present. These events, although of considerable geo-physical interest, are not of auroral origin and are therefore beyond the scope of this review.

To set the framework for a discussion of auroral particle precipitation observations, a summary of the current understanding of the geomagnetic field topology is presented along with a brief description of the terms most commonly used in the presentation of these data. Auroral optical observa-tions are then discussed and the gross characteristics of particle precipitation inferred from these measurements are outlined. The significant direct observations of auroral precipitation are summarized in the next sections and the interpretation of these data in terms of energization and precipitation mechanisms and source regions appears in the final section.

II. MAGNETOSPHERIC TOPOLOGY

The topology of the geomagnetic field, as well as is known at this time, is shown in Figure 4. Represented here is a model of the magnetosphere, the region of space surrounding the earth in which the geomagnetic field is confined, and adjacent regions of geophysical interest. At large distances from the earth, greater than 20 or 30 earth radii (R_e) on the dayside, one observes the unperturbed solar wind.

1. Solar wind

The solar wind, which results from the hydrodynamic expansion of the solar corona into interplanetary space, is observed in the vicinity of earth orbit (1 AU) to be a supersonic plasma streaming away from the sun (see Parker and

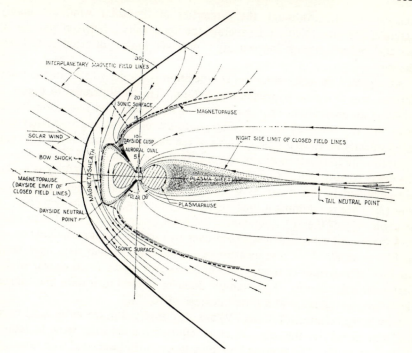

FIGURE 4   Model of the magnetosphere.

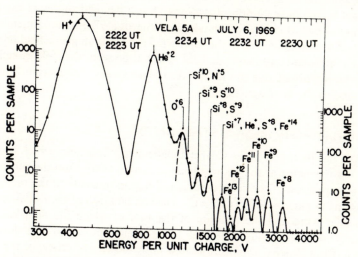

FIGURE 5   Ion composition of the solar wind (Bame et al., 1970).

Ferraro (1971)). Although the properties of the solar wind are highly variable, it is found that on the average the ion density is between one and 10 per cm^3 and the average ion velocity is near 400 km/sec directed away from the sun.

As expected the solar wind ion composition is similar to that of the solar corona, that is mostly hydrogen and helium. This is illustrated in Figure 5 which is an energy/unit charge analysis of the solar wind ions from Bame et al. (1970). Since all the ion species are travelling at approximately the same velocity an electrostatic analysis of the energy distribution of the beam separates the ions into peaks in the energy per unit charge (E/Q) spectrum, the peaks corresponding to different mass per unit charge (M/Q). The peaks are identified in this figure as to their most likely ion species. Hydrogen ions (H^+) are the dominant ions followed by doubly ionized helium (He^{++}) at about 4% of the H^+ abundance, the ratio being highly variable in time and space. All other ion species are over two orders of magnitude down in density from He^{++}. These measurements and other similar data have been used to infer various characteristics of the source region for the solar wind, that is the solar corona. For example Bame et al. used their data to estimate the source temperature at a little in excess of 10^6 °K.

Low energy electrons ($\sim$ few eV) are also observed in the solar wind with the same density as the ions, which is required for the solar wind to retain charge neutrality. The electron and ion temperatures associated with random thermal motions are of the order of 10^5 °K at 1 AU, the average electron temperature being somewhat higher than the ion temperature, therefore, most of the solar wind energy and momentum in the earth's frame of reference is carried by the bulk motion of the ions. For a detailed survey of the solar wind the review article by Kavanagh and Schardt (1970) is recommended.

Carried by and imbedded in the solar wind is a weak magnetic field referred to as the interplanetary magnetic field. As will be seen, this field is believed to play an important role in the transfer of energy and momentum from the solar wind to the geomagnetic field.

2.   Interplanetary magnetic field (IMF)

The IMF has its origin in the solar magnetic field. This field, which is of the order of a few gauss at the photosphere, is carried away from the sun by the outward expansion of the solar corona and has an intensity of the order of a few gamma (1 gauss $\sim 10^5$ γ) at 1 AU. The combination of the radial motion of the solar wind and the rotation of the sun, to which the field lines are ultimately connected, causes the IMF to be formed into a spiral as shown in Figure 6 (from Wilcox, 1968). A second observed characteristic of the IMF shown in Figure 6 is the sector structure. It has been found that the field can

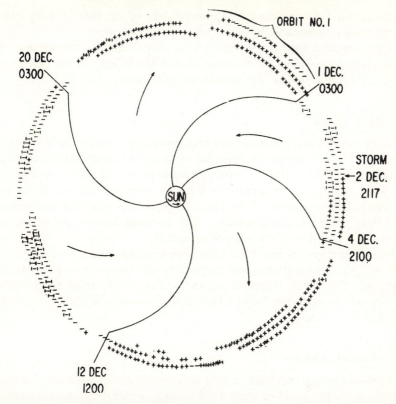

FIGURE 6 Schematic diagram of the interplanetary magnetic field showing the "garden hose" effect and sector structure (Wilcox, 1968).

be devided into sectors where the field is directed either away or toward the sun and this pattern is observed to corotate with the sun. The angle between the IMF and the earth-sun line at the earth's orbit is normally of the order of 45°. This is often called the "garden hose" angle because of the similarity of the spiral pattern shown in Figure 6 and that generated by a common garden variety water sprinkler.

It is believed that when the IMF lines come in contact with the geomagnetic field the two can become connected as indicated schematically in Figure 4. This interconnection of field lines originating in the polar caps and the interplanetary field has been suggested as a cause of various important geophysical phenomena. For example interconnection may be the mechanism responsible for the transfer of energy from the solar wind to the magnetosphere where it appears in various forms such as large scale convective motion and auroral particle kinetic energy. Although there is not yet sufficient experimental data to determine the exact role of the IMF in geophysical phenomena it appears likely that it will be a crucial one.

When the solar wind encounters an obstacle such as the earth the resulting interaction depends on the characteristics of the object. In the case of the earth an upstream shock, the Bow Shock, is formed.

3. Bow shock

Various theoretical arguments have been presented to justify the assumption that, for certain calculations, the solar wind may be treated as a continuum fluid (e.g. Levy et al. 1964). With this assumption and the knowledge that the solar wind gas is in supersonic flow, the formation of a collision free standing shock upstream of the magnetosphere was predicted theoretically and later observed experimentally. The standoff distance for this shock shown in Figure 4 is of the order of several R_e.

In the vicinity of the shock the solar wind flow changes from supersonic to subsonic, the particle density and temperature and the magnetic field strength increase. The shock therefore converts most of the solar wind kinetic energy, which was in the form of bulk motion outside the shock, to random thermal motion inside.

4. Magnetosheath

The region between the shock and the compressed geomagnetic field, which is populated by the shocked solar wind plasma, is called the magnetosheath. Near the nose of the magnetosphere, magnetosheath ions and electrons have typical energies of 500 eV and 100 eV respectively and the bulk flow velocities are low compared to the random (thermal) motion. As the plasma flows from the nose the random motion is converted into bulk motion and the plasma once again passes into supersonic flow. The position of the transition region from subsonic to supersonic flow, shown in Figure 4 as the "sonic surface", is not well known but is expected to occur near the region indicated.

The earthward boundary of the magnetosheath or equivalently the outer boundary of the magnetosphere is called the magnetopause (see Figure 4).

5. Magnetopause

The measured position and shape of the magnetopause are in reasonable agreement with theoretical predictions based on a continuum gas-dynamics approach (Spreiter and Alksne, 1969). In this approximation the magnetopause is defined as the surface where the magnetosheath particle pressure P is balanced by the geomagnetic field pressure $B^2/8\pi$ (in c.g.s. units), B being the magnetic field strength on the earthward side of the magnetopause. This approximation is justified since the magnetic field pressure in the magneto-sheath and the particle pressure inside the magnetopause are usually small compared to the above terms.

Although this approach is generally successful in predicting the position it does not yield any information on the detailed structure of the magnetopause. For example it implies that there is no interconnection between the inter-planetary and geomagnetic fields which, as mentioned previously, is probably of prime importance. The determination of the small scale magnetopause structure requires a much more sophisticated approach to the problem. A reasonably complete review of the present theoretical understanding of the magnetopause was presented by Willis (1972).

6. The plasma sheet and neutral sheet

The interaction of the solar wind with the geomagnetic field causes the mag-netosphere to be distorted into a comet like shape with an extended down-stream tail (Figure 4). The core of the geomagnetic tail is populated by an energetic plasma, the Plasma Sheet, with typical energies of the order of keV's. This plasma is observed near the equatorial plane and extends typically from $\sim 8\ R_e$ to large distances in the midnight sector and from $\sim 5\ R_e$ to the magnetopause in the noon sector (Vasyliunas, 1968). It will be shown in the following that the particles in the plasma sheet are similar to auroral primaries and are probably the auroral particle source.

The low magnetic field region in the tail which separates the oppositely directed magnetic field lines is referred to as the neutral sheet and is the region in which the cross-tail currents are concentrated producing the extended tail configuration.

Earthward of and overlapping the inner portion of the plasma sheet, the geomagnetic field is relatively undistorted and stable. It is in this region that the Van Allen radiation belts are found.

7. The radiation belts

The radiation belts are formed by very energetic particles, typically tens to hundreds of keV and higher, which are trapped in the dipole field. Most of these particles are thought to have diffused radially inward from the plasma sheet and to have been energized in the process. Another possible source for these particles is the so-called CRAND (Cosmic Ray Albedo Neutron Decay) process which might supply a significant fraction of the very energetic inner radiation belt particles.

Since this region of the magnetosphere is reasonably well shielded from solar wind induced perturbations, trapped particles have long lifetimes. Some inner radiation belt particles have lifetimes of the order of many years. A convenient review of radiation belt characteristics is given by Roederer (1970).

A brief summary of particle motions in a magnetic field is given below.

If a particle is injected into a static uniform magnetic field $(\bar{B})$ at an angle α (pitch angle) to the field it will execute a helical motion with a radius of curvature of gyroradius ρ given by

$$\rho = \frac{P \sin \alpha}{Bq} = \frac{P_\perp}{Bq}$$

where m and q are the mass and charge of the particle and p is the momentum. The angular frequency of the motion ω_c, or the cyclotron angular frequency is

$$\omega_c = \frac{Bq}{m}.$$

The magnetic moment μ, sometimes referred to as the first adiabatic invariant, defined as $\mu = P_\perp^2/2mB$ is obviously a constant of motion under these conditions. It can be shown that μ is also conserved in a time varying non-uniform field provided spatial variations in B are small over one gyro-radius ρ and temporal fluctuations are slow compared to ω_c.

Assuming the μ is conserved and that no external forces act on the particle, meaning that the kinetic energy is constant, a particle's velocity parallel to $\bar{B}$ ($v_\parallel$) is given by

$$v_\parallel = v \left[1 - \frac{B_f \sin^2 \alpha_i}{B_i} \right]^{\frac{1}{2}}$$

where i and f refer to the initial and final values of the parameters. Therefore as a particle moves towards a region of increasing B, $v_\parallel$ decreases and eventually the particle will reverse its motion at a value of B given by B_m (the value of B at the mirror point) where

$$B_m = \frac{B_i}{\sin^2 \alpha_i}.$$

Now consider charged particle motions in the earth's geomagnetic field which is approximately dipolar in the region occupied by the radiation belts (Figure 7). As the particle approaches the earth it will encounter an increasing B and will eventually mirror at a point M. It will then travel back

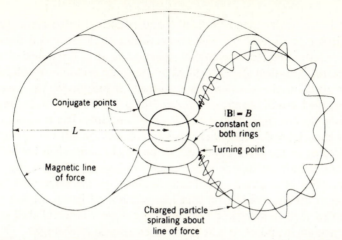

FIGURE 7   Motion of a charged particle in the geomagnetic field (Space Science, C. A. Londquist ed., 1966, McGraw-Hill, N. Y.).

through the equatorial plane along approximately the same field line and mirror at the conjugate point M^*. Thus the particle will bounce between mirror points, the time required to complete one cycle being the bounce period (T_b).

If a particle is injected into the radiation belts such that one of its mirror points is below the atmosphere, usually assumed to be 100 km above the earth's surface, the particle will be lost from the trapping region when it strikes the atmosphere. The pitch angle associated with a mirror point at 100 km is therefore called the loss cone angle (α_{LC}) and particles with angles less than α_{LC} are considered to be in the loss cone.

The quantity J associated with this bounce motion and defined as

$$J = \int_M^{M^*} p_{||} \, ds$$

where the integral is taken along the magnetic field lines between mirror points (M and M^*), is also an invariant of motion as long as the magnetic field changes are small over one bounce period. This quantity, the second adiabatic invariant, is related to the more intuitive quantity $I=J/P$ which has units of length and a value approximately equal to the distance between mirror points measured along the field line.

If there are gradients in the magnetic field, such as in the geomagnetic field, first order drifts in the particle guiding centres will result (see e.g. Roederer). For radiation belt particles the dominant drift is caused by the gradient in the scaler magnetic field B and resulting drift is in a direction perpendicular to $\vec{\nabla}_B$ and $\vec{B}$. The conservation of the first and second invariants requires that the trajectory of a particle lie on a closed shell surrounding the earth. This drift shell is defined by magnetic field lines which join two irregular rings in the northern and southern hemispheres, the ring being defined by the locus of the mirror points of the particle as it drifts in longitude, and satisfy the condition that I and B_M are constant. The trajectories of a group of particles with different energies and masses initially on the same field and mirroring at the same point will all lie on the same shell but will have bounce and drift times that depend on mass and energy. On the other hand, the trajectories of a group of particles initially on the same field line but mirroring at different points on that line will not all lie on the same shell; this is called shell splitting and implies that a slightly different shell is swept out for each mirror point on a field line or equatorial pitch angle.

McIlwain (1961) has shown that for the earth's field, in the region of the radiation belts, it is possible to define (by analogy with the dipole case) a useful shell parameter

$$L = f(B, I)$$

which by definition is constant along the path followed by a particle mirror point as it drifts in longitude and also is approximately constant (to within about 1 %) along field lines. At distances from the earth less than about 4 earth radii the shell parameter L and the field strength B have proved to be extremely useful in orgainizing radiation belt intensity measurements obtained at different locations in the geomagnetic field; such measurements are typically displayed as intensity contours on a B-L plot.

Also by analogy with a dipole field another parameter, Λ, equivalent to L, called the invariant latitude is defined by

$$\cos^2\Lambda = \frac{1}{L}.$$

This parameter has proved to be useful in organizing data from low altitude polar orbiting satellites. For a dipole field, L is the distance from the centre of the dipole to the equatorial crossing of the field line in question and Λ is the magnetic latitude. These are also useful intuitive meanings in the case of the geomagnetic field which is approximately dipolar.

If variations in the geomagnetic field in one drift period (T_D) are small then a third adiabatic invariant, the flux invariant ϕ is also conserved. The flux invariant is defined as the total magnetic flux through the surface bounded by a magnetic drift shell. This conservation law is used extensively in radial diffusion calculations for radiation belt particles but is of little consequence to auroral theories since the time scales for auroral events are short compared to particle drift periods.

At high latitudes geophysical phenomena are highly dependent on local time therefore a third coordinate, magnetic local time (*MLT*) is used. *MLT* is defined by the angle between two planes both containing the geomagnetic axis, one plane passing through the sun and the other through the observation point.

8. The plasmasphere and plasmapause

A dense low energy plasma occupies approximately the same region of space as the radiation belt particles. This plasma has typical energies of the order of a few electron volts and an ion composition representative of the ionosphere, the dominant ions being H^+ and He^+ at altitudes greater than ~ 1000 km. In typical measurements of plasma densities, latitudinal profiles show a sharp drop in density at L values between 4 and 10 R_e. This region defined by the large radial gradients in density is the plasmapause. The region between the plasmapause and the ionosphere, which is filled with low energy plasma originating in the ionosphere, is called the plasmasphere.

The processes involved in the formation of the plasmapause are still a matter of debate, however, there is reason to associate this surface with the inner termination of large scale electric fields which are induced by the solar wind and penetrate deep into the magnetosphere.

The relationship, if any exists, between auroral precipitation and the plasmasphere is also largely uncertain. Some authors have suggested that the low energy plasma inside the magnetosphere may play a role in particle precipitation and have presented some data which support their claim. These subjects will be discussed in more detail in Section VI, 3.

This completes the very brief survey of the plasma regimes found in the near earth space environment. In the following sections of this report, an attempt will be made to relate this general picture of the magnetosphere to the indirect (ground based) and direct (in situ) observation of auroral particle precipitation.

III. THE OPTICAL AURORA

The optical aurora is only one manifestation of magnetospheric processes associated with high latitude geophysical phenomena. Other observable effects include geomagnetic field pertubations due to currents flowing in the ionosphere and magnetosphere, electromagnetic wave emission and absorption over the spectral range from x-rays to ULF waves, particle energization and precipitation and many others. It has been found that the optical aurora forms a convenient framework for the discussion of these phenomena, therefore this aspect will be presented in some detail.

For a detailed summary of the field of ground based auroral optical observations the reader is referred to the publication by Akasofu et al. (1966). Some of the more relevant observations which form an important background to the particle observations appear in the following section.

1. Dimensions of auroral forms

The most commonly observed auroras have a ribbon or drapery-like appearance (see Figure 1) and have a maximum luminosity in the 100 to 130 km altitude range in the nightside auroral oval. These forms which will be referred to as arcs, typically extend from horizon to horizon in the nominal east-west direction, have vertical heights of the order of tens of kilometers and north-south thicknesses as small as hundreds of meters (Maggs and Davis, 1968). It is believed that these arcs may at times be continuous over most of the oval. Other terms such as bands, corona, patches, etc. are used to describe the appearance of other less common auroral forms, however, it is not clear how these classifications are related to processes causing the emissions. Therefore this aspect of auroral morphology will not be discussed.

Evidence of the continuous nature of the oval has recently been derived from a scanning photometer system in the ISIS-II satellite. A picture of the distribution of auroras may be derived from this sensor during one pass of the satellite over the pole. An example (courtesy of Dr. C. D. Anger) of the results from one such pass appears in Figure 8. Shown here is a slightly distorted picture of the 5577 Å emission which would be seen by an observer stationed above the geomagnetic pole. The venetian blind effect is an artifice of the scanner system and should therefore be ignored. The noon-midnight meridian passes approximately through the centre of the oval in the vertical direction. The sunlit hemisphere not shown in this photograph would appear at the top of the figure. A continuous oval shaped emission region extending completely around the geomagnetic pole is shown and embedded in it are structured regions of enhanced emission. Near the midnight sector of the oval a region of enhanced emission is observed extending to higher latitude This poleward bulge is a typical feature associated with the occurrence of an auroral disturbance referred to as a substorm (see Section III, 5).

Typical arc dimensions place severe restrictions on any auroral precipitation theories since the mechanism must be capable of producing forms with widths of the order of one hundred electron gyroradii (or a few proton gyroradii)

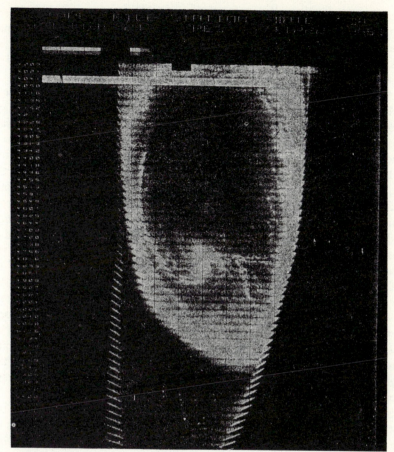

FIGURE 8 An example of the ISIS-II scanning photometer photographs of the aurora showing the complete auroral oval (courtesy Dr. C. D. Anger).

for typical auroral particle energies (~ 10 keV) and at the same time the mechanism must involve collective effects which have correlation distances of the order of tens of thousands of kilometers.

As well as determining the gross spatial distribution of the incident particles, optical observation can also be used to infer something about their energy distribution.

2. Auroral spectroscopy and the primary auroral particle spectrum

Measurements of height profiles of various auroral emissions have led to some insight into the energy distribution of primary auroral particles responsible for the emissions. As mentioned previously, since the maximum luminosity occurs in the 100 to 130 km altitude range in the nightside oval, the primary electrons responsible for emission must have energies in the 1 to 10 keV range.

Auroral spectroscopists also hope that, when the atmospheric response is well enough understood, it will be possible to infer the primary particle spectrum by measuring a sufficient number of emission parameters. To date this remote sensing approach has met with only a limited amount of success. The intensity ratios of various emission lines have been shown to be sensitive to the electron energy spectrum. For instance there is some evidence that the ratios of the intensities of 6300 Å emission to other common auroral lines such as 4278 Å and 5577 Å increases with decreasing average electron energy. This correlation presumably exists because 6300 Å is a high altitude emission and is therefore excited mostly by low energy electrons which are absorbed at this altitude. The 4278 Å and 5577 Å intensities correspond more to the total energy input (see e.g. Heikkila et al. 1972).

Confirmation of the relationship between 6300 Å emissions and low energy ($E < 1$ keV) electron precipitation was presented by McEwen and Sivjee (1972). They launched a sounding rocket into a type A aurora, identified by an increased 6300 Å/5577 Å emission ratio, and found that an unusually soft electron energy spectrum was responsible for the emissions. The average electron energy was found to be near 500 eV. Later in the flight the intensity ratio approached the normal value and electron spectra typical for these latitutdes and local times were observed.

Shepherd (1971) showed how rotational temperature measurements of the optical emission lines can be related to the altitude of the emissions. Thus by measuring the rotational temperature in time varying auroral forms (pulsating auroras) he was able to follow the temporal development of the optical pulsations. He concluded that during pulsations the energetic electrons arrived before the more numerous low energy electrons. This time delay was interpreted as a dispersion effect due to a distant source, the source distance being of the order of 10^5 km.

Although these remote sensing techniques show promise of producing valuable data on primary electron energy spectra, the processes are not well enough understood at this time to make the measurements definitive.

3. Proton aurora

Doppler shifted hydrogen Balmer lines (H_α, H_β and H_γ) in auroral spectra have been observed and correctly interpreted for some time as resulting from the incidence of energetic protons on the upper atmosphere. Measurements of

H_α and H_β have been used to define the spatial distribution of the proton precipitation region and the magnitude of the Doppler shift suggested a minimum proton energy of several keV. It has been found that the precipitation is confined to a relatively narrow latitudinal region, referred to as the proton auroral oval, which is approximately coincident with the electron excited oval.

A survey of the average proton auroral oval as indicated by H_α measurements is shown in Figure 9 which is taken from Wiens and Vallance Jones (1969). As described in that article the proton aurora occurs in the vicinity of the electron oval but the peak intensity is generally found a few degrees equatorward of the electron precipitation in the early evening sector. The two ovals either overlap or the proton oval is slightly poleward in the post midnight sector. No auroral proton emissions are observable at latitudes poleward of the auroral oval, that is in the polar cap.

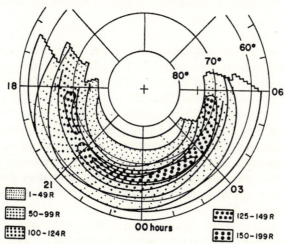

FIGURE 9   Average H_α intensity distributions (Wiens and Vallance Jones, 1969).

Unlike the electron excited auroral forms, virtually no small scale temporal ($\sim$ minutes) or spatial ($\sim$ kilometers) structure is observed in proton emissions. The lack of spatial structure is easily explained in terms of the expected interaction between the atmosphere and the beam of auroral protons. Consider a spatially confined beam of protons with an isotropic pitch angle distribution (the usual distribution, Section IV, 1, d) which is incident on the upper atmosphere. Charged protons will spiral along a particular field line until they interact with the neutral atmosphere, pick up an electron, and become electrically neutral. These energetic neutrals are no longer constrained by the geomagnetic field and can travel across field lines over large

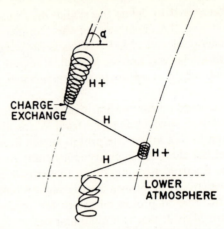

FIGURE 10   Trajectories of low-energy protons entering the atmosphere (Davidson, 1965).

distances before charge exchanging once more. This effect is illustrated in Figure 10 which is from Davidson (1965). As Davidson showed this process is repeated several times before the proton reaches the lower atmosphere. Charge exchange reactions then become so frequent that the proton starts spiraling once more but with a gyroradius consistent with a particle with an effective charge less than unity. Using a Monte Carlo method Davidson showed that the spread in the incident proton beam can be as large as 600 km. Hence, the lack of spatial structure in proton emissions results from atmospheric interactions with the proton beam and does not reflect the spatial character of the incident beam. (For a further discussion of this point see Section V, 1.)

As mentioned above, no fluctuations are observed in proton emission on time scales less than a few minutes (see e.g. Eather, 1967). This result cannot be explained in terms of atmospheric interactions and is presumably therefore a source effect.

These optical measurements have proven to be useful in defining the overall proton precipitation patterns however no conclusive measurements of the primary proton energy spectrum can be derived from these data.

4. Polar cap aurora

The morphology of electron auroras in the polar cap is quite different from those commonly observed in the auroral zone. Polar cap auroral arcs which appear in the region between the instantaneous auroral oval and the geomagnetic pole, tend to align themselves with the earth-sun line and

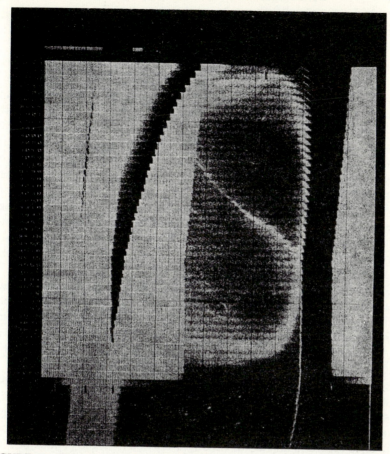

FIGURE 11 Example of a polar cap electron aurora as observed by the ISIS-II scanning photometer (courtesy Dr. C. D. Anger). The arc appears to pass continuously from the dayside (upper left) to the nightside auroral oval.

some studies indicate that they anticorrelate with overall geomagnetic activity, their occurrence in the polar cap being a maximum during periods of low activity. A remarkable example of one such polar cap auroral arc appears in Figure 11 (courtesy of Dr. C. D. Anger) which is from the ISIS-II scanning photometer. The bright vertical strips appearing in this picture are due to sunlight scattering into the photometer. The feature of interest in this photograph is the bright form near the centre of the diffuse oval region. This arc appears to connect to the dayside oval on the right hand side and to the nightside oval in the upper left. This arc was aligned in the earth-sun direction and passed over the polar cap near the geomagnetic pole. A survey of ground based observations of the polar cap auroras was given by Lassen (1969).

5. Geomagnetic storms and substorms

The magnetosphere will be defined as being in one of two possible states, "quiet" or "disturbed", although it is not clear how meaningful it is to make this distinction. The state of the magnetosphere is monitored by a worldwide network of ground based magnetometers which respond to current systems flowing in the ionosphere and magnetosphere. Various indices dependent on the average magnitudes of the deviations of the magnetometers from their quiet day readings are calculated and used by geophysicists as an indicator of the degree of geomagnetic disturbance. Two of the more commonly used indices are K_p, which combines the magnetic activity observed at certain standard midlatitude stations and A_E which uses auroral zone stations and therefore responds mostly to auroral current systems, mainly the auroral electrojet.

Geomagnetic storms produce measurable effects on a world-wide scale. Equatorial magnetograms generally show a depression in the geomagnetic field strength which is attributed to the growth of a large high altitude current system near the equatorial plane. This current system is referred to as the Ring Current and results from the injection of energetic ions deep within the magnetosphere to radial distances of the order of several R_e.

Auroral zone magnetograms are also active at these disturbed periods indicating the presence of large ionospheric current systems. A close inspection of the magnetograms shows that the activity tends to be concentrated into short bursts (lasting 15 mins to 1 hr) which are followed by periods of relative magnetic quiescence (lasting a few hrs.). These periods of high auroral zone activity are referred to as Substorms.

The morphology of auroral substorms is still under investigation, and while there is agreement on some of the more important features there is still considerable disagreement on exactly how to determine the onset of a substorm (see Section VIII, 3). In the quiet periods before the onset of the substorm electron produced auroral arcs are observed throughout most of the oval. These arcs are relatively stable in shape and intensity, are aligned with the oval and may be continuous over lengths comparagle with the oval. No large, short period ($\sim$ minutes) fluctuations in the local geomagnetic field are observed in association with these forms.

At the onset of the substorm the arcs brighten near the mignight meridian, become active and there is a sudden enhancement in the east-west ionospheric current (the auroral electrojet) in the vicinity of the arcs. The arcs expand poleward and, in a less spectacular fashion, equatorward. This expansion can cause the auroral forms to lose their continuity resulting in a "break-up" condition. The continuing expansion in the midnight sector forms a bulge in the oval (see Figure 8) which propagates east and west eventually activating the complete auroral oval in the case of large substorms.

During the recovery phase of the substorm the activity subsides, the electrojet current decreases and the arcs return to the presubstorm condition. A complete description of the auroral substorm is given in the article by Akasofu et al. (1966).

IV. AURORAL PARTICLE ENERGY SPECTRA AND PITCH-ANGLE DISTRIBUTIONS

In this section a discussion of the direct observations of the large scale spatial and velocity dependent part of the energetic auroral particle distribution functions will be presented. The time dependence and the small scale spatial dependence of these functions will appear in the next section. This division into various parts is not meant to imply that these functions are mathematically separable but is done only to simplify the presentation.

Particle observations indicate that the parameters characterizing the precipitation in and near the auroral oval are local time dependent. The observations will therefore be separated into four quadrants centred on local midnight, dawn, noon and dusk. In most cases these characteristics change smoothly between each sector and abrupt changes in characteristics at any particular local time are not usually observed.

1. Dusk sector

At times corresponding to local sunset and early evening, under quiet conditions, relatively stable single or multiple, auroral arcs, reflecting the general pattern of electron precipitation, are usually observed lying along the auroral oval. A few degrees equatorward of the arcs hydrogen line emissions resulting from energetic proton precipitation occur over a broad latitudinal range ($\sim 5°$). Figure 12 which is from Ackerson and Frank (1972) clearly shows this separation into two zones. Energy versus time spectrograms, in which the measured intensity is coloured-coded in accordance with the calibration strip shown on the right hand side of the figure is used. Particle energies are plotted logarithmically on the ordinate and universal time (UT) on the abscissa. Magnetic latitude (Λ), local magnetic field strength (B) and magnetic local time (MLT) are also shown. The authors identified the intense burst of electrons at $\sim 70°$ invariant latitude with a simultaneously observed auroral arc, the electrons responsible for the arc being in the keV energy range.

Proton precipitation was observed a few degrees equatorward of the electron precipitation with an average energy of several keV.

With the onset of substorm activity in the midnight sector the arcs may become activated to a small degree or in the case of large substorms the auroral bulge, expanding near local midnight may propagate into the evening sector, that is, form a "Westward Travelling Surge" and produce visual effects similar to those observed in the midnight sector.

a) *Electron energy spectrum* Typical auroral electron energy spectra measured over an early evening auroral arc are shown in Figure 13, which is from Choy et al. (1971). Also shown in this figure are spectra observed simultaneously in the equatorial plane by the geostationary satellite ATS 5. The rocket is normally defined as being over an auroral arc when it is on geomagnetic field lines which pass through the auroral form, this being based on the assumption that the primary electrons spiral along field lines until they strike the atmosphere and produce the aurora. The maximum energy flux, as opposed to number flux, for these electron spectra was carried by electrons in the 1 to 10 keV energy range. This result is in agreement with the auroral height measurements which suggested electrons of this energy as the excitation source.

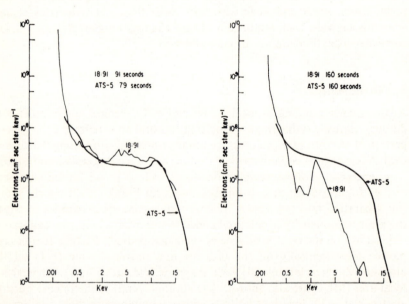

FIGURE 13   Precipitating electron energy spectra (18:91) observed in an early evening electron aurora (Choy et al., 1971).

It should be noted that at low altitudes (less than ~300 km) the primary electron energy spectrum below 100 eV can be contaminated by secondary electrons produced by the impact of the primary beam on the atmosphere. Therefore the sharp increase in intensity below 100 eV could be due in part to these secondaries.

Another obvious conclusion is that these electrons have been significantly energized since in the two possible source regions, the solar wind (the most probable source) and the ionosphere, the electron energies are of the order of electron volts rather than kilovolts. One of the more interesting unanswered questions concerns the auroral particle energization mechanisms.

Another observation which bears on the question of particle energization is the shape of the energy spectrum. For comparison purposes a Maxwellian distribution with a temperature consistent with typical auroral particle energies ($kT = 10$ keV) is shown in Figure 14. Visual inspection of this curve and the measured distributions shown in Figure 13 confirms that the primary electron distribution function is far from Maxwellian. This leads to the conclusion that the source mechanisms must involve non-stochastic processes.

Direct observations (e.g. see Westerlund, 1969) also indicate that the bright auroral arcs are embedded in a diffuse region of low energy electron precipitation. As an example an electron spectrum measured just poleward of the auroral arcs, produced by electrons with spectra shown in Figure 13,

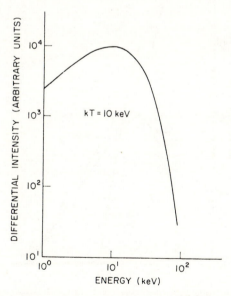

FIGURE 14   A Maxwellian intensity distribution for a particle population with $kT = 10$ keV.

appears in Figure 15. Also shown are typical electron spectra observed in the tail at 18 R_e by the Vela satellites. These electrons have characteristic energies of the order of hundreds of electron volts or less. This hardening, relative increase in energetic particles over the lower energy component, in auroral forms is also evident in satellite observations (Figure 12). These observations weigh heavily against one of the original auroral arc models. In this model auroral particles were envisioned to originate in a population trapped near the equatorial plane with a latitudinal extent large compared to the auroral forms. It was proposed that the arcs resulted when increased pitch-angle scattering from some unspecified source, on field lines connecting the equatorial population with the auroral forms filled the loss-cone with particles. Increased particle intensities in auroral forms but with similar energy spectra inside and in regions adjacent to arcs would result, which is

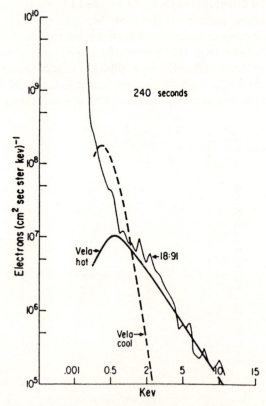

FIGURE 15   Precipitating electron energy spectrum (18:91) observed to the north of an early evening auroral arc. Also shown are typical electron spectra observed in the geomagnetic tail at 18 R_e by the Vela satellite (Choy et al., 1971).

not consistent with observations. A pitch-angle diffusion mechanism strongly dependent on energy must be introduced to overcome this difficulty.

b) *Electron angular distributions* To specify the distribution function of a group of particles the angular part of the function must also be described. A convenient local coordinate system to use defines the magnetic field direction as the pole of a spherical coordinate system with the angle between the field direction and the particle velocity vector being called the pitch angle (α). Since, to first order, auroral particles spiral about magnetic field lines the distribution functions are azimuthally symmetric. Therefore, only the pitch-angle distribution of particles is presented and inherent in this is the assumption that the function has rotational symmetry about the magnetic field direction. This symmetry has of course been confirmed observationally.

Typical auroral electron pitch-angle distributions at various energies are shown in Figure 16 which is from Whalen and McDiarmid (1972a) and refer

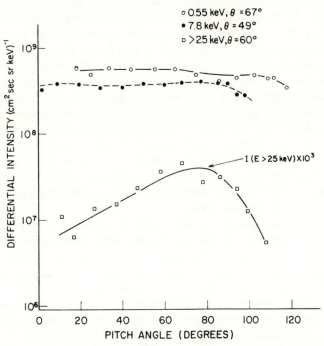

FIGURE 16 Electron pitch-angle distributions observed over an early evening aurora arc. The scale refers to all detectors, however, the units for the integral energy detector ($E > 25$ keV) are cm^{-2} sec^{-1} ster^{-1} (Whalen and McDiarmid, 1972a).

to an early evening auroral arc. These curves illustrate the two most common-ly observed distributions, isotropic over the upper hemisphere or anisotropic with a maximum intensity near $\alpha = 90°$. Electron intensities returning from the atmosphere ($\alpha > 90°$) are dependent on the pitch-angle and energy distri-bution of the incident beam however this return flux is usually of the order of 10% of the primary flux (see e.g. Choy et al.) which is consistent with that expected from atmospheric backscattering.

At high altitudes there may be a significant amount of magnetic field line convergence between the observation point and the region where the electrons are stopped by the atmosphere. This increasing magnetic field strength causes some electrons to be magnetically reflected or mirror below the rocket. Therefore one would expect that, for an incident isotropic distribution, the return flux (i.e. particles with $\alpha \geq 90°$) would be isotropic out to the loss cone angle α_{LC}. Beyond this angle the intensity should drop to a value consistent with atmospheric loss.

The loss cone angle may be calculated by using the conservation of the first invariant and assuming the atmosphere has a sharp onset at 100 km (Section II, 7). Since atmospheric absorption occurs over a finite altitude interval the resulting cutoff will, in fact, be spread over a finite pitch-angle interval near α_{LC}. This effect is illustrated in Figure 16 where $\alpha_{LC} \simeq 90°$.

The pitch-angle distributions shown in this figure are consistent with the generally accepted view that the source of auroral particles is the distant magnetosphere. An isotropic distribution is expected in any thermalized plasma and in particular if the source is at great distances from the ionos-phere since in this region the precipitating particles are confined to a very small loss cone. For example, if the source were in a region with a 10 γ ($1\gamma = 10^{-5}$ gauss) magnetic field strength, which is typical of equatorial fields at this latitude, the loss cone angle would be of the order of a few degrees. A source which could produce significant intensity changes over a few degrees in pitch angle is difficult to visualize although some have been postulated (e.g. see Speiser, 1969).

Distributions peaked near $\alpha = 90°$ (Figure 16) are easily explained in terms of the mirror geometry of the geomagnetic field. The decrease at $\alpha \geq 90°$ is a local loss cone effect as discussed above whereas the decrease at small pitch angles results from a loss cone in the opposite hemisphere. Electrons with pitch angles near zero degrees must have been scattered into this pitch angle range by collisions with other particles (e.g. the atmosphere in the opposite hemisphere) or by some other pitch angle scattering mechanism. Therefore observations of pitch-angle distributions peaked near $\alpha = 90°$ are usually interpreted as indicating that the particles are on field lines which allow some trapping, that is, that allow particles to execute at least one complete bounce cycle. These field lines are referred to as "closed". It is believed that some high latitude field lines do not terminate in the opposite hemisphere but in fact are connected to the interplanetary field (see Figure 4) and these are called "open" field lines.

A third class of pitch-angle distributions is also observed in auroral precipitation, although less frequently. These distributions peak at small pitch-angles. The occurrence of such distributions indicate a streaming or bulk flow of particles and may be related to currents flowing parallel to the local magnetic field lines which in turn play an important role in most auroral theories.

An example (Whalen and McDiarmid, 1972a) of field-aligned electron pitch-angle distributions observed in the early evening auroral oval appears in Figure 17. These data were taken at low altitudes near the poleward boundary

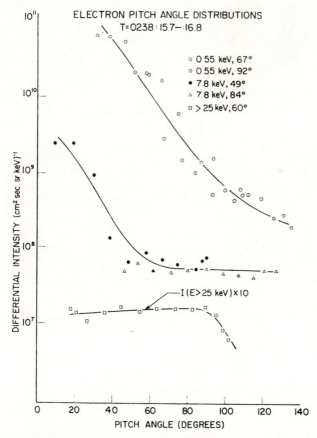

FIGURE 17   Field-aligned electron distributions observed at the poleward boundary of a series parallel auroral arcs (Whalen and McDiarmid, 1972a).

of a series of parallel east-west oriented arcs, the electrons in the arcs having isotropic angular distributions as shown in Figure 16. These field-alinged distributions were found to be confined to electrons with energies less than 10 keV and to occur over a narrow latitudinal interval.

One possible explanation for field-aligned distributions, discussed by the authors, requires the existence of a quasi-static electric field parallel to the local magnetic field ($\xi_{\parallel}$). Low energy electrons injected isotropically into this

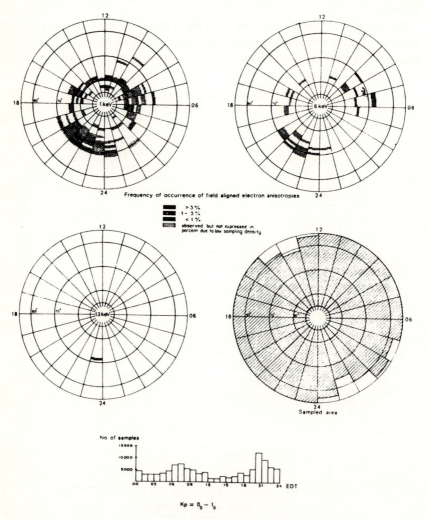

FIGURE 18 Frequency of occurrence of field-aligned electron precipitation for periods of low geomagnetic activity (Holmgren and Aparicio, 1972).

region will be accelerated parallel to $\bar{B}$. If the convergence of B over the acceleration region is small, the electrons will emerge from the electric field region with a pitch-angle distribution peaked along $\bar{B}$. The authors showed that by injecting the low energy electrons at various altitudes and electrostatic potentials into a field with a total potential drop greater than 7.8 keV, distributions consistent with those in Figure 17 are predicted. The electric field region was required to be within 1 R_e of the observation point otherwise the diverging effect of the geomagnetic field, that is motion of particles to

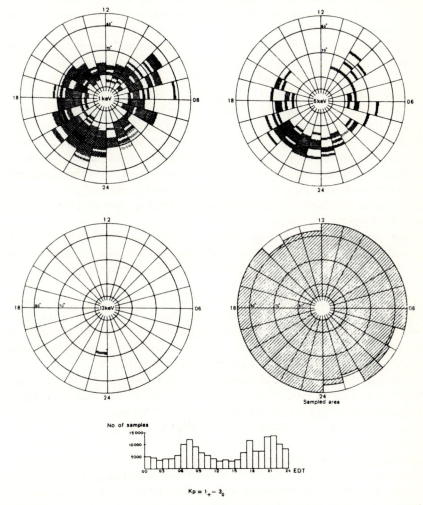

FIGURE 19 Frequency of occurrence of field-aligned electron distributions for periods of moderate activity, $1 +\leq k_p \leq 3_0$ (Holmgren and Aparicio, 1972).

larger α's with increasing B, would tend to overcome the converging effect of $\xi_\parallel$ as the particles approached the ionosphere.

The location and frequency of occurrence of low altitude field-aligned distributions has been investigated by Holmgren and Aparicio (1972) using data from the ESRO 1A satellite. Their frequency of occurrence plots for three different electron energies, (1, 6 and 13 keV) during periods of low geomagnetic activity $(k_p \leq 1_o)$ appear in Figure 18.

As described by the authors, field-aligned electron distributions occur in the region poleward of the auroral oval and are strongly dependent on energy, the frequency of occurrence increasing with decreasing energy. The distributions are local time dependent with maxima in occurrence in the early evening and near local noon. It was also found that the occurrence frequency, which is always low ($\leq 3\%$), increased with geomagnetic activity. To illustrate this effect Figure 19 shows the frequency pattern for periods of high geomagnetic activity $(1_+ \leq k_p \leq 3_o)$.

There have been a number of reports of rocket measurements (Chappell, 1968; Reasoner and Chappell, 1973; Arnoldy and Choy, 1973) in which large fluxes of low energy (typically less than a few keV) electrons were observed to travel up field lines (i.e. pitch angle distributions peaked at angles greater than 90°). The fluxes observed were in excess of those expected from Coulomb scattering of the incident beam in the atmosphere and it has been argued that these measurements are evidence for low altitude acceleration mechanisms.

The significance of electron energy and pitch-angle distributions and possible inferences concerning auroral particle acceleration mechanisms and current systems are discussed in more detail in Section VIII, 2.

c) *Proton energy spectrum*   A spectrum typical of those measured in early evening hydrogen auroras is displayed in Figure 20 which is taken from Whalen et al. (1971b). These data were gathered from a sounding rocket at altitudes in excess of 400 km. Measurements below this altitude are effected by charge exchange interactions described in Section III, 3.

The spectrum is observed to peak, or at least to have an energy flux peak, near 6 keV with an intensity of between 10^5 and 10^6 $(cm^2\ sec\ sr\ keV)^{-1}$. A comparison of Figures 20 and 14 indicates that the shape of the spectrum closely approximates a Maxwellian distribution with a $kT \simeq 6$ keV. The average proton energy is more than an order of magnitude higher than in the solar wind or magnetosheath and many orders of magnitude higher than ionospheric ion energies, therefore, as was found with the electrons, auroral protons have also been energized between the source region and the auroral zone.

Using simultaneous observations of H_β emissions and the primary proton

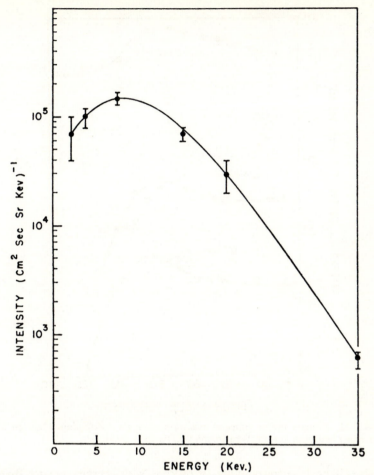

FIGURE 20   Proton spectrum obtained in an early evening hydrogen aurora (Whalen et al., 1971b).

flux (Miller and Shepherd, 1969) were able to show that the predicted H_β height profiles and intensities were in agreement with measured primary proton flux. Thus the validity of the use of hydrogen Balmer emissions to determine proton precipitation patterns was verified.

d) *Proton pitch-angle distributions*   An example of pitch-angle distributions observed in an early evening hydrogen aurora appears in Figure 21. These data, which are from Whalen and McDiarmid (1969), refer to observations near apogee ($\sim$380 km) of a sounding rocket which was launched into a hydrogen aurora. Shown here are the results from three integral proton

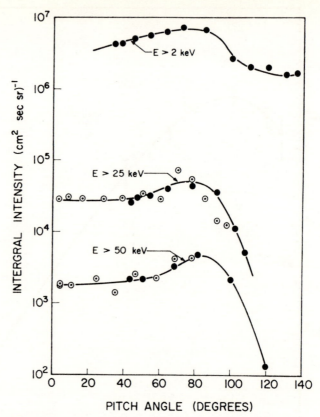

FIGURE 21   Primary proton (integral energy) pitch-angle distributions observed in an early evening hydrogen aurora (Whalen and McDiarmid, 1969).

intensity channels corresponding to protons with energies greater than 2, 25 and 50 keV. As was the case with electron precipitation, with very few exceptions the primary proton pitch-angle distributions are observed to be either isotropic or anisotropic peaked near 90°. Using the same arguments presented earlier, one concludes that the precipitation was occurring on closed geomagnetic field lines and that proton pitch-angle scattering was occurring somewhere along the trapped particle trajectories.

Observations of field-aligned proton pitch-angle distributions are rare. One of the most comprehensive observations was reported by Hultqvist et al. (1971). Using two similar proton detectors, on satellites ESRO 1A and ESRO 1B, with peak energy responses near 6 keV, one looking up the field line at a 10° pitch angle and the second looking at 80°, they were able to monitor the ratios of protons at two pitch angles. Count rate profiles observed

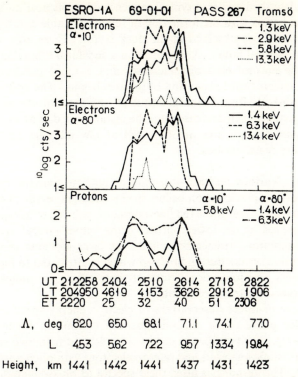

FIGURE 22   Count rate profiles observed from ESRO 1A satellite showing regions of field aligned proton precipitation (Hultqvist et al., 1971).

in typical auroral zone passes when field-aligned protons were present appear in Figure 22. Displayed in the top two boxes of this figure are the electron count rates at various pitch angles.   The proton count rates (bottom box) at 5.8 keV and 6.3 keV with pitch angles (α) of 10° and 80° respectively are seen to be comparable at low and high latitudes whereas they differ by a factor of about 50 near the middle of the precipitation zone. The authors show that the most reasonable interpretation is that this difference resulted from a pitch-angle distribution strongly peaked near zero degrees. These distributions occurred fairly frequently in the auroral zone, and tended to increase in frequency with increasing latitude.   The authors speculated that the field aligned ion distributions may be associated with a parallel electric field which accelerated the ions towards the atmosphere.

2.   Midnight sector

The most visually spectacular of all auroral displays occur in the midnight sector. As described earlier, a typical sequence of events near midnight in the auroral oval begins with the formation of relatively stable east-west oriented arcs similar to those observed in the dusk sector. These arcs may remain quiescent for several hours. At the onset of a substorm ("breakup") the arcs brighten, the auroral electrojet increases and a rapid poleward and less spectacular equatorward expansion of the arcs begins. The poleward expansion may propagate from its initial latitude ($\sim 65°$) to latitudes as high as 80° invariant. Activity then decreases and the disturbance dies away. Some time later the quiet auroral arcs are observed at low latitudes and the initial conditions are re-established.

a) *Electrons*   Various indirect observations indicate that the most intense energetic electron precipitation occurs in the midnight sector in conjunction with substorms. Low altitude red line emissions, referred to as type B aurora, is often observed at the lower border of the active auroral arcs. Cosmic radio noise monitors, riometers, indicate the large amounts of ionospheric radio absorption occurs during breakup which is attributed to increased low altitude (80–90 km) ionization. Both of these effects are indicative of enhanced energetic electron precipitation.

An example of direct observations of electron energy spectra made after the expansive phase of an auroral substorm appears in Figure 23 (from Rearwin, 1971). The two spectra shown were derived from two separate but similar electrostatic analysers and refer to electrons with pitch-angles between 60° and 65°. The author has fitted these spectra to Maxwellian distributions and gets a good fit for $kT \simeq 3$ keV. As expected the intensity of energetic electrons ($E > 20$ keV) is much higher than observed in the quiet early evening arcs. Substorm intensities near 20 keV are typically of the order of 10^5 to 10^6 (cm^2 sec sr keV)$^{-1}$ whereas in the dusk sector the energetic electrons spectrum typically cuts off sharply above 10 keV (Figure 13). Whalen and McDiarmid (1970) showed that the most intense energetic electron precipitation occurred in the leading edge of the northward propagating auroral arcs during the expansive phase of a substorm.

Polar orbiting satellites often encounter intense energetic electron precipitation in the midnight sector of the auroral oval. An example of these count rate profiles appears in Figure 24 (McDiarmid and Burrows, 1965). The intensity of electrons with energy greater than 40 keV is plotted here as a function of latitude. Precipitation at low altitudes is from the outer radiation belt whereas the intense peak at high latitudes ($\Lambda \simeq 71°$) is undoubtedly associated with electron injection during a substorm. McDiarmid and Burrows plotted the position in latitude and local time of these electron "spikes" and the result appears in Figure 25. This figure clearly illustrates that these injection events are restricted to local times near midnight and

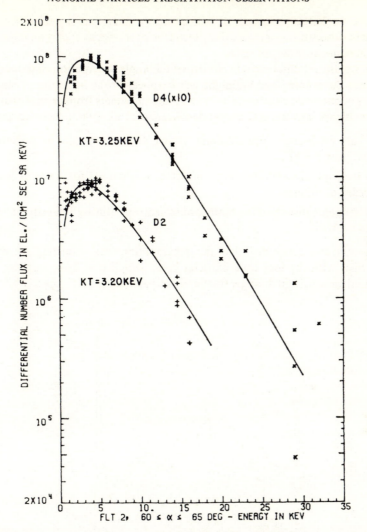

FIGURE 23   Low-energy electron energy distributions observed in the midnight sector during an auroral substorm. The continuous line drawn through the data points is a best fit Maxwellian distribution with kT as shown (Rearwin, 1971).

occur on the poleward boundary of the outer radiation zone.

These observations and those of Rearwin (1971) Fig. 23) suggest that the production of intense energetic populations at high latitudes, in association with auroral substorms, is limited to the midnight sector and, unlike the case discussed in Section IV, 1 for quiet arcs, that the source mechanisms

involve stochastic processes. It will be shown later that several other independent observations indicate that stochastic processes may play an important
role in some auroral processes.

No significant differences in electron pitch-angle distribution characteristics
between the evening and midnight sector appear in the literature. Using the
results of energetic electron ($E > 25$ keV) measurements from several sounding
rocket flights into auroral events McDiarmid et al. (1967) reported that:

i) The pitch-angle distributions varied from isotropic to anisotropic
peaked near $\alpha = 90°$.

ii) Isotropic distributions are sometimes maintained for times at least of
the order of minutes.

iii) Angular distributions tend toward isotropy during intensity enhancements.

The authors concluded that the precipitation was occurring on closed
field lines with the loss cone particles being supplied at least partially by a
mechanism which pitch-angle scattered the particles from trapped orbits into

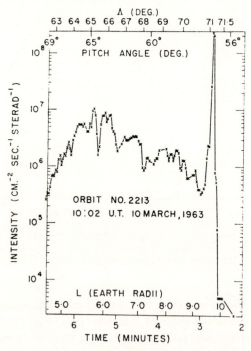

FIGURE 24   Count rate profile for electrons with energies greater than 40 keV. Note
the intense precipitation "spike" observed near $\Lambda = 71°$ (McDiarmid and Burrows, 1965).

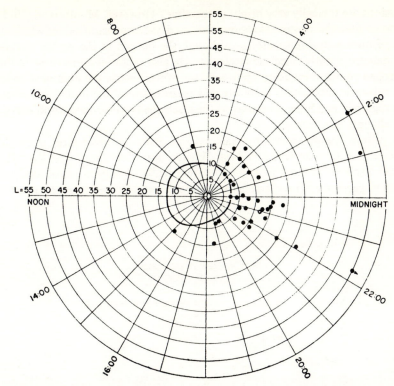

FIGURE 25 Distribution of high latitude precipitation "spikes". The smooth curve is the average high latitude boundary of the outer radiation zone (McDiarmid and Burrows, 1965).

the loss cone. Observation ii) requires an acceleration mechanism to supply particles at approximately the same rate as they are being lost to the atmosphere since, with no particle injection and an isotropic loss cone distribution, the trapped intensity would exhibit an exponential decay with a time constant of the order of hundreds of seconds (see Section VIII, 2).

The third observation suggests that the pitch-angle diffusion mechanism may be intensity dependent with larger diffusion coefficients occuring with increased intensities.

b) *Protons*   Since hydrogen line emissions from auroral proton precipitation are very weak they tend to be masked by the stronger electron excited auroral emissions when the two regions overlap, as is the case in the midnight sector (Wiens and Vallance Jones, 1969). The response of the proton emissions during substorms is even more difficult to observe since the electron emissions are typically very bright at this time. However, Montbriand (1971) reported that the proton emissions followed the expansion of the electron excited aurora during substorms although the motion of the proton aurora was delayed slightly from the electrons. He also reported an increase in the hydrogen emission intensity up to a factor of four inside the expanded zone. The substorm mechanism therefore affects the extent of the proton precipitation region in much the same way as the electrons however the proton energy flux changes less dramatically.

A proton energy spectrum measured in a substorm near local midnight appears in Figure 26 which is from Whalen and McDiarmid (1972b). The results displayed here are from two different sensors, an electrostatic analyzer (proton spectrometer) and an energetic ion mass spectrometer (α_1, α_2 and α_3).

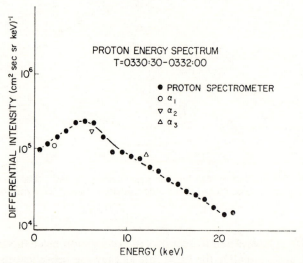

FIGURE 26   Proton energy spectrum observed in an auroral substorm near local midnight (Whalen and McDiarmid, 1972b).

As expected from the discussion above, there are no large differences between this spectrum and the quiet early evening spectrum (Figure 20). The same conclusion was reached by Miller et al. (1973) on the basis of a more statistically significant analysis of many sounding rocket flights into different auroral events. The reported hydrogen line enhancements are therefore most likely to be associated with a flux increase rather than a spectral change. This lack of significant spectral response to substorms places stringent requirements on proposed acceleration processes (Section VIII, 3).

Strikingly different proton spectra from those shown above, have been reported. As an example Figure 27 is a spectrum from Bernstein et al. (1969). These authors found the proton spectrum to be exponential in shape, with intensities near 10^8 (cm^2 sec sr keV)$^{-1}$ at 1 keV. Bernstein (1973) recently suggested a possible explanation for this discrepancy other than instrumental error. His analysis of the occurrence frequency of spectra like those in Figures 26 and 27 indicate that the intense exponential proton distributions (Figure 27) are a post-midnight phenomenon.

Auroral proton pitch-angle distributions measured during substorms are not found to be significantly different from those reported in Section IV, 1.

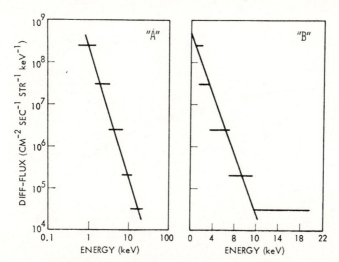

FIGURE 27   Proton energy spectrum in the midnight sector obtained by Bernstein et al. (1969).

3. Dawn sector

Early morning auroras usually appear as patchy diffuse regions of enhanced emission superimposed on a general background glow. Various morphological features in this sector are related to substorms initiated near the midnight meridian such as the appearance of pulsating auroras in the post-breakup time period. Photometric and cosmic noise absorption measurements (see e.g. Hartz and Brice, 1967) indicate that the morning sector is a region of relatively hard electron precipitation.

Proton induced hydrogen line emissions are also detected in the morning sector at approximately the same intensity as near midnight. The position of the proton precipitation region with respect to the electron aurora is not certain however several reports in the literature indicate that the two regions overlap (Eather, 1967).

Satellite observations in the dawn sector normally show broad unstructured electron intensity profiles in the auroral oval which is consistent with the precipitation pattern derived from ground based observations. A typical dawn sector electron energy spectrum, as observed by the energetic particle detectors on the ISIS-II satellite is displayed in Figure 28 (McDiarmid, unpublished). Although not very intense the spectrum is relatively hard. The energetic electron intensity at energies greater than 10 keV are comparable to those observed near midnight whereas the low energy component of the spectrum is one to two orders of magnitude lower.

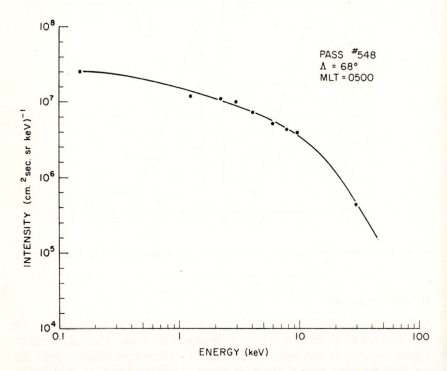

FIGURE 28   Dawn sector electron energy spectrum observed from the ISIS-II spacecraft (McDiarmid, unpublished).

Electron pitch-angle distribution observations in the dawn sector are similar to those observed near dusk and midnight with one possible exception. As Figures 18 and 19 indicate, the frequency of occurrence of high latitude field-aligned electron distributions has a minimum in the early morning hours.

4. Noon sector

In the noon sector the visual auroral oval is situated in the 75° and 80° invariant latitude interval (see Figure 3). Non-structured red line emissions (6300 Å), several degrees wide is commonly observed in the region of the oval (Shepherd and Thirkettle, 1973). The particles responsible for this emission are probably electrons with energies in 100 eV range (see e.g. Heikkila et al., 1972). Less frequently, structure relatively stable arcs are also observed in this region aligned along the oval (Anger, 1973, private communication). The particles responsible for the arcs are probably electrons with energies of the order of a kilovolt.

Hydrogen line emissions (H_β) with a smaller Doppler shift than observed in the night hemisphere have also been detected in this region generally overlapping the electron excited arcs. Thus the presence of proton precipitation in the dayside oval with energies considerably lower than observed near midnight is indicated.

The correlation of the auroral oval position with worldwide geomagnetic activity is well documented, the oval moving equatorward with increasing activity. However, responses of the dayside oval to individual substorms is not well known although there are clearly some substorm related effects. For example, Akasofu (1969) indicated that the dayside auroral oval becomes activated approximately 15 to 30 minutes after the onset of a substorm near midnight. Other responses such as the equatorward motion of the dayside oval coincident with the onset of substorms have also been reported by Akasofu (1972a). It would appear however that no definitive investigation of the detailed correlation between dayside auroral activity and substorms has been published.

No simultaneous auroral optical observations and dayside particle measurements are available at this time however some particle data which were plausibly associated with this aurora were published by Winningham and Heikkila (1973). They showed that electron precipitation in the dayside oval occurs over a latitudinal interval of about 5° near local noon. Some electron energy spectra at various pitch angles (α_p), observed in this region, appear in Figure 29. The average electron energy was found to be of the order of one hundred electron volts which is in accordance with the photometric observations mentioned previously. However, embedded in the low energy electron precipitation is occasional higher energy structured precipitation which is presumably responsible for the dayside arcs.

Coincident with the electron aurora is a region of low energy proton precipitation. An example of proton energy spectra appears in Figure 30 which is from Heikkila and Winningham (1971). Spectra were observed to peak near 400 electron volts with a peak intensity of in excess of 10^7 (cm^2 sec sr keV)$^{-1}$.

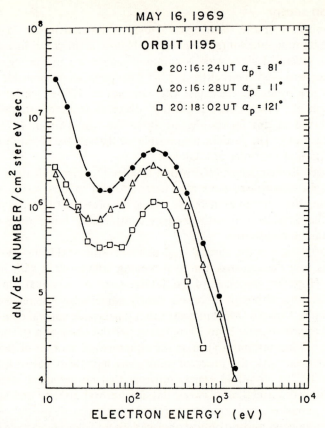

FIGURE 29   Electron energy spectra observed in the dayside auroral oval at three different pitch angles (a_p) from Winningham and Heikkila (1973).

These particle spectra are markedly different from those observed in other sectors of the auroral oval but, as Heikkila and Winningham commented, are similar to spectra observed in the magnetosheath (IMP 4 curve in Figure 30). Various mechanisms which could account for the presence of magneto-sheath-like plasma on field lines connecting with the dayside auroral oval are discussed in Section VII.

Frank (1971b) has also published data pertaining to the dayside auroral oval but his results were obtained at higher altitudes than those of Heikkila and Winningham. Spectra similar to those at low altitudes were observed, however his data suggest that the ion energy spectra differ from those in the magnetosheath. Figure 31 shows Frank's observations in the magnetosheath at 10.6 R_e, in the dayside auroral region, here referred to as the "Polar Cusp", at 5.35 R_e and a typical distant plasma sheet spectrum (IMP 4). Frank's

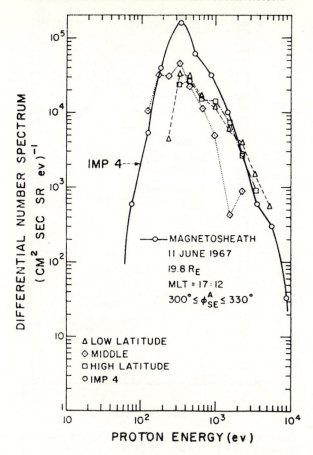

FIGURE 30   Proton energy spectra observed at low altitudes and three latitudes by the ISIS-II spacecraft and typical IMP 4 magnetosheath spectrum (Heikkila and Winningham, 1971).

data indicate a significant difference between polar cusp and magneto-sheath ion spectra.

These observations also suggest that the electron and ion precipitation regions were not coincident as claimed by Heikkila and Winningham but separated into two adjacent sheets with the electrons equatorward of the ions. Confirmation of this observation was given by Frank and Ackerson (1971), using data from the low altitude polar orbiting satellite Injun 5. Winningham (1972) attempted to resolve some of the apparent disagreement with some success, however, the structure of the dayside precipitation remains a matter of some disagreement.

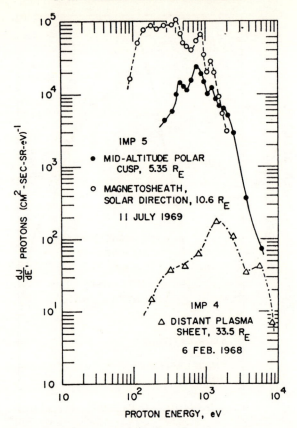

FIGURE 31   Proton energy spectra observed in the midaltitude polar cusp, the magneto-sheath and in the distant plasma sheet (Frank, 1971b).

A separate and distinct class of precipitation occurs equatorward of the late morning and noon auroral oval. Photometric and direct satellite observations (see e.g. Heikkila et al. 1972) show a relatively unstructured region of energetic electron precipitation extending over ~5° in latitude equatorward of the auroral oval. It is the incidence of these particles which causes the late morning cosmic noise absorption events discussed by Hartz and Brice (1967). Jelly and Brice (1967) showed that the auroral absorption in this sector frequently had its onset shortly after the beginning of a substorm near local midnight.

Several sounding rockets have been launched into late morning auroral absorption events and some typical results appear in Figure 32.

The data marked Chase (1968) were derived from a flight near local noon and the Budzinski (1968) results are from a flight at 0800 hrs LT. Both flights were from Churchill Research Range, invariant latitude ~ 70°, which is a few degrees equatorward of the auroral oval at these local times. No significant differences between these spectra and the general morning sector precipitation (Figure 28) is evident, thus, a common origin for these electrons is suggested.

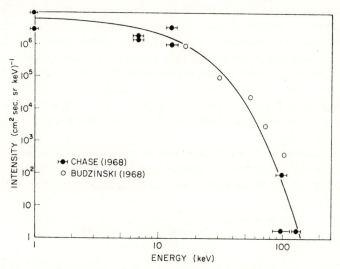

FIGURE 32   Sounding rocket observations of electron energy spectra in late morning cosmic noise absorption events.

5.   Polar cap electron auroras

Structured earth-sun oriented visual auroral forms (Figure 11), the polar cap auroras, are often observed in the region poleward of the auroral oval (see Lassen, 1969). Auroral height measurements indicate that these auroras typically occur at higher altitudes than auroral oval forms in the midnight sector. A lower average energy for polar cap auroras is therefore indicated.

Only one direct particle measurement with simultaneous ground based observations to confirm the presence of a polar cap aurora appears on the literature that being a publication by Whalen et al. (1971c). They launched a sounding rocket from a site near Resolute Bay, NWT (invariant latitude 84.5°) over a polar-cap aurora. Energy spectra measured while the rocket was over the earth-sun oriented auroral form appear in Figure 33.

Electron spectra were observed to be peaked in the 1.5- to 2.0- keV energy range with a peak intensity of ~ 10^7 (cm^2 sec sr keV)$^{-1}$. These values of average energy and intensity are consistent with the ground based observations. The peak energy of polar cap aurora, therefore lies between typical dayside and nightside auroral oval energies of 100 eV and 10 keV respectively. Pitch-angle distributions of precipitating electrons were also measured and found to be isotropic at all energies as indicated by Figure 33. No proton precipitation was observed in this flight and an upper limit on the intensity of primary proton precipitation was set near 10^5 $(cm^2 \ sec \ sr \ keV)^{-1}$.

Winningham and Heikkila (1973) have reported observations of polar cap precipitation, as measured from the ISIS-I satellite, which they refer to as polar "squalls" or "showers". Spectra in these events, shown in Figure 34, are similar although more intense than the data of Whalen et al., and leads one to believe that both belong to the same class of auroral events. This figure also illustrates what Heikkila and Winningham describe as typical pitch-angle distributions observed at high altitudes. Their measurements indicate that the most energetic electron spectral distributions occurred at small pitch angles.

These authors also reported that electron precipitation with energy spectra similar to dayside auroral oval particles (energies ~ 100 eV) but several orders of magnitude down in intensity is sometimes present over most of the polar

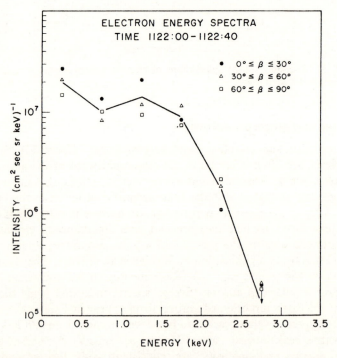

FIGURE 33 Electron energy spectra observed at three different pitch-angles (β) over an earth-sun oriented polar cap aurora (Whalen et al., 1971c).

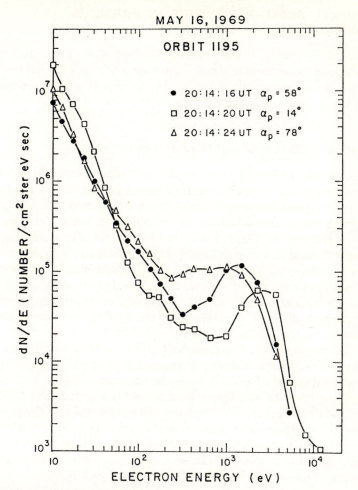

FIGURE 34   Electron energy spectra measured in the polar cap from the ISIS-I space-craft (Winningham and Heikkila, 1973).

cap. They associated this precipitation with the diffuse subvisual 6300 Å auroral emissions occasionally observed in the polar cap (Sandford, 1964). Proton and energetic electron intensities in the polar cap and near structured polar cap auroras have been below the limit of detectability in all cases.

This completes the summary of the gross energy and pitch-angle dependence of high latitude auroral particle precipitation. In the next section the small scale spatial and temporal structure observations will be presented and some of the implications of these data concerning auroral source mechanisms discussed.

V. SMALL SCALE STRUCTURE IN PARTICLE PRECIPITATION

The preceding section was devoted to the description of the gross spatial and temporal characteristics of auroral particle precipitation. Outlined in the following are measurements of the detailed features of auroral precipitation and some of the consequences of these observations.

1. Small scale spatial and temporal structure

The observations of small scale spatial and temporal structure in auroral precipitation provide valuable information on particle source mechanism. As will be shown in this section spatial distribution measurements of particle precipitation place severe restrictions on source mechanisms and measurements of temporal fluctuations may be used to infer the source distances. Photographic and photometric records of the optical aurora define the gross pattern of auroral particle precipitation and give some insight into the energy flux spatial distribution and at times a crude indication of the energy spectrum of the incident particles. Direct observations of the primary beam above the perturbing effects of the atmosphere are required however if complete information on the detailed structure is to be obtained.

In the interpretation of measurements from probes borne by sounding rockets and satellites it is generally impossible to unambiguously separate temporal and spatial effects. A particular response will be defined as temporal if it is observed in a sensor fixed in invariant latitude and local time and spatial if the simultaneous responses of two identical spatially separated sensors differ. These definitions of course only apply to non-relativistic effects but are satisfactory for most auroral particle phenomena. Since the particle sensors are invariably in motion in this coordinate system the separation of temporal and spatial effects becomes very difficult.

The spatial extent of auroral electron precipitation as determined by photographic techniques has north-south dimensions as small as a few hundred meters and an east-west extend of the order of thousands of kilometers (see e.g. Maggs and Davis, 1968). Where sufficiently accurate simultaneous photometric and direct electron precipitation measurements have been made reasonable agreement between the two techniques has been found. For example in a sounding rocket flight over several auroral arcs reasonable agreement between the total electron energy flux at energies greater than 100 eV (Whalen and McDiarmid, 1972a) and the optical emissions measured below the rocket by an on-board optical scanning system were obtained (C. D. Anger, private communication).

Although the exact correspondence of primary electron precipitation and auroral emissions is still to be established (see e.g. Evans 1971) it is likely that the characteristic widths of auroral precipitation can be as small as a few hundred meters. This corresponds to several tens of gyroradii for the auroral electrons responsible for the emissions or of the order of or less than one auroral proton gyrodiameter. An energy of 10 keV is assumed for both particle species.

The detection of small scale spatial effects in proton precipitation using hydrogen line emissions is severely hampered by the interaction of the atmosphere with the incident charged protons. Theoretical calculations (Davidson, 1965) indicate that a proton beam will charge exchange with ambient atmospheric constituents, become electrically neutral and travel freely across geomagnetic field lines (see Section III, 3). Whalen et al. (1971a) reported observing significant charge exchange effects starting near 400 kms altitude. As Davidson showed, charge exchange processes will cause a well confined isotropic beam of incident protons to spread over an area ~ 300 km in radius before they lose any significant portion of their energy and emit their characteristic hydrogen lines. Therefore no significant small scale spatial structure is expected in hydrogen line emissions independent of the structure in the primary beam. Photometric observations are in agreement with this prediction. For a recent review of these observations see Eather (1967).

Similarly any temporal modulations of the incident proton beam would not be observed in the hydrogen emissions unless the fluctuations were coherent over distances comparable with the spreading ($\sim$ several hundred kilometers). This effect may be partially responsible for the lack of significant ($\sim 50\%$) temporal fluctuations in auroral hydrogen emissions over time scales of the order of minutes (Eather, 1967). Assuming a magnetic field strength of the order of 30 γ as the equatorial value on the field lines concerned, this time period is of the order of 100 proton gyro-periods in the equatorial plane.

With few exceptions direct proton measurements from sounding rockets at altitudes less than ~ 350 km support the photometric observations. As an example Figure 35 shows data from Miller et al. (1973) which refer to a sounding rocket flight launched near local midnight fromChurchill Research Range (inv. lat. = 70°) into an auroral "breakup". Plotted here are data from five different energy channels which cover most of the range of interest. The channel marked neutral protons responded to all neutrals with energy greater than 1 keV. As explained by the authors, response in this channel was due almost exclusively to energetic neutrals. No large fluctuations in count rates over time periods $\sim$ several minutes or distances $\sim$ tens of kilometers appear in the data with the exception of the rapid decline in count rates near the end of the flight which was due to atmospheric absorption

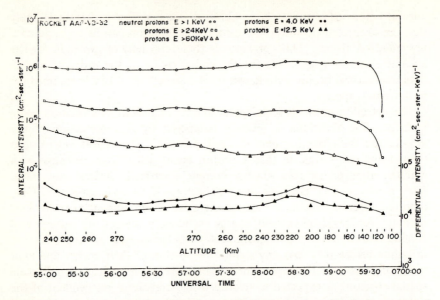

FIGURE 35 Proton intensity versus time profiles observed at low altitudes in an auroral "break-up" showing the typical lack of high frequency fluctuations (Miller et al., 1973).

of the protons as the rocket re-entered the atmosphere.

Few measurements at altitudes above 400 km where small scale effects not present at low altitudes might be observed appear in the literature. Miller et al. presented proton data from a high altitude sounding rocket flight in an auroral substorm and their results appear in Figure 36. Detectors similar to those described in the preceding paragraph were flown in this rocket. Also shown in this figure is the 10.1 keV electron intensity. At low altitudes (<400 km) they observed the usual featureless proton profiles however at higher altitudes large intensity modulations appeared in all charged particle channels. The lack of modulation in the neutral proton detector was interpreted as indicating that the fluctuations in the charged particle detectors were of a spatial rather than a temporal origin. This conclusion follows since any purely temporal fluctuation would be expected to produce corresponding intensity modulations in the energetic neutral flux, the neutrals presumably resulting from charge exchange of the primary beam above the rocket.

These measurements indicate that, at times, spatial structure in the primary proton beam is present above the atmosphere. Using the known horizontal rocket velocity and typical time scales for the fluctuations the authors estimated the scale lengths for the structures to be of the order of 5 kms which is several tens of local auroral proton gyroradii.

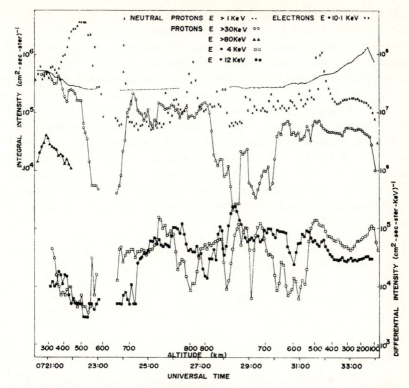

FIGURE 36   Proton and electron count rate profiles observed at high altitudes during an auroral "breakup" (Miller et al., 1973).

The most commonly observed variation in auroral electron precipitation results from the motion or deformation of the total or portions of auroral arcs. This would be classified as a combination of temporal and spatial effects. Purely temporal auroral phenomena are observed and appear as regions of enhanced auroral emissions fluctuating in intensity, at times with marked periodicities while retaining a fixed spatial structure. Such names as "pulsating", "flaming", and "flickering" aurora, based on the physical appearance, have been applied to this general class of temporal effect. Estimates of modulation frequencies associated with these effects place the fluctuations in the 0.1 to 10 Hz range.

Large electron count rate fluctuations are usually observed in and near auroral forms. Although several reports of quasi-periodic short lived pulsations appear in the literature (see e.g. Evans 1971) when the data are subjected to a detailed frequency analysis no sustained periodicities in the data have been found (e.g. Arnoldy 1970). Direct particle measurements have been made in events with well defined temporal structure and this structure has been observed in the electron precipitation. Figure 37 shows particle measurements from a sounding rocket flight into a pulsating aurora and the simultaneously observed optical pulsations (Whalen et al. 1971a). Plotted here is the output from a ground based 3914 Å photometer pointed at the rocket location and the simultaneously measured energetic ($E > 27$ keV) electron intensity. The high degree of correlation between these data is obvious. The authors concluded that these optical pulsations, with periods of the order of 10 seconds, resulted from the modulation of the energetic ($E > 15$ keV) component of the incident electron beam and were not accompanied by similar modulations in the primary proton beam. They also presented data which indicated that the pulsations resulted from periodic enhancements in the pitch angle diffusion rate somewhere along the flux tube above the observation point.

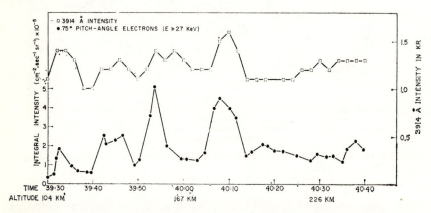

FIGURE 37 Energetic electron ($E > 27$ keV) intensities and simultaneously observed 3914 Å emissions in a pulsating aurora (Whalen et al., 1971a).

Bryant et al. (1971) made measurements in a similar event and their results appear in Figures 38 and 39. They observed a similar coherence in electron and 3914 Å intensities (Figure 38) at three different energies, 4 keV, 6 keV and >22 keV. With simultaneous low and high energy data they were also able to infer the source distance for the modulation by using a cross correlation technique (Figure 39) and measuring the delay in arrival times between the electrons with different energies. By making the reasonable assumption that the modulation occurred simultaneously at all electron energies producing the precipitation, the source distance was determined to be $\sim 4 \times 10^4$ km distant, which is the approximate distance from the observation point to the equatorial plane measured along the geomagnetic field lines. Thus they

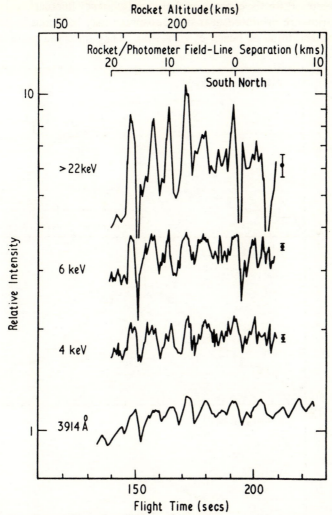

FIGURE 38   Electron and 3914 Å emission intensities observed in a pulsating aurora (Bryant et al., 1971).

concluded that the source of these pulsations was situated near the equatorial plane.

Other less convincing observations of dispersion in energetic electron arrival times have also been published (Evans, 1971). These data typically refer to high frequency (~ 10 Hz) fluctuations. The lack of any measurable dispersion has been interpreted as indicating a local source, that is a source within a few thousand kilometers of the ionosphere, for these modulations.

Some support for the contention that high frequency fluctuations in auroral precipitation are produced near the ionosphere may be found in the photometric observations of Beach et al. (1968). They observed periodic 10 Hz

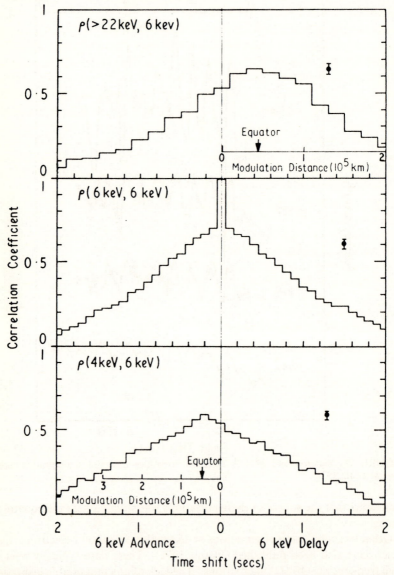

FIGURE 39   Correlation coefficients between 4-, 6-, and > 22 keV electrons. Delays in arrival times were attributed to the source region being at a distance of $\sim 4 \times 10^4$ km (Bryant et al., 1971).

pulsations in active auroras which they referred to as "flickering". If it is assumed that the electrons have an energy spread similar to other auroral particles, then the requirement for coherence in the 10 Hz modulation sets an upper limit on the source distance comparable to that mentioned above.

A note of caution should be injected at this time concerning the interpretation of these temporal effects. Although the modulations may have originated locally the initial energization need not be local. As an example one could imagine that the modulation resulted from low altitude pitch-angle scattering of a group of energized and trapped particles. Thus although the last perturbation may have occurred near the ionosphere, no information on the energization source distance is gained.

In summary, the minimum observed time scales for coherent temporal modulations for electrons is of the order of 1/10 sec and for protons of the order of minutes. Assuming a value of about 30 gamma for the equatorial magnetic field strength on auroral field lines it is found that these time periods are of the order of hundreds of equatorial gyroperiods for both particle species. Similarly spatial structure as small as several tens of gyrodiameters has been observed in electron and proton precipitation. Dispersion measurements associated with temporal structure in electron precipitation has convincingly located the source of long periods (~ 10 sec) pulsations at or near the equatorial plane. Less definitive measurements and some plausible arguments suggest a local source for the short period ($\sim 1/10$ sec) modulations.

2. Electron-proton correlations

Except for the gross spatial correlation of the electron and proton auroral ovals and the long term substorm related temporal effects, small scale temporal and spatial correlations of electron and proton precipitation are not typically observed. Attempts to detect such correlations by monitoring electron and proton auroral emissions have been reported (see Eather, 1967) however these attempts have met with little success. Some isolated reports (see e.g. Johnstone, 1971) of direct measurements of electron-proton correlations do appear in the literature however these observations would appear to be the exception rather than the rule in auroral phenomena.

An example of simultaneous proton and electron precipitation observations appears in Figure 36 which is from Miller (1973). Shown here are the intensities of protons in the 1- to 100 keV energy range and 10.1 keV electrons all measured from a high altitude sounding rocket flight launched into an auroral substorm from Churchill Research Range (70° inv. lat.). Modulations similar (synchronous) to those shown for the 10.1 keV electrons were observed in all low energy electron channels. It can be seen by inspection that although large modulations occur in both electron and proton intensities no systematic correlation exists between count rates recorded for each species. As noted by the authors this lack of correlation is a general feature observed in auroral precipitation.

These results obviously weigh heavily against any auroral arc theory which invokes a large electric field parallel to the geomagnetic field ($\xi_\parallel$) as the main mechanism to accelerate auroral particles. However, there is evidence that parallel electric fields do exist (see Section IV, 1 and VI, 1) but their role in the acceleration of auroral particles is probably secondary.

VI. OTHER OBSERVATIONS RELATED TO AURORAL PARTICLE PRECIPITATION

The requirement for a magnetospheric acceleration mechanism has been amply demonstrated by particle observations. To produce this energization electric fields are required. The existence of such fields of sufficient magnitude to produce the required acceleration has been established by many direct and indirect observations. These electric field observations and their relationship to auroral precipitation will be examined in the following. The presence of a highly conducting plasma in the vicinity of these electric fields suggests that currents will flow in the manetosphere. Such magnetospheric currents have long been inferred from magnetometer observations on the earth's surface and recently from satellite-borne probes. Although the exact function of these, currents in auroral processes in unclear at this time, their strong correlation with auroral precipitation suggests they may be of prime importance.

The dynamic nature of magnetospheric phenomena and in particular auroral processes indicates that transient electric and magnetic fields, and electromagnetic waves may play an important role in auroral phenomena. As discussed previously the presence of electromagnetic turbulence was involked to explain the strong pitch-angle scattering observed in auroral precipitation. A very brief discussion of electromagnetic waves and turbulence will therefore be included in this section.

1. Quasis-static electric fields

The magnetospheric electric field ξ will be divided into two parts $\xi_\perp$ and $\xi_\parallel$ where $\xi_\perp$ is the component of ξ perpendicular to the local geomagnetic field ($\bar{B}$) and $\xi_\parallel$ is the parallel component. Mozer (1973) has given an extensive review of techniques used to measure $\xi_\perp$ and $\xi_\parallel$ both in the ionosphere and in the magnetosphere. As a result of the high electron mobility along field lines it might be expected that $\xi_\parallel$ would in general be close to zero

(Rees et al., 1971). However, theoretical predictions (see e.g. Swift, 1965) and laboratory experiments (Hamberger and Jancarik, 1970) have shown that under certain circumstances large parallel resistances can develop as a result of a plasma instability and thus allow non-negligible values of $\xi_{\parallel}$. Direct probe measurements of large values of $\xi_{\parallel}$ (~ 20 mv/m) have been made by Mozer and Bruston (1967), Mozer and Fahleson (1970) and Kelley et al. (1971). However, there is still some controversy about the existence of large parallel electric fields in the magnetosphere. For instance Mozer (1973) states ". . . that parallel electric field measurements by the double probe technique are reliable and sufficiently sensitive to measure the large parallel fields that might often to present in the turbulent auroral ionosphere. . ." while Heppner (1972) states "in total, there are various grounds for skepticism regarding the validity of the parallel fields reported. . .".

Since it is beyond the scope of this paper to discuss electric fields in detail we shall assume that $\xi_{\parallel}$ is zero in the following discussion of the large scale effects of electric fields in the magnetosphere.

A charged particle in a uniform magnetic field ($\bar{B}$) acted upon by a non-zero $\bar{\xi}$ will drift at a velocity of $\bar{v}_c$, referred to as the convection velocity, given by

$$\bar{v}_c = \frac{\bar{\xi} \times \bar{B}}{B^2}$$

That is the particle will drift in a direction perpendicular to $\bar{\xi}$ and $\bar{B}$ at a speed given by $\xi_{\perp}/B$. For a review of the adiabatic motion of charged particles the reader is referred to Northrop (1963). In equivalent magneto-hydrodynamic terms the magnetic field lines are visualized as being in motion at a velocity v_c and the particles are "convected" along by the field. A particle in this moving frame of reference sees no electric field. In discussing magnetospheric phenomena (see e.g. Axford, 1969) it is usually easier to treat these fields in a self-consistent manner by using the concept of magnetic field line motion rather than electric fields.

Quasi-static convective electric fields have been observed in the auroral zones and over the polar caps by various techniques including direct measurement by satellite and sounding rocket-borne probes, detection of ionospheric and artificially injected ion cloud motions, and more recently by measuring the local charged particle distribution functions. All these techniques lead to the determination of the bulk flow or convection velocity of the local plasma.

A highly idealized plasma flow pattern which summarizes most of the low altitude convection measurements appears in Figure 40. The flow directions, indicated by small arrows, are plotted as a function of local time and invariant latitude. In the polar cap the plasma is observed to be convecting in the antisolar direction. At low latitudes on the midnight side this flow is diverted, the premidnight flow being towards the dusk meridian and the postmidnight towards the dawn, thus forming a return flow to the dayside of the magnetosphere. At lower latitudes the plasma is observed to be more or less corotating with the earth. As will be discussed in Section VIII, 1 it is believed that the high latitude twin vortex convection pattern is induced by solar wind interactions whereas the low latitude flux tubes, which are shielded from solar wind effects, are set in motion by the earth's rotation. The region dominated by corotational electric fields is usually associated with the plasmasphere.

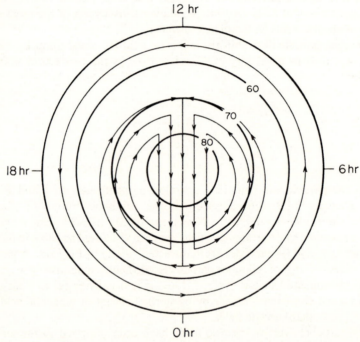

FIGURE 40   A highly idealized plasma flow pattern normally observed at high latitudes·

High latitude electric fields have been measured using most of the techniques listed above and are generally consistent with the overall pattern shown in Figure 40. Typical electric fields are found to be of the order of tens of millivolts per meter (mv/m) over the polar caps and at times ~100 mv/m near the instantaneous auroral oval. It is found that the auroral oval lies near the reversal region between sunward and anti-sunward flow in the dawn and dusk sectors and in the region of east-west flow in the midnight and noon sectors. Although some evidence for the auroral oval being on field lines convecting in the antisolar direction has been published (see Gurnett, 1972) most of the observational evidence suggests that the oval is coincident with or equatorward of the demarcation line between antisolar convecion and convection paths associated with the return flow to the dayside of the magnetosphere.

The relative positions of the twin vortex convection pattern and the auroral oval is of considerable importance since, current magnetospheric models associate antisunward convection with "open" field lines and sunward convection with "closed" field lines. These observations therefore are related directly to the question of whether auroras occur on open or closed field lines.

Detailed measurements of correlations between $\xi_\perp$ and individual auroral arcs have been attempted by several groups with somewhat contradictory results. Aggson (1969) found that the amplitude of $\xi_\perp$ decreased by as much as an order of magnitude inside the arc. This decrease in $\xi_\perp$ was attributed to enhanced conductivity inside the arc due to the increased electron precipitation which tended to short out the electric field. On the other hand the Berkeley group (see e.g. Kelley et al., 1971) have reported no decrease of electric fields inside arc structures.

2. Current systems

As expected the large electric fields observed at ionospheric heights in and near the auroral oval produce currents in the ionosphere. The interaction of the convecting plasma with the relatively stationary neutral atmosphere causes currents to flow at altitudes lower than ~ 150 km in a direction perpendicular to $\xi_\perp$ (Hall currents) and parallel to $\xi_\perp$ (Pederson currents). At altitudes above 150 km both electrons and ions drift at the same velocity in the $\xi \times \bar{B}$ direction. Below 150 km ion-neutral collisions cause the ions to move in the $\xi \times \bar{B}$ direction and in the $\xi_\perp$ directions and at lower altitudes (~ 120 km) even electron-neutral collisions become important and Pederson currents dominate. Significant currents directed along the geomagnetic field lines may also be present.

The most easily observed high latitude current system is the auroral electrojet which can cause several thousand gamma deflections on ground magnetograms. The electrojet which occurs in conjunction with the expansive phase of auroral substorms has been shown to flow near and approximately parallel to auroral arcs in the 100 to 150 km altitude range. The electrojet is believed to be mostly a Hall current, which is consistent with theoretical predictions. Pederson currents are also believed to flow to the ionosphere however the presence of these currents is difficult to infer unambiguously from ground-based magnetometer measurements. These currents are of vital importance to many auroral arc theories which require Pederson currents to serve as a return path for field-aligned currents.

Magnetic perturbations from field-aligned currents which flow between the distant magnetosphere and the ionosphere have been detected by satellite-borne probes. Zmuda et al. (1970) reported observing high latitude transverse magnetic perturbations as large as 1000 gammas in and near the auroral oval. Using direct observation of energetic electron precipitation and high altitude magnetic field perturbations, Cloutier (1971) concluded that a significant portion of the upward directed field-aligned (Birkeland) current was carried by the energetic electrons producing a visible auroral arc whereas Whalen and McDiarmid (1972a) and others on the basis of particle observations alone, showed that field-aligned energetic electron precipitation, of sufficient intensity to produce the magnetic perturbations reported by Zmuda et al. occurred at the poleward boundary of the auroral arcs and that the pitch-angle distributions (Section IV, 1) of these electrons suggested a parallel electric field as a current source.

None of the above reports can claim to have unambiguously identified the particles carrying the field-aligned currents. To do this requires a measurement of the total distribution function of both electrons and ions at all energies. These publications reported only on the energetic component of the plasma and said nothing about the ambient ionospheric plasma flow, which could be the actual current source. More detailed measurements of the thermal ionsopheric plasma are required before the ultimate source of field aligned currents can be determined.

The function of field-aligned currents in magnetospheric phenomena is unclear. Various models associate these currents with the auroral electrojet, the vertical Birkeland currents forming the source and sink for the horizontal electrojet system. An example of these current systems is shown in Figure 41 which is from Akasofu and Meng (1969). They showed that the distribution of magnetic disturbance observed on the earth's surface during an auroral substorm could be accounted for by the three dimensional current system shown, in which the Birkeland currents completed the circuit involving the electrojet and a partial ring current. A similar but more complex theoretical current system was presented by Schield et al. (1969). More recently it has been suggested that the closure of the current system in the deep magnetosphere occurs via the tail currents rather than the asymmetric ring currents. Birkeland currents have also been associated with currents flowing on the surface of the magnetopause. Whatever the outcome of present investigations, it is likely that Birkeland currents will play a key role in auroral precipitation mechanisms.

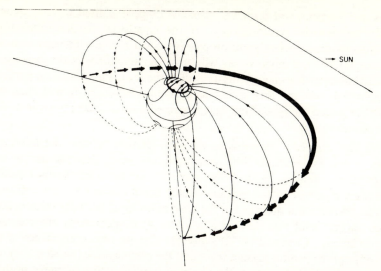

FIGURE 41   Model current system showing the proposed relationship between the auroral electrojet, Birkeland currents and the asymmetric ring current (Akasofu and Meng, 1969).

3. Electromagnetic waves and turbulence

It is beyond the scope of this report to review the observations of electromagnetic waves in the magnetosphere. Only a few examples indicating the possible function of these waves in auroral phenomena will be presented. For recent reviews of this subject the reader is referred to Kennel (1969) and Roberts (1969).

The relevance of plasma waves to auroral particle precipitation has not been clearly established, however their presence has been invoked to explain many of the characteristics observed in auroral particles. Kennel and Petschek (1966) for instance predicted that turbulent electromagnetic (whistler-mode) waves, resulting from a plasma instability fed by anisotropies in particle pitch-angle distributions, could occur near the equatorial plane. They then showed that these waves could produce the pitch angle diffusion effects normally observed in auroral precipitation (Section IV). The mechanism involves a resonant interaction between trapped particles and waves propagating in the whistler-mode with a frequency ω. Particles with a gyrofrequency Ω will interact strongly with these waves when

$$\omega - k_{\parallel} v_{\parallel} = \Omega$$

where $k_{\parallel}$ is the particle wave number and $v_{\parallel}$ is the component of the particle velocity parallel to the magnetic field lines. This equation states that the waves and particles will resonate when the Doppler-shifted wave frequency equals the particle gyrofrequency. If the equatorial particle pitch-angle distributions are sufficiently anisotopic (with a maximum intensity near 90° pitch angle) whistler mode waves will grow (be amplified) at the expense of perpendicular particle energy. As a result of the resonant interaction particles are scattered into the loss cone. Kennel and Petschek showed that this instability placed an upper limit on trapped equatorial fluxes consistent with some observations.

Since the value of $k_\parallel$ depends on the total particle density, the minimum energy which can resonate with whistler-mode waves depends on the low energy plasma density. Brice (1969) pointed out that cold plasma injection from the ionosphere in the morning sector could account for the enhanced electron precipitation observed in the later morning at latitudes equatorward of the auroral oval (see Section IV, 3 and 4) by decreasing the resonant electron energy. Cornwall (1972) suggested that the near coincidence of the inner edge of the proton plasma sheet and the plasmapause (Section VII) was indicative of the same process, the resonant particles being energetic protons and the low energy plasma being of plasma-spheric origin. Both authors have proposed that artificial injection of low energy plasma into the equatorial plane could cause enhanced (measurable) precipitation and thus provide direct evidence for this instability.

It should be noted that although plasma waves have been detected by ground based and satellite borne receivers whistler-mode waves have not been observed with the intensity required by the Kennel and Petschek mechanism and consequently it is not clear if the mechanism is responsible for auroral particle pitch-angle scattering.

Electrostatic waves of sufficient intensity to cause strong pitch-angle scattering of auroral electrons have been observed on field lines connecting with the auroral oval (Scarf et al. 1970). Using pitch-angle distribution measurements in the loss cone Whalen and McDiarmid (1973) measured the pitch-angle diffusion coefficients of low energy auroral electrons and found that such a mechanism was consistent with their observations.

Several authors have also suggested that plasma turbulence from various instabilities could produce significant auroral particle acceleration. For example turbulence could produce anomalously low conductivity along field lines (Bostrom, 1972). This would then allow significant electric fields parallel to the geomagnetic field to develop with finite Birkeland currents. As mentioned previously $\xi_\parallel$'s are a possible energization source for auroral particles. Stochastic acceleration of auroral particles by plasma turbulence was suggested by particle observations (Section IV, 2). Perkins (1968) theorized that the ionosphere might be the site of the turbulence and as such the source for energetic auroral electrons. The data on high frequency fluctuations in auroral precipitation (Section V) give some support to this contention. However, since waves of the intensity required by these theories have not been detected in or near the ionosphere, the significance of plasma turbulence in auroral energization remains unclear.

VII. AURORAL PARTICLE SOURCE REGIONS

Previously presented results and some new observations will be discussed in this section in terms of current magnetospheric models and theories. It will be shown that except for possibly a few unusual cases, the energy and particles for auroral precipitation originate in the solar wind. Plausible modes of particle entry and transport through the magnetosphere will be discussed and some problem areas delineated.

The first step in auroral source region determination involves particle trajectory tracing from the auroral oval to the source region. This is done by assuming that to the first approximation auroral particles follow geomagnetic field lines. The result of such a projection of the auroral oval, as determined by Feldstein (1966), along field lines is shown in Figure 42, which is from Fairfield (1968). He found, using the best geomagnetic field model available,

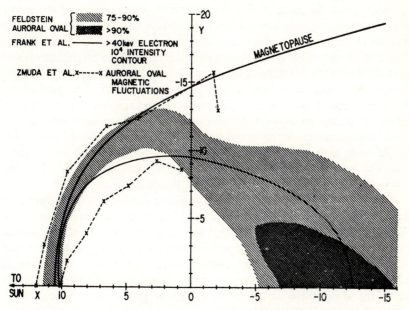

FIGURE 42 Auroral oval and associated low altitude phenomena projected along field lines to the equatorial plane using a geomagnetic field model (Fairfield, 1968).

that in the dark hemisphere the auroral oval projected into the equatorial plane whereas on the dayside the oval was found to lie on field lines which pass near the magnetopause, that is near the boundary between open and closed field lines (see Figure 4). Fairfield concluded that the nightside aurora occurred on closed field lines and that the dayside aurora originated in the magnetopause.

Energetic particles also serve as reliable tracers of geomagnetic field lines. On a typical low altitude satellite pass the high count rate observed in an energetic-electron detector (~ 30- to 40 keV) as the satellite passes through the outer radiation belt and the auroral oval will eventually approach the cosmic ray background level. McDiarmid et al. (1972) showed that the latitude at which the count rate fell to the cosmic ray background level corresponded to the low-latitude limit of closed field lines. Evans and Stone (1972) came to much the same conclusion. McDiarmid and Burrows (1968) showed that the "limit of closed field lines" was coincident with the high latitude edge of Feldstein's oval at all local times, thus indicating that all of the oval was on closed field lines.

The similarities of energy spectral shapes of protons and electrons observed in the dayside oval and the magnetosheath along with the near coincidence of the oval with the limit of closed field lines, led Heikkila and Winningham (1971) to suggest that these particles originate in the magnetosheath. They speculated that the magnetosheath was able to penetrate to low altitudes on the dayside through the dayside neutral points, some times referred to as the dayside or polar cusps. This interpretation has been questioned by Burrows et al. (1972) who showed that the regions defined by these magnetosheath-like particles are often overlapped by energetic ($E > 36$ keV) particles, measured on the same spacecraft, which have pitch-angle distributions peaked near 90° (i.e. trapped particle distributions). These authors suggest the precipitation occurs on closed field lines and that the entry mechanism may be radial diffusion across the magnetopause or local convective entry. They would only associate the narrow features at the high latitude boundary observed by Frank and Ackerson (1971) (see Section IV, 4) with direct entry of magnetosheath particles through the dayside cusps.

The problem of plasma confinement by a cusp configuration as found on the dayside magnetosphere has been treated by many authors. In particular Willis (1969) has recently shown that the cusp configuration allows particles to penetrate directly into the magnetosphere, the region of penetration being confined to dimensions of the order of an ion gyroradius. Once into the magnetosphere these particles can spiral down the field lines until they strike the atmosphere or mirror in the geomagnetic field. Therefore it is expected that magnetosheath plasma can gain access to the magnetosphere through the dayside cusps and will precipitate into the atmosphere over a very narrow latitudinal range coincident with the last closed field line. The narrow feature observed by Frank and Ackerson is more in keeping with this expectation than the relatively broad precipitation region spreading to lower latitudes described by Heikkila and Winningham. This controversy about the significance of dayside entry of magnetosheath plasma and its relation to dayside auroral precipitation remains unresolved at this time.

Most observations indicate that the nightside auroral oval is on closed field lines. As mentioned previously the survey by McDiarmid and Burrows and the projection of the auroral oval into the equatorial plane by Fairfield are in agreement on this point. Furthermore particle pitch-angle distribution measurements (Section IV, 2) also show quite conclusively that these auroras normally occur on geomagnetic field lines which support some trapping. If this conclusion is correct similar particle distributions would be expected to be observed in the equatorial plane, that is in the plasma sheet, and at low altitudes in the auroral zone. Such observations have been reported by several groups.

Typical auroral electron energy spectra when compared with those observed in the plasma sheet have been found to be similar (see e.g. Chase, 1969). As an example an electron spectrum observed in the equatorial plane at 9.8 R_e near local midnight, reported by Schield and Frank (1970), appears in Figure 43. No striking differences between this equatorial spectrum and those typically observed at the same local time in the auroral oval (Figure 23) are apparent.

A more direct comparison of low altitude and equatorial electron energy spectra appears in Figure 13. As previously discussed the data marked 18:91 refer to low altitude sounding rocket observations over an auroral arc. The data marked ATS-5 refer to near simultaneous data taken in the equatorial plane at 6.6 R_e by the geostationary satellite ATS-5 which was stationed on field lines passing near but to the south of the sounding rocket trajectory. The similarity of the spectra at times is obvious. These data clearly show that nightside auroral electron energy spectra do exist near the equatorial plane, that is in the plasma sheet.

Vasyliunas (1970) using simultaneously gathered auroral and midaltitude particle observations was able to demonstrate the coincidence of the low altitude projection of the plasma sheet and the instantaneous auroral oval. His suggestion that the oval was simply the low altitude projection of the plasma sheet is in full agreement with the preceding discussion.

Combining the field line tracing and energy spectrum results the source of the nightside auroral oval precipitation is identified as the plasma sheet. Furthermore, these results suggest that the plasma sheet is on closed field lines and that no significant electron energization occurs between the equatorial plane and the ionosphere.

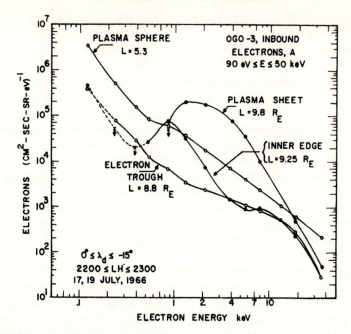

FIGURE 43   Electron energy spectra observed in and near the earthward edge of the plasma sheet (Schield and Frank, 1970).

Some examples of possible low altitude acceleration of auroral particles by local electric fields were presented in Sections IV and V. However, these results probably refer to acceleration mechanisms of a limited spatial and temporal extent and may not be related to the gross pattern of auroral precipitation. High latitude observations of field-aligned auroral-electron precipitation by Ackerson and Frank (1972) were interpreted as indicating that this arc was on open field lines and resulted from the direct acceleration of magnetosheath electrons by a low altitude $\xi_\parallel$. Sharber and Heikkila (1972) also observed field-aligned distributions on the nightside at high latitudes but they suggested the particles are accelerated on closed field lines by a Fermi type process (see Section VIII, 2). However Haskell and Southwood (1972) have questioned this interpretation on the grounds that any reasonable convective electric and magnetic field configuration could not produce the observed anisotropies at low altitudes (i.e. in such a confined equatorial pitch angle range). Hence, low altitude parallel electric fields may still be the most likely way to produce field-aligned distributions whether on open or closed field lines. Although the source region for auroral precipitation at or near the high latitude boundary of the nightside auroral oval still remains unsettled, the weight of evidence suggests that all the oval is on closed field lines.

Proton energy spectrum and spatial distribution measurements in the plasma sheet and at low altitudes in the auroral oval also suggest a plasma sheet origin for nightside auroral particles. Photometric observations of the relative positions of the proton and electron auroral ovals (Section III, 3) indicate that the electron precipitation is located a few degrees poleward of the proton aurora in the premidnight sector whereas in the postmidnight sector the two regions tend to overlap. An equivalent distribution of equatorial plasma sheet particles was found by Frank (1971a). A schematic summary of the particle spatial distributions in the equatorial plane is shown in Figure 44 where the plasma sheet electron energy density is shown as the plasma sheet and the ion energy density as the extraterrestrial ring current. Also indicated on this figure are the plasmapause, the energetic particle trapping boundaries and the earthward edge of the electron distribution in the plasma sheet which Vasyliunas (1970) showed projected down to the low latitude edge of the electron oval. Precipitation from this pattern in the equatorial plane would produce proton aurora equatorward of the electron aurora in the early evening and overlapping patterns after midnight which is consistent with the generally observed low altitude pattern.

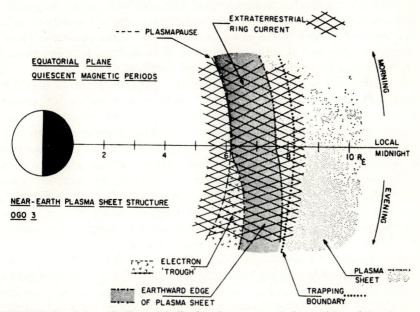

FIGURE 44 Schematic diagram indicating the average relative position of the plasmapause, and the inner edge of the electron and proton (extraterrestrial ring current) plasma sheet in the equatorial plane (Frank, 1971a).

Similar ion spectra are also observed in the equatorial plane and at low altitudes in the auroral oval. Figure 45 shows proton energy spectra from Pizzella and Frank (1971), measured near the equatorial plane on field lines connecting to the proton auroral oval. These spectra are quite similar to but more intense than the normal proton spectra observed at low altitudes (Figure 20). A comparison of an auroral ion spectrum measured from a sounding rocket during an auroral substorm and simultaneously observed equatorial ion distributions at approximately the same local time but at slightly lower latitude ($\sim 2°$ lower) appears in Figure 46. The equatorial intensities (ATS-5) (courtesy Dr. S. E. DeForest) were observed to be an order of magnitude higher than the low altitude data (IVB-25, Whalen, unpublished). Aside from possible latitudinal effects, the differences in intensity at energies > 3 keV could most easily represent a pitch-angle effect. Remembering that the plasma sheet is on closed field lines the equatorial distribution might be expected to exhibit the normal trapped angular distribution, that is a peaking near 90°. Therefore, since the ATS-5 intensities refer to large pitch angles and the IVB-25 data refer to loss cone intensities with equatorial pitch angles less than a few degrees, the intensity difference may only reflect the expected equatorial angular distribution. It is not clear at this time if the large discrepancy at energies less than 3 keV is an instrumental effect or a significant observational difference between the two populations. It is clear however that a population with sufficient intensity to serve as a source for auroral ions exists in the vicinity of the equatorial plane. Hence both ion and electron observations are consistent with the plasma sheet being the source for nightside auroral oval particle precipitation.

It has been shown that the most probably immediate source of nightside auroral precipitation is the plasma sheet, however the question still remains as to the source of plasma sheet particles. Two regions could serve as the source, the solar wind or the ionosphere. Electrons originating in either of these two regions are of course indistinguishable. However the composition of solar wind ions is quite different from that in the ionosphere. Axford (1970) pointed out that since the solar winds major ion constituents are hydrogen ions (H^+) and doubly ionized helium ions (He^{++}) (see Figure 5) and the high altitude ionosphere is composed of mostly H^+ and singly ionized helium He^+, a measurement of the helium ion charge state in auroral precipitation should determine the ion source region.

Measurements of auroral ion composition have been successfully made and the dominant helium ion was found to be He^{++}. An example of such a measurement from a sounding rocket launched into an auroral breakup was reported by Whalen and McDiarmid (1972b). The output from one channel of their energetic ion mass spectrometer, which was designed to measure the mass distribution in the energy range pertinent to auroral ions (2–20 keV),

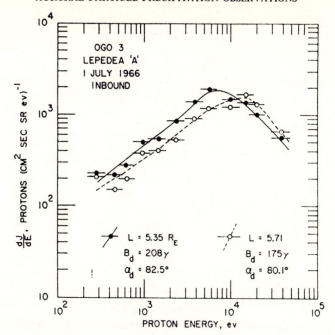

FIGURE 45 Proton energy spectrum measured in the equatorial plane on field lines passing near the proton auroral oval (Pizzella and Frank, 1971).

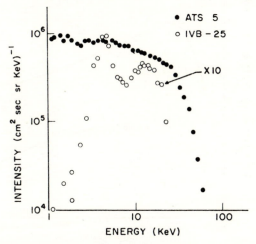

FIGURE 46 Simultaneously observed low altitude proton precipitation (IVB-25) and equatorial intensities (ATS 5 data courtesy Dr. S. E. Deforest).

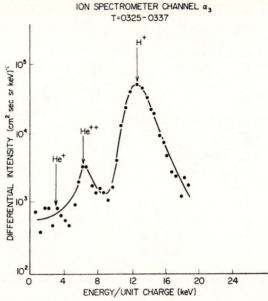

FIGURE 47 Energetic auroral ion mass composition measurement (Whalen and Mc-Diarmid, 1972b).

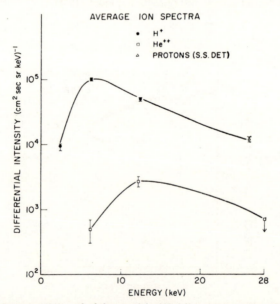

FIGURE 48 Proton (H^+) and alpha particle (He^{++}) energy spectra measured during an auroral substorm (Whalen and McDiarmid, 1972b).

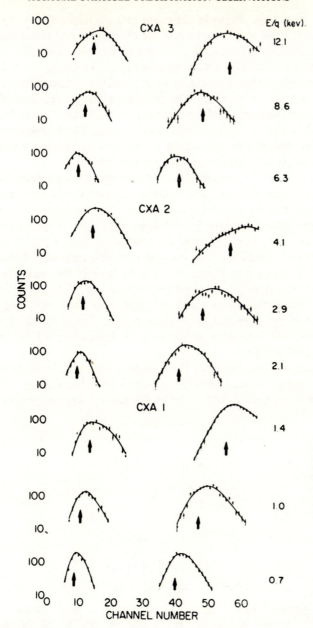

FIGURE 49   Mass distribution of energetic ion precipitation at nine different energies per unit charge (E/Q). The arrows indicate the centroid position for $M/Q=1$ (right) and $M/Q=16$ (left) predicted from laboratory calibrations (Shelley et al., 1972).

appears in Figure 47. Peaks in the count rate corresponding to various ion species are expected to be observed in the energy per unit charge scan at the positions indicated by the arrows. As this figure shows only H^+ and He^{++} were found in statistically significant numbers. Combining the results from all four channels of their instrument an H^+ and He^{++} spectrum was derived and is presented in Figure 48. The integrated flux of H^+ and He^{++} was calculated from these data and the flux ratio of He^{++} to H^+ was found to be $\sim 3\%$ which is similar to typical solar wind values. These measurements and other similar observations have been interpreted as strong evidence for a solar wind origin of nightside auroral precipitation.

Shelley et al. (1972) reported ion composition measurements from a satellite-borne spectrometer which under normal conditions agreed with those quoted above. However, during large geomagnetic storms intense fluxes of energetic heavy ($m/Q \sim 16$) ions, presumed to be singly ionized oxygen atoms O^+, were observed to precipitate over a broad latitudinal interval. Some examples of the mass distribution measured during one of these periods appears in Figure 49. Shown here is the mass distribution observed at nine different energies per unit charge (E/Q). The arrows indicate the centre channel in which H^+ (the right hand side) and O^+ (the left hand side) are expected to produce peaks. The presence of energetic massive ions with intensities comparable to those observed for protons is indicated at all energies.

The occurrence frequencies of these massive ions was found to be related to the geomagnetic index k_p as illustrated in Figure 50. Plotted in the upper curve is the three-hour k_p index and in the lower a number proportional to

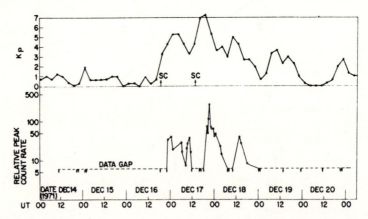

FIGURE 50 Peak energetic heavy-ion (O^+) number flux and geomagnetic disturbance index k_p plotted as a function of UT during the geomagnetic storm of December 17-18, 1971 (Shelley et al., 1972).

the peak flux of energetic heavy ions observed during one satellite pass. It is seen that during December 17 and 18, a period of high geomagnetic activity, large fluxes of heavy ions were observed whereas at other times the intensity was below the limit of detectability. The authors suggested that during geomagnetic storms a low altitude acceleration mechanism is operative, in the altitude range where O^+ is the dominant ion species, which accelerates ambient ionospheric ions to energies of the order of 10 keV. They further suggested that these ions could make a significant contribution to the extra-terrestrial ring current and may even be responsible for the production of low altitude SAR arcs.

On the basis of these rather sparse data on energetic ion mass composition, one may conclude that the source of the nightside auroral oval precipitation, the plasma sheet, is populated by solar wind particles. It is also possible that at times of large geomagnetic storms ionospheric ions may be accelerated and could compete with the solar wind as a energetic ion source.

VIII. AURORAL PARTICLE ENERGIZATION MECHANISMS

Examples of some of the currently popular magnetospheric theories and acceleration mechanisms which we consider to be most compatible with observations will be presented in this section. References to detailed discussions of these and other theories will be given where relevant. The energization mechanisms will de divided into two categories, the quiescent or steady state and the disturbed or substorm associated mechanisms. The continuous presence of auroral precipitation, large scale convective electric fields, ionospheric currents and other energy dissipative magnetospheric phenomena require a more or less continuous exchange of energy between the solar wind, the presumed energy source, and the magnetosphere. The mechanisms believed responsible for particle acceleration associated with this quasi-steady state condition will be discussed in the second part of this section. The sudden energy release which occurs during auroral substorms appears to be associated with an internal reconfiguration of the magnetosphere and will be discussed as a separate phenomenon although one which is intimately associated with the quiescent mechanism.

1. Solar wind—magnetosphere energy exchange

It is generally accepted that the energy source for high latitude phenomena is associated with large scale quasi-static convective electric fields induced in the magnetosphere by the solar wind (Axford, 1969). Two mechanisms for momentum transferral between the solar wind and the magnetosphere have been suggested. Axford and Hines (1961) proposed that a viscous interaction in the vicinity of the magnetopause could cause magnetospheric flux tubes to be set in motion whereas Dungey (1961) suggested that convection was driven by a connection of interplanetary and geomagnetic field lines. A large body of data favouring reconnection as the dominant source now appears in the literature. A convenient summary of these observations was given by Axford (1969).

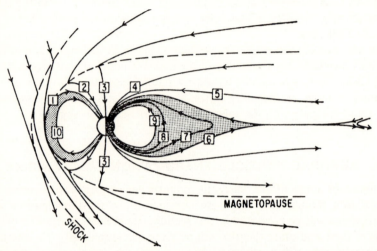

FIGURE 51 Magnetospheric model showing a possible source of magnetospheric convection. The numbers indicate successive positions of geomagnetic field lines with reconnection occurring on field lines 1 and 6 (Axford, 1969).

An idealized model of Dungey's mechanism was presented by Axford and appears in Figure 51. Shown here is a cross section of the magnetosphere in the noon-midnight meridian and the north-south component of the interplanetary magnetic field which is assumed to be pointed in the southerly direction. In this model the interplanetary field lines are considered to be "frozen in" the solar wind plasma and are carried away from the sun to the nose of the magnetosphere where field line cutting occurs resulting in the connection of the interplanetary and geomagnetic field lines (line 1). Field line 1 is then carried over the polar caps by the solar wind, the progressive positions being indicated by increasing numbers. The feet of these field lines are able to slip freely across the poles at ionospheric heights because of the presence of the insultaing atmosphere (Gold, 1959). Eventually this field line reaches a position indicated by line 6 where reconnection once again occurs. This completely internal field line then moves toward the earth (position 7, 8 and 9), slips around the earth to position 10 and proceeds radially outward to position 1 completing the cycle.

The motion of the feet of these field lines in the ionosphere corresponding to positions 1 through 6 is over the polar cap in the antisolar direction and to lower latitudes in the same direction for positions 7, 8 and 9. The motion from positions 9 to 10 describes an approximately constant latitude line and the motion from 10 to 1 is once again toward the pole. With a little though it is possible to convince oneself that this field line motion produces the observed low altitude convection pattern shown in Figure 40.

A similar low altitude convection pattern is predicted by the Axford and Hines theory. The differences in the two theories is more easily observed in the equatorial convection patterns. Figure 52 shows the equatorial pattern predicted by Axford and Hines. The antisolar flow in this model which is driven by a viscous interaction is along the flanks of the magnetosphere.

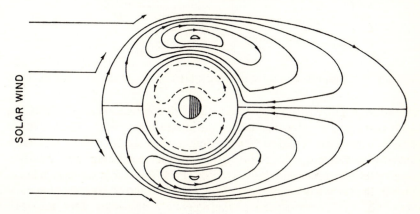

FIGURE 52   Model of equatorial convection pattern predicted if magnetospheric convection is driven by a viscous interaction between the solar wind and the magnetosphere (Axford and Hines, 1961).

In Dungey's model this part of the flow does not appear in an equatorial cross section, since the field lines undergoing antisunward flow do not pass through the equatorial plane. The sunward flow in the equatorial plane simply terminates at the magnetopause as shown in Figure 53 which is from Brice (1967).

A considerable amount of indirect evidence favouring field line reconnection as opposed to viscous drag as the convection source has been published. Most of these studies have taken the form of correlations of the interplanetary field direction with various observable geophysical phenomena. No other solar wind parameter such as dynamic pressure, density or velocity seem to correlate with geophysical activity to the same degree. As an example Nishida and Maezawa (1971) showed that the worldwide current systems

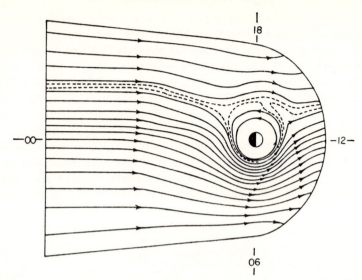

FIGURE 53   Equatorial convection pattern predicted by the Dungey model (Brice, 1967).

referred to as DP2, which they associated with convection electric fields, are highly correlated with the direction of the interplanetary field. The amplitude of DP2 currents are found to be a maximum when the IMF has a maximum southerly component, which corresponds to the orientation most advantageous to reconnection. Similarly Arnoldy (1971) found a positive correlation between the southerly component of the IMF and the occurrence of auroral substorms, substorms occurring approximately one hour after the IMF turned south.

More direct evidence for the interconnection of the IMF with polar cap field lines is derived from the study of energetic solar particles entry into the magnetosphere. During solar disturbances energetic ($\sim$1 MeV) electrons and protons are emitted by the sun and impinge on the magnetosphere. The lower magnetic rigidity particles ($<$1 MeV) will tend to follow magnetic field lines near the earth and in the absence of any diffusion across field lines or interconnection will be excluded from the magnetosphere. It is found however that these particles appear in the polar caps more or less simultaneous with their arrival at the magnetopause (see e.g. Anderson, 1970) indicating they have direct access to the polar caps down to latitudes defined by the limit of closed field lines (e.g. McDiarmid et al., 1972). These results might be expected from rapid diffusion of these particles across the geomagnetic tail. However the more likely interpretation is that these high latitude field lines are connected to the interplanetary magnetic field (shown in Figure 54 from Anderson) and that the solar particles simply spiral along the IMF

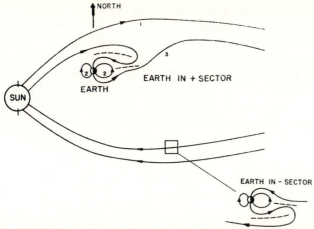

FIGURE 54   Field line topology of open magnetosphere indicating how polar cap field lines could be connected to interplanetary field lines and eventually to the sun (Anderson, 1970).

until they arrive in the polar cap and are detected. Thus a large body of observational evidence supports Dungey's contention that interconnection of geomagnetic and IMF lines does occur and is the major driving force for magnetospheric convection. With this view of the gross magnetospheric configuration we now turn to a discussion of particle motions and resulting energization.

2.   Quasis-static acceleration mechanisms

Magnetosheath (solar wind) plasma may enter the magnetosphere in the vicinity of the two regions of weak magnetic field shown in Figure 51 as x-type neutral points. Plasma entering near the dayside neutral point at the nose of the magnetosphere will be convected over the polar cap and into the tail. In the tail this plasma may be lost to the interplanetary medium or become trapped on closed field lines after the flux tube has reconnected. This plasma plus that entering from the downstream flanks in the vicinity of the tail neutral point will be convected deep into the magnetosphere on the contracting closed geomagnetic field lines (lines 7, 8 and 9 in Figure 51). The plasma trapped on these contracting field lines is compressed and therefore energized as the volume of the flux tubes decrease. This adiabatic heating can be thought of as resulting from the combined effects of Fermi and betatron acceleration (Axford, 1969).

Fermi acceleration refers to the increase in particle energy as a result of a collision with a moving magnetic field. As an example consider a particle moving in the x direction with a speed v_o. If the particle encounters a magnetic field moving in the $-x$ direction with a speed v and is elastically reflected in the $-x$ direction the resulting particle speed will be increased by $2v$. Particles trapped on contracting field lines will observe their mirror points in each hemisphere approaching each other and on each reflection will gain energy parallel to the field lines. Using the conservation of the second invariant (Section II, 7) the ratio of the initial (T_i) to the final (T_f) kinetic energy is approximately related to the initial (l_i) and final (l_f) field line lengths by the equation:

$$\frac{T_i}{T_f} = \left(\frac{l_f}{l_i}\right)^2$$

Using a current magnetic field model Sharber and Heikkila (1972) estimated a maximum energy ratio of $T_f/T_i \simeq 20$ which is more than an order of magnitude increase in parallel kinetic energy. These authors suggested that their measurements of field-aligned electron precipitation at the poleward edge of the auroral zone are consistent with Fermi acceleration of these particles. However it is not clear that the Fermi mechanism can produce the required distribution in the equatorial plane (the particles must be aligned to within a few degrees of the field line) with any reasonably magnetospheric model (see Section VII).

As particles are convected into the magnetosphere they also encounter an increasing magnetic field strength and as a result experience betatron acceleration. Using the conservation of the first invariant (Section II, 7) the increase in kinetic energy (ΔT) perpendicular to the magnetic field is related to the increase in magnetic field (ΔB) seen by the particle by the equation

$$\frac{\Delta T}{T} = \frac{\Delta B}{B}$$

where B and T are the initial field strength and kinetic energy. Using current magnetic field models, one calculates that an equatorial particle in moving from a point in the distant tail to typical auroral latitudes will have its transverse energy increased by at least an order of magnitude. Combined Fermi and betatron acceleration therefore can increase particle energies by several orders of magnitude.

In this model the magnetosheath plasma is increasingly energized as it is convected radially inward on the nightside. It then flows around the earth at lower latitudes to the dayside where it is convected radially outward and decompressed (de-energized). A decreasing temperature (or average energy) with increasing latitude at more or less all local times is therefore predicted. This trend has been reported in many statistical studies of auroral particle precipitation (e.g. Sharp and Johnson, 1971) and lends support to the convection theory of particle energization.

There seems to be observational evidence and theoretical justification for attributing auroral particle energization to adiabatic compression of magneto-sheath plasma by solar wind driven convection. It is not likely however that this mechanism alone could explain the fine scale structure observed in auroral precipitation. As an example this model would predict rather featureless latitudinal profiles of particle precipitation with particle energies increasing monotonically with decreasing latitudes rather than the highly structured profiles normally observed in association with auroral forms. Various mechanisms have been suggested which might account for the small scale structure.

It is not unreasonable to suggest that particles trapped on contracting flux tubes in the nightside magnetosphere will rapidly develop a depleted loss cone distribution due to betatron acceleration. These particles could be convected deep into the magnetosphere without serious atmospheric losses. The structured low altitude precipitation could then be explained by invoking a spatially confined region of strong pitch-angle diffusion near the equatorial plane which would enhance the local precipitation rate (Kennel, 1969). The presence of strong pitch-angle diffusion on auroral field lines has been inferred from many observations (Section IV). Whalen and McDiarmid (1973) showed that the exponential decay observed in electron intensities following an auroral substorm was consistent with their being lost from a trapping region at the maximum rate possible from pitch-angle diffusion. The equivalent minimum lifetime (T_M) was estimated by Kennel (1969) using the following reasoning. Consider a trapped particle population near the equatorial plane in strong diffusion. Under these conditions the pitch angle distribution is completely isotropic, even in the loss cone. Therefore the probability that a particle will find itself in the loss cone is $\Delta\Omega/2\pi$ where $\Delta\Omega$ is the loss cone solid angle. If α_{LC} is the loss cone angle and T_B is the travel time from the equatorial plane to the ionosphere, or equivalently the one quarter bounce period, then the minimum lifetime for a particle is given by

$$T_M = 2T_B/(\alpha_{LC})^2$$

Since auroral particles have been observed with lifetimes comparable to T_M and pitch-angle distributions in the loss cone indicate pitch-angle diffusion coefficients high enough to cause strong pitch-angle scattering (Whalen and McDiarmid) it may be concluded that sufficient pitch angle scattering occurs on auroral field lines to produce some of the spatial structure observed in auroral precipitation.

Although the spatial structure of auroral arcs may be explained by enhanced pitch angle diffusion the increase in average electron energy observed inside auroral arcs (Section IV, 1, a) cannot be accounted for by this simple mechanism. An energy selective pitch-angle diffusion mechanism could be invoked to explain this observation, the region of enhanced energy selective diffusion being confined to distances in the equatorial plane of the order of a few tens of gyroradii to produce the narrow arcs. However, this seems a rather unlikely solution to the problem of the formation of discrete electron arcs.

It has been postulated that a dynamic interaction between the ionosphere and the magnetosphere could cause the formation of discrete auroral arcs. Atkinson (1970) showed that increased ionization in the ionosphere resulting from energetic auroral electron precipitation could cause polarization electric fields to develop in the ionosphere which in turn could produce enhanced precipitation, thus establishing a positive feedback system between the ionosphere and the deep magnetosphere. With appropriate assumptions, he was able to show that precipitation regions consistent with auroral arcs will result. Electron energization on field lines connecting with auroral forms does not appear explicitly in the model. A large $\xi_\parallel$ probably could be incorporated to produce the particle precipitation.

Coroniti and Kennel (1972) also invoked ionosphere-magnetosphere interactions to show that large parallel electric fields might be formed to accelerate auroral particles. This model and Atkinson's requires a significant component of the convective electric field to be parallel to the auroral arcs.

Some direct observations of auroral particles have led to speculation that large quasi-static electric fields parallel to magnetic field lines ($\xi_\parallel$) could be responsible for auroral particle energization. Electron energy distributions peaked in the 1- to 10 keV energy range and pitch-angle distributions peaked along field lines (Section IV, 1, a and b) are suggestive of such an acceleration mechanism. Some authors have suggested that large field-aligned currents could lead to substantial values of $\xi_\parallel$ by causing potential double layers (sheaths) to form in the ionosphere or by causing plasma instabilities which drastically decrease the parallel conductivity. These mechanisms are discussed in a recent article by Block (1972). However, since no convincing evidence exists which indicates that large $\xi_\parallel$'s are the main mechanism responsible for the formation of auroral arcs, the validity of all the above models remains to be demonstrated.

Neutral point acceleration has also been suggested as a possible source for auroral particle energization. As indicated in Figure 51 neutral or x-type neutral points may exist in the magnetosphere in the vicinity of the last closed field line (lines 1 and 6) which have been shown to lie near the high latitude edge of the auroral oval. Assuming a geomagnetic tail geometry similar to that shown in Figure 51 and the existence of a dawn to dusk electric field across the neutral sheet, which is predicted by the convection model discussed previously, Speiser (1967) calculated the trajectories of particles in the vicinity of the neutral sheet. Some typical trajectories in the geomagnetic tail are illustrated in Figure 55. Speiser showed that particles injected near the flanks of the tail will oscillate about the neutral sheet and be accelerated by the electric field as they move across the tail. A small magnetic field perpendicular to the sheet will cause these energized particles to be ejected from the tail along field lines passing near the neutral point with a pitch-angle distribution strongly peaked near 0°. Field-aligned electron distributions which may be associated with such a mechanism have been detected in the tail by Hones et al. (1971a).

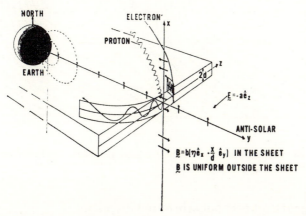

FIGURE 55   Particle trajectories in the vicinity of the tail neutral sheet assuming a cross-tail convective electric field. Particles injected in the flanks of the tail oscillate about the sheet and accelerate as they cross the tail and are eventually ejected from tail toward the earth by a small perpendicular component of the tail field (Speiser, 1967).

Evidence for the existence of multiple neutral points in the tail was recently presented by Schindler and Ness (1972). Figure 56 is a model presented by the authors which was found to be in qualitative agreement with their observations. Hruska (1973) has associated the single and multiple enhancements in electron precipitation observed in low altitude latitudinal profiles, presumably auroral arcs, with neutral point or multiple neutral point acceleration. It should be noted that such a magnetic field configuration allows neutral point acceleration to occur equatorward of (inside) closed field lines.

The existence of neutral points in the magnetosphere seems likely, however the dynamics of field line reconnection, the formation of x-type neutral points and the acceleration of particles in the vicinity is not properly understood

(Axford, 1969). Therefore the relationship between neutral points and auroral particle energization and precipitation is still highly conjectoral.

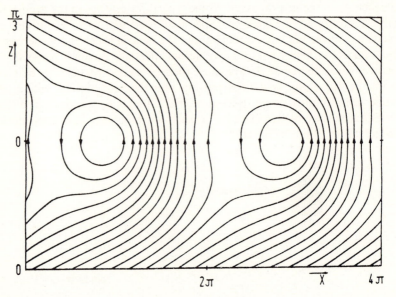

FIGURE 56　Model of tail magnetic field indicating field line configuration with multiple neutral points (Schindler and Ness, 1972).

3. Magnetospheric substorm mechanisms

The causes and effects of magnetospheric substorms are not well understood and are currently the subject of intensive investigation. A summary of the ground based observations of auroral substorms appears in Section III, 5. Briefly the explosive or expansive phase is characterized by:

i) The sudden brightening and activation of auroral arcs in the oval near midnight followed by a rapid poleward and less spectacular equatorward expansion of the oval.

ii) A rapid growth of the auroral electrojet and a corresponding increase of field alinged currents, and

iii) A greatly enhanced energetic (E > 10 keV) electron precipitation in the midnight sector and later in the dawn and dusk sectors.

Direct low altitude measurements of auroral precipitation also indicate that energetic electrons are injected near midnight at the commencement of the substorm. Changes in electron spectra during substorms have not been well documented however by comparing Figures 13 and 23, which represent quiet or presubstorm and substorm spectra respectively, some properties can be inferred. No increase in electron average energy is apparent rather a redistribution to a more Maxwellian shape is suggested. The sudden appearance of energetic (E > 10 keV) electrons is therefore likely to be associated with the thermalization of the pre-existing nonthermal electron distribution. That is these data suggest that the energetic particle injection during a substorm is associated with a stochastic acceleration mechanism. Typical loss cone pitch-angle distributions obtained during substorms also suggest stochastic processes, in this case, pitch-angle diffusion is the result.

The characteristics of proton precipitation are relatively unaffected by substorm, (Section IV, 2, b). Ground based photometric observations show no strong correlation with substorm activity aside from the expansion of the proton precipitation region near midnight (Section IV, 2, b). Direct measurements of auroral proton precipitation reveal no substorm correlations with possible exception of the appearance of small scale spatial structure which has only been observed during substorms (Section V, 1).

Some of the most consistent features observed near the equatorial plane in association with auroral substorms are the equatorward motion of the inner edge of the plasma sheet and simultaneous injection of plasma deep into the magnetosphere, the thinning and later expansion of the tail plasma sheet and the recovery of the geomagnetic field from a tail like (pre-substorm) to a more dipole-like (post-substorm) configuration.

Mende et al. (1972) presented observations which illustrated most of these characteristics. The substorm studied occurred after a period of low geomagnetic activity and was thus well isolated from preceding substorms. The onset of the expansive phase was determined from ground based photometric and magnetometer data to have occurred near 0530 UT on February 13, 1970. Observations in the solar wind from HEOS 1 showed that the IMF which had a northerly component previously switched southward at about 0330 UT, 2 hours before the substorm onset. At approximately 0430 UT measurements in the plasma sheet at 18.5 R_e from Vela 5B, situated near the tail neutral sheet, showed a gradual decrease in ion density and pressure (Figure 57). This decrease continued and was followed by a plasma drop out at 0510 (~20 minutes before the substorm onset). These characteristics in the Vela particle data have been shown to be associated with a plasma sheet thinning, the satellite being inside the plasma sheet originally and outside in the low plasma density high latitude tail, after thinning (Hones et al., 1971b).

Data acquired simultaneously by sensors on the geostationary satellite ATS 5 (Figure 57) showed that the plasma sheet thinning was accompanied by an increase in plasma density at 6.6 R_e in the evening sector. This effect

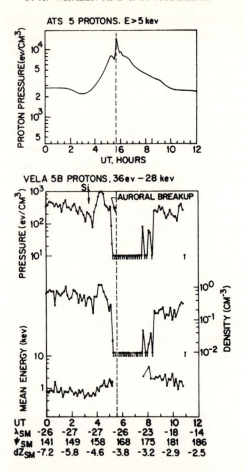

FIGURE 57   Simultaneous ATS 5 (equatorial at 6.6 R_e) and Vela 5B (tail measurements at 18,5 R_e) proton measurements during an auroral substorm (Mende et al., 1972).

was presumed to be due to the equatorward motion of the inner edge of the plasma sheet and the subsequent envelopment of the satellite, which at quiet times is located equatorward of the plasma sheet, by plasma sheet particles (Shelley et al., 1971).

These effects have been attriubted to increased reconnection at the dayside neutral point resulting from the southward turning of the IMF. This process has the net effect of transporting dayside flux tubes into the tail. In the absence of a corresponding increase in convection to the dayside magneto-pause an increased tail flux and an inward motion of the magnetopause will result (see Figure 58, from Kennel and Coroniti, 1971). Such an erosion of the dayside magnetopause was observed by Aubry et al. (1970).

The increasing tail magnetic field is accompanied by an increasing tail current and the resulting deformation of high latitude field lines into a more

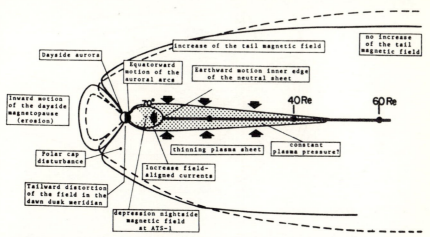

FIGURE 58 Model of magnetospheric changes during the growth phase of an auroral substorm. The dashed line represents the shape of the magnetopause at the end of the growth phase (Kennel and Coroniti, 1971).

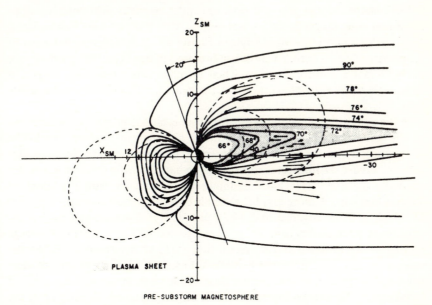

FIGURE 59 Pre-substorm model magnetic field configuration (Fairfield and Ness, 1970).

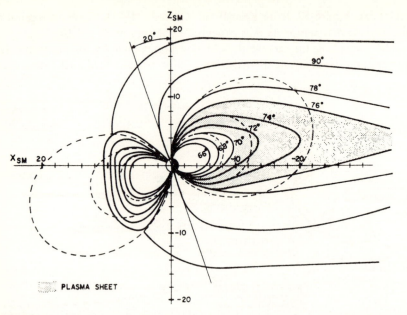

FIGURE 60   Quiet post-substorm model magnetic field configuration (Fairfield and Ness, 1970).

tail-like shape. Model pre- and post-substorm magnetic field configurations consistent with magnetic field observations were presented by Fairfield and Ness (1970) and are shown in Figures 59 and 60. It is believed that most of the energy released during a substorm is stored in this expanded geomagnetic tail and that a substorm results when an unknown mechanism triggers a large scale internal reconfiguration of the geomagnetic field to a more stable form.

In the equatorial plane the substorm onset is characterized by the recovery of the geomagnetic field at synchronous altitudes to a more dipole-like configuration and the injection of energetic plasma deep into the magnetosphere. Figure 57 illustrates the sudden enhancement in proton intensities normally observed near local midnight in coincidence with the onset of auroral breakup. A more graphic example of this injection is shown in Figure 61 which is from Deforest and McIlwain (1971). Electron and proton intensities at pitch angles near 90° are represented by a grey scale with energy plotted as the ordinate and universal time (UT) as the abscissa. When the satellite is near local midnight (~0600 UT) simultaneous injections of electrons and protons at all energies are observed in coincidence with the onset of substorms. This plasma injection into 6.6 R_e is presumed to result from the sudden equatorward motion of the plasma sheet. The rising and falling

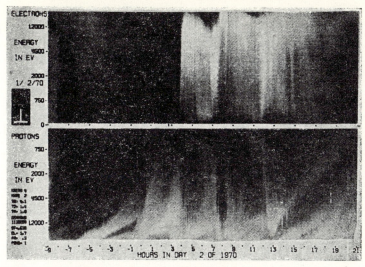

FIGURE 61 Spectrogram of the perpendicular ($\beta \simeq 90°$) proton and electron fluxes observed by ATS 5 particle spectrometers (Deforest and McIlwain, 1971).

traces observed at other local times result when particles are injected near midnight and then drift in the combined electric (convective) and magnetic fields. These traces represent dispersion curves resulting from the energy, mass and charge dependence of the particle drift-velocities. Convective electric field patterns (McIlwain, private communication) near the synchronous orbit were derived by fitting these dispersion profiles to model fields and the results of this fitting procedure appear in Figure 62. Clearly this pattern is similar to the theoretical patterns shown in Figures 52 and 53. These direct observations of plasma injection are consistent with the contention that the extraterrestrial ring current results from repeated particle injections into the ring current in association with substorms (Akasofu, 1969).

In the tail the onset of the substorm is followed by the recovery or thickening of the plasma sheet which is illustrated in Figure 57 by the reappearance of plasma sheet particles at Vela 5B, in this case, approximately two hours after the substorm. This is a little longer than typical recovery times (Hones et al., 1971b).

The probable coincidence of the poleward boundary of the oval and the last closed field line coupled with the poleward expansion of the oval is strongly suggestive of field line closure or tail reconnection during substorms. McDiarmid and Hruska (1972) used energetic electron latitudinal intensity profiles to show that geomagnetic field lines become temporarily closed up to 80° invariant latitude during substorms, which is 10° higher than normal.

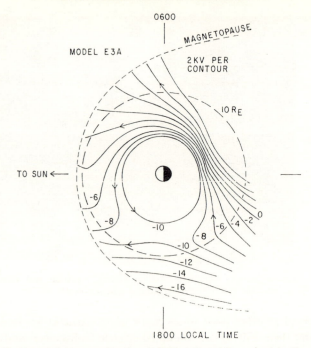

FIGURE 62 Equatorial convective electric fields inferred from particle dispersion curves observed at 6.6 R_e (McIlwain, private communication).

Therefore, plasma sheet thickening observed by the Vela satellites may not be due solely to a recovery of the geomagnetic field but may also reflect the increased extent of closed field lines in the night hemisphere. Rapid reconnection of tail field lines has been suggested as a possible substorm energy source (Axford, 1969).

On the dayside various substorm responses have been reported from activation of auroral arcs tens of minutes after the onset (Akasofu, 1969) to an equatorward motion of auroral arcs coincident with the onset of substorms near local midnight (Akasofu, 1972). The equatorward motion of the dayside aurora has been related to an equatorward shift of the boundary of closed field lines or the inward motion of the dayside magnetopause resulting from erosion of dayside flux tubes. This interpretation is in agreement with most substorm models (e.g. Figure 58) except on the important question of timing. Previous results suggested that the erosion of the dayside magnetopause and the subsequent increase in tail flux, presumably due to an increased southward component of the IMP, generally preceded the onset of auroral substorms and was therefore part of the substorm growth phase. This apparent conflict illustrates one of the continuing controversies concerning auroral substorms, this being whether measurable effects associated with a substorm growth phase systematically occur before the onset of auroral substorms.

McPherron (1970) proposed that the existence of a substorm growth phase could be inferred from auroral zone magnetograms and the equatorial particle observations summarized previously are strongly suggestive of a growth phase. Mozer and Manka (1971) using balloon-borne electric field probes claimed that increased convective electric fields appear in the polar cap as a precursor to substorms although these observations have not been confirmed by satellite-borne probes (Heppner, 1972b). Akasofu (1972) has attacked the concept of a growth phase mostly on the basis of timing and showed that many of phenomena were in fact associated with a substorm expansive phase. This conflict may partly result from the difficulty in determining the exact onset time of substorms from any set of ground station data now available. The exact temporal correlation of the vast body of ground based substorm observations in the night hemisphere with the dayside aurora and satellite measurements is unresolved at this time.

A complete summary of all auroral particle observations and theories has not been attempted in this report. Rather we presented those observations, speculations, models and theories which we believe best represent the current status of the understanding of high latitude auroral phenomena. We have discussed important competing interpretations where they existed and referenced these works. Undoubtedly some and possibly most of the concepts discussed here will change with the introduction of new measurements and theories. It is with considerable interest that we look forward to these advances and to the resolution of the outstanding problems of auroral physics.

References

Ackerson, K. L. and L. A. Frank (1972) *J. Geophys. Res.* **77**, 1128.

Aggson, T. L. (1969) in B. M. McCormac and A. Omholt (eds.), *Atmospheric Emissions*, Van Nostrand Reinhold, Co., N. Y.

Akasofu, S. -I., S. Chapman and A. B. Meinel (1966) *Handbuch der Physik*, **49**, 1.

Akasofu, S. -I. (1969) in B. M. McCormac and A. Omholt (eds.), *Atmospheric Emissions*, Van Nostrand Reinhold, Co., N. Y.

Akasofu, S. -I. and C. -I. Meng (1969) *J. Geophys. Res.* **74**, 293.

Akasofu, S. -I. (1972a) *J. Geophys. Res.* **77**, 2303.

Akasofu, S. -I. (1972b) *J. Geophys. Res.* **77**, 6275.

Anderson, K. A. (1970) in B. M. McCormac (ed.), *Particles and Fields in the Magnetosphere*, D. Reidel, Dordrecht.

Arnoldy, R. L. (1970) *J. Geophys. Res.* **75**, 228.

Arnoldy, R. L. (1971) *J. Geophys. Res.* **76**, 5189.

Arnoldy, R. L. and L. W. Choy (1973) *J. Geophys. Res.* **78**, 2187.

Atkinson, G. (1970) *J. Geophys. Res.* **75**, 4746.

Aubry, M. P., C. T. Russel and M. G. Kivelson (1970) *J. Geophys. Res.* **75**, 7018.

Axford, W. I. and C. O. Hines (1961) *Can. J. Phys.* **39**, 1433.

Axford, W. I. (1969) *Revs. Geophys.* **7,** 421.
Axford, W. I. (1970) in B. M. McCormac, *Particles and Fields in the Magnetosphere,* D. Reidel, N. Y.
Bame, S. J., J. R. Asbridge, A. J. Hundhausen and M. D. Montgomery (1970) *J. Geophys. Res.* **75,** 6360.
Beach, R., G. R. Crosswell, T. N. Davis, T. J. Hallinan and L. R. Sweet (1968) *Planet. Space Sci.* **16,** 1525.
Bernstein, W., G. T. Inouye, N. L. Sanders and R. L. Wax (1969) *J. Geophys. Res.* **74,** 3601.
Bernstein, W. (1973) private communication.
Block, L. P. (1972) *Cosmic Electrodynamics* **3,** 349.
Bostrom, R. (1972) in E. R. Dyer (ed.), *Critical Problems of Magnetospheric Physics,* Proceedings of the Joint COSPAR/IAGA/URSI Symposium, MADRID.
Brice, N. M. (1967) *J. Geophys. Res.* **72,** 5193.
Brice, N. M. (1969) The morning maximum in energetic electron precipitation and its relation to convection of the magnetosphere, presented at Conference on Electric Fields in the Magnetosphere, Rice University, Houston, Texas.
Bryant, D. A., G. M. Courtier and G. Bennett (1971) in B. M. McCormac (ed.), *The Radiating Atmosphere,* D. Reidel, Dordrecht.
Burrows, J. R., I. B. McDiarmid and M. D. Wilson (1972) in B. M. McCormac (ed.), *Earth's Magnetospheric Processes,* D. Reidel, Dordrecht.
Chappell, C. R. (1968) The interaction of auroral electrons with the atmosphere, Ph.D. Thesis, Rice University, Houston, Texas.
Chase, L. M. (1969) *J. Geophys. Res.* **74,** 348.
Choy, L. W., R. L. Arnoldy, W. Potter, P. Kintner and L. J. Cahill, Jr. (1971) *J. Geophys. Res.* **76,** 8279.
Cloutier, P. A. (1971) *Revs. Geophys. and Space Phys.* **9,** 987.
Cornwall, J. M., F. V. Coroniti and R. M. Thorne (1971) *J. Geophys. Res.* **76,** 4428.
Cornwall, J. M. (1972) *Revs. Geophys. and Space Phys.* **10,** 993.
Coroniti, F. V. and C. F. Kennel (1972) *J. Geophys. Res.* **77,** 2835.
Davidson, G. T. (1965) *J. Geophys. Res.* **70,** 1061.
Deforest, S. E. and C. E. McIlwain (1971) *J. Geophys. Res.* **76,** 3587.
Dungey, J. W. (1961) *Phys. Rev. Letts.* **6,** 47.
Eather, R. H. (1967) *Revs. Geophys.* **5,** 207.
Evans, D. S. (1971) in B. M. McCormac (ed.), *The Radiating Atmosphere,* D. Reidel, Dordrecht.
Evans, L. C. and E. C. Stone (1972) *J. Geophys. Res.* **77,** 5580.
Fairfield, D. H. (1968) *J. Geophys. Res.* **73,** 7329.
Fairfield, D. H. and N. F. Ness (1970) *J. Geophys. Res.* **75,** 7032.
Feldstein, Y. I. (1966) *Planet Space Sci.* **14,** 121.
Frank, L. A. (1971a) *J. Geophys. Res.* **76,** 2265.
Frank, L. A. (1971b) *J. Geophys. Res.* **76,** 5202.
Frank, L. A. and K. L. Ackerson (1971) *J. Geophys. Res.* **76,** 3612.
Gold, T. (1959) *J. Geophys. Res.* **64,** 1219.
Gurnett, D. A. (1972) in B. M. McCormac (ed.), *Earth's Magnetospheric Processes,* D. Reidel, Dordrecht.
Hamberger, S. M. and J. Jancarik (1970) *Phys. Rev. Letters* **25,** 999.
Hartz, T. R. and N. M. Brice (1967) *Planet. Space Sci.* **15,** 301.
Haskell, G. P. and D. J. Southwood (1972) *J. Geophys. Res.* **77,** 6926.
Heikkila, W. J. and J. D. Winningham (1971) *J. Geophys. Res.* **76,** 883.
Heikkila, W. J., J. D. Winningham, R. H. Eather and S. -I. Akasofu (1972) *J. Geophys. Res.* **77,** 4100.
Heppner, J. P. (1972a) in E. R. Dyer (ed.), *Critical Problems of Magnetospheric Physics,* Proceedings of the Joint COSPAR/IAGA/URSI Symposium.
Heppner, J. P. (1972b) *J. Geophys. Res.* **77,** 4877.
Holmgren, L. and B. Aparicio (1972) preprint, Kiruna Geophysical Observatory, No. 72:302.

Hones, E. W., J. R. Asbridge, S. J. Bame and S. Singer (1971a) *J. Geophys. Res.* **76**, 4663.
Hones, E. W., J. R. Asbridge and S. J. Bame (1971b) *J. Geophys. Res.* **76**, 4402.
Hruska, A. (1973) *J. Geophys. Res.* **78**, 7509.
Hultqvist, B., H. Borg, W. Riedler and P. Christopherson (1971) *Planet. Space Sci.* **19**, 279.
Jelly, D. and N. M. Brice (1967) *J. Geophys. Res.* **72**, 5919.
Johnstone, A. H. (1971) *J. Geophys. Res.* **76**, 5259.
Kavanagh, L. D. and A. W. Schardt (1970) *Revs. Geophys. and Space Phys.* **8**, 389.
Kelley, M. C., F. S. Mozer and U. V. Fahleson (1971) *J. Geophys. Res.* **76**, 6054.
Kennel, C. F. and H. E. Petschek (1966) *J. Geophys. Res.* **71**, 1.
Kennel, C. F. (1969) *Revs. Geophys.* **7**, 379.
Kennel, C. F. and F. V. Coroniti (1971) presented at IUGG Meeting, Moscow.
Lassen, K. (1969) in B. M. McCormac and A. Omholt (eds.), *Atmospheric Emissions*, Van Nostrand Reinhold, N. Y.
Levy, R. H., H. E. Petschek and G. L. Siscoe (1964) *AIAA Journal* **2**, 2065.
Maggs, J. E. and T. N. Davis (1968) *Planet. Space Sci.* **16**, 205.
McDiarmid, I. B. and J. R. Burrows (1965) *Can. J. Phys.* **70**, 3031.
McDiarmid, I. B., E. E. Budzinski, B. A. Whalen and N. Sckopke (1967) *Can. J. Phys.* **45**, 1755.
McDiarmid, I. B. and J. R. Burrows (1968) *Can. J. Phys.* **46**, 49.
McDiarmid, I. B., J. R. Burrows and M. D. Wilson (1972) *J. Geophys. Res.* **77**, 1103.
McDiarmid, I. B. and A. Hruska (1972) *J. Geophys. Res.* **77**, 3377.
McEwen, D. J. and G. G. Sivjee (1972) *J. Geophys. Res.* **77**, 5523.
McIlwain, C. E. (1961) *J. Geophys. Res.* **66**, 3681.
McPherron, R. L. (1970) *J. Geophys. Res.* **75**, 5592.
Mende, S. B., R. D. Sharp, E. G. Shelley, G. Haerendel and E. W. Hones (1972) *J. Geophys. Res.* **77**, 4682.
Miller, J. R. and G. G. Shepherd (1969) *J. Geophys. Res.* **74**, 4987.
Miller et al. (1973) in publication.
Montbriand, L. E. (1971) in B. M. McCormac (ed.), *The Radiating Atmosphere*, D. Reidel, Dordrecht.
Mozer, F. S. and P. Bruston (1967) *J. Geophys. Res.* **72**, 1109.
Mozer, F. S. and U. V. Fahleson (1970) *Planet. Space Sci.* **18**, 1563.
Mozer, F. S. and R. H. Manka (1971) *J. Geophys. Res.* **76**, 1697.
Mozer, F. S. (1973) *Space Sci. Revs.* **14**, 272.
Nishida, A. and K. Maezawa (1971) *J. Geophys. Res.* **76**, 2254.
Northrop, T. G. (1963) in *The Adiabatic Motion of Charged Particles*, Interscience Publishers, N. Y.
Parker, E. N. and V. C. A. Ferraro (1971) *Handbuch der Physik* **49**, 131.
Perkins, F. W. (1968) *J. Geophys. Res.* **73**, 6631.
Pizzella, G. and L. A. Frank (1971) *J. Geophys. Res.* **76**, 88.
Rearwin, S. (1971) *J. Geophys. Res.* **76**, 4505.
Reasoner, D. L. and C. R. Chappell (1973) *J. Geophys. Res.* **78**, 2176.
Rees, M. H., R. A. Jones and J. C. G. Walker (1971) *Planet Space Sci.* **19**, 313.
Roberts, C. S. (1969) *Revs. of Geophys.* **7**, 305.
Roederer, J. G. (1970) in J. G. Roederer and J. Zahringer (eds.), *Physics and Chemistry in Space*, Vol. 2.
Sandford, B. P. (1964) *J. Atmospheric Terrestrial Phys.* **26**, 749.
Scarf, F. L., C. F. Kennel, R. W. Fredericks, I. M. Green and G. M. Crook (1970) in B. M. McCormac (ed.), *Particles and Fields in the Magnetosphere*, D. Reidel, Dordrecht.
Schield, M. A., J. W. Freeman and A. J. Dessler (1969) *J. Geophys. Res.* **74**, 247.
Schield, M. A. and L. A. Frank (1970) *J. Geophys. Res.* **75**, 5401.
Schindler, K. and N. F. Ness (1972) *J. Geophys. Res.* **77**, 91.
Sharber, J. R. and W. J. Heikkila (1972) *J. Geophys. Res.* **77**, 3397.
Sharp, R. D. and R. G. Johnson (1971) in B. M. McCormac (ed.), *The Radiating Atmosphere*, D. Reidel, Dordrecht.
Shelley, E. G., R. G. Johnson and R. D. Sharp (1971) *Radio Sci.* **6**, 6092.

Shelly, E. G., R. G. Johnson and R. D. Sharp (1972) *J. Geophys. Res.* **77**, 6104.
Shepherd, G. G. (1971) in B. M. McCormac (ed.), *The Radiating Atmosphere*, D. Reidel, Dordrecht.
Shepherd, G. G. and F. W. Thirkettle (1973) *Science* **180**, 737.
Speiser, T. W. (1967) *J. Geophys. Res.* **72**, 3919.
Speiser, T. W. (1969) in B. M. McCormac and A. Omholt (eds.), *Atmospheric Emissions*, Van Nostrand Reinhold, N. Y.
Spreiter, J. R. and A. Y. Alksne (1969) *Rev. Geophys. Space Phys*, **7**, 11.
Swift, D. W. (1965) *J. Geophys. Res.* **70**, 3061.
Vasyliunas, V. M. (1968) *J. Geophys. Res.* **73**, 2839.
Vasyliunas, V. M. (1970) *E.O.S. Trans. A.G.U.* **51**, 814.
Westerlund, L. H. (1969) *J. Geophys. Res.* **74**, 351.
Whalen, B. A. and I. B. McDiarmid (1969) in B. M. McCormac and A. Omholt (eds.), *Atmospheric Emissions*, Van Nostrand Reinhold, N. Y.
Whalen, B. A. and I. B. McDiarmid (1970) *J. Geophys. Res.* **75**, 123.
Whalen, B. A., J. R. Miller and I. B. McDiarmid (1971a) *J. Geophys. Res.* **76**, 978.
Whalen, B. A., J. R. Miller and I. B. McDiarmid (1971b) *J. Geophys. Res.* **76**, 2406.
Whalen, B. A., J. R. Miller and I. B. McDiarmid (1971c) *J. Geophys. Res.* **76**, 6847.
Whalen, B. A. and I. B. McDiarmid (1972a) *J. Geophys. Res.* **77**, 191.
Whalen, B. A. and I. B. McDiarmid (1972b) *J. Geophys. Res.* **77**, 1306.
Whalen, B. A. and I. B. McDiarmid (1973) *J. Geophys. Res.* **78**, 1608.
Wiens, R. H. and A. Vallance Jones (1969) *Can. J. Phys.* **47**, 1493.
Wilcox, J. M. (1968) *Space Sci. Rev.* **8**, 258.
Willis, D. M. (1969) *Planet. Space Sci.* **17**, 339.
Willis, D. M. (1972) in E. R. Dyer (ed.), *Critical Problems of Magnetospheric Physics*, Proc. of the Joint COSPAR/IAGA/URSI Symposium. MADRID.
Winningham, J. D. (1972) in B. M. McCormac (ed.), *Earth's Magnetospheric Processes*, D. Reidel, Dordrecht.
Winningham, J. D. and W. J. Heikkila (1974) *J. Geophys. Res.* **79**, 949.
Zmuda, A. J., J. C. Armstrong and F. T. Heuring (1970) *J. Geophys. Res.* **75**, 4757.

Polar-Cap Absorption—
Observations and Theory

GEORGE C. REID

National Oceanic and Atmospheric Administration, Boulder, Colorado 80302, U.S.A.

1 INTRODUCTION

The term 'polar-cap absorption', introduced as a convenient description for a class of events observed with riometers (see below) at high magnetic latitudes, is now generally employed to describe the entire range of ionospheric effects caused by energetic solar particles. Many of these effects have little to do with absorption, and some of them are not even confined to the polar caps. For the sake of convenience, however, we shall adopt the general usage here, defining the scope of the paper as a review of the effects of energetic solar particles (i.e., protons and heavy nuclei with energies in the Mev range) on the earth's ionosphere. The possible role of solar electrons will be discussed in its context.

We shall not be concerned here with the properties and behavior of the solar particles themselves, as revealed by the wealth of information that has been derived from direct satellite observations within the past decade, but simply with the observation and interpretation of the ionospheric effects. In a sense, we can think of the ionosphere as a vast ionization chamber pervaded by a magnetic field, and we can use ground-based radio observations to infer some properties of the gas in the chamber (i.e., the upper atmosphere) and of the portion of the magnetic field that governs the trajectories of the particles (i.e., the outer magnetosphere). The ionospheric and magnetospheric aspetcs of PCA events will be treated separately in what follows, though it must be borne in mind that these aspects are not as clearly separated in actual events, giving rise to some difficulty in interpretation. The review will begin with brief surveys of the historical development of the area and of the observational techniques commonly employed.

2 HISTORICAL SURVEY

The first recorded outburst from the sun of particles with cosmic-ray energies occurred on 28 February 1942 in conjunction with an intense solar flare that also provided the first evidence of solar radio-noise emission through its effect on wartir   radars operating in England. In the three decades since their birthdate, the twin disciplines of solar cosmic rays and solar radio astronomy have flourished, adding greatly to our understanding of solar-terrestrial relationships, and producing new sub-disciplines as they progressed. One such offshoot, the study of polar-cap absorption, began when certain anomalous ionospheric effects were observed following the major solar-cosmic-ray event of 23 February 1956. These effects took the form of a loss of signal on high-latitude VHF communications circuits (Bailey, 1957), strong absorption of cosmic radio noise penetrating the icnosphere from above (Little and Leinbach, 1958), and sudden disturbances of the amplitude and phase of LF and VLF radio signals (Allan et al., 1957; Belrose et al., 1956; Ellison and Reid, 1956; Pierce, 1956). The effects were interpreted by several authors as the result of atmospheric ionization produced by the solar cosmic ray particles, but Bailey (1957, 1959) was the first to put this suggestion on a quantitative basis and to show the importance of particles with energies well below those detectable by ground-based cosmic-ray instrumentation, if such particles did indeed exist.

The International Geophysical Year of 1957-58 provided an impetus for geophysical observations at high geomagnetic latitudes, and led to the location of a number of riometers (see below) at several such sites in Alaska, Canada, and Scandinavia. It was soon found that events similar to that of 23 February 1956 were quite frequent--of the order of one per month at this period--but with the important difference that no detectable increases in ground-level cosmic-ray flux could be observed in the great majority of cases. These events were quickly interpreted as the ionospheric effects produced by out-bursts of particles from the sun with energies considerably less than those needed to penetrate the entire atmosphere (Hultqvist and Ortner, 1959; Leinbach and Reid, 1959; Reid and Collins, 1959), leading to the inference that such events were much more common than the rare outbursts of particles of cosmic-ray energy. The particles were assumed to be mainly protons, and the energies required were of the order of millions of electron volts (Mev), in contrast to the billion-electron-volt (Bev) energies of the cosmic-ray protons, and also to the kilovolt energies of the particles associated with auroras and geomagnetic storms. The development of polar-cap effects in the ionosphere prior to certain major geomagnetic storms was also described at about the same time by Japanese workers (Hakura et al., 1958) on the basis of blackouts recorded by ionospheric sounding equipment.

Independently of these ionospheric observations, the first direct detection of solar protons in the earth's atmosphere was made on 22 August 1958 by Anderson (1958), using balloon-borne equipment at Churchill, Canada. This observation coincided with one of the ionospheric events described above, and provided the needed confirmation of the interpretation placed on these events. Thus an important gap in the energy spectrum of solar particles was filled, revealing an entirely new aspect of solar-terrestrial relations. The first list of PCA events (Reid and Leinbach, 1959) in which the term 'polar-cap absorption' was first introduced, described 24 events that occurred between May 1957 and July 1959. Since then the list has been extended both forward in time as new events were detected, and backward in time through the interpretation of existing high-latitude ionospheric sounding records (Collins et al., 1961), until the known events now total several hundreds.

In the early development, a considerable amount of effort was devoted to attempts to infer the fluxes and spectra of the solar particles themselves through quantitative interpretation of the PCA effects. Following the first satellite detection of these particles (Rothwell and McIlwain, 1959), however, satellite observations have supplanted ionospheric observations as a direct source of information on the particles. As mentioned above, we shall not be concerned with direct observations of the particles here, other than to mention two results that have a bearing on the interpretation of the ionospheric effects. The first is the discovery that the solar-particle fluxes contain appreciable number of alpha-particles (Freier et al., 1959) and heavier nuclei (Fichtel and Guss, 1961), and the second is the discovery of associated fluxes of energetic electrons in the relativistic (Meyer and Vogt, 1962) and non-relativistic (Van Allen and Krimigis, 1965; Anderson and Lin, 1966) energy ranges.

Within recent years, the advent of accurate knowledge of the particle fluxes and spectra from satellite observations has in principle allowed the original problem to be inverted − instead of using our sketchy and incomplete knowledge of the behavior of the lower ionosphere to infer the properties of the particle fluxes, we can now use the detailed knowledge of the particle fluxes to improve our understanding of the ionosphere. A certain amount of progress in this direction has been made, but has been hampered by the fact that the ground-based techniques for observing PCA usually integrate their effects through a large range of altitudes, leading to a considerable amount of uncertainty in interpretation. Recently a few altitude profiles of electron densities and ion composition have been obtained with rockets during PCA events, and the exploitation of this approach should accelerate the rate of progress.

Apart from this potential for exploring the lower ionosphere, ground-based studies of PCA have certain inherent advantages over satellite observations for investigating the magnetic-field topology of the outer magnetosphere; this region is responsible for controlling the access of the solar particles to the ionosphere, and ground-based observations at high latitudes can provide essentially continuous information on the variability of the pattern of precipitation. These continuous observations, combined with the 'cuts' across the polar caps provided by polar-orbiting satellites, can yield a great deal of important information on the behavior of the outer magnetosphere and on its relationship to the behavior of the solar wind.

3 OBSERVATIONAL TECHNIQUES

As we shall see in the next section, the major effects of solar-proton bombardment occur at altitudes below 100 kilometers (we use the term 'solar-proton' here for convenience, bearing in mind that the particles are not exclusively protons). In the conventional ionospheric terminology, PCA is thus a phenomenon of the D region, where the collision frequency between free-electrons and neutral atmospheric constituents is relatively high, and where radio-wave energy is most strongly absorbed. As a result, the most widely used techniques for observing PCA events have made use of the absorption suffered by radio waves traversing the ionosphere.

At frequencies higher than the penetration frequency of the ionosphere, extraterrestrial signals can be received at the earth, and the transmission properties of the ionosphere can be monitored continuously by simple measurements of signal power. This is the principle employed by the riometer, alluded to above, which records cosmic radio noise at some frequency that is usually chosen to be high enough to avoid most terrestrial interference, and low enough that the cosmic-noise signal (which varies with frequency as $f^{-2.7}$) is readily measurable. The riometer (Little and Leinbach, 1959) is basically a highly stable noise receiver that avoids problems of gain variations by continuously comparing the incoming noise signal with the noise output of a local source and adjusting the latter to maintain equality. The output of the local source is then recorded on strip chart, and the integrated ionospheric absorption is measured as the decrease of this level below that recorded at the same sidereal time before and after the event.

While the riometer does suffer from certain disadvantages (for example, weak PCA events of long duration are difficult to detect), its simplicity has led to widespread use. Frequencies in the vicinity of 30 MHz have generally been employed, and the absorption in decibels recorded by a 30-MHz riometer has become a widely used measure for the intensity of a PCA event. The receiving antennas used have generally had fairly wide beams (of the order of $60°$ between half-power points); the absorption recorded in this case is substantially more than would be recorded by a pencil beam pointing vertically upward, since the noise that is received from off-vertical directions suffers more absorption by traversing a slanting path through the absorbing region.

In principle, absorption measurements should be converted to equivalent vertical absorption values in order the standardize the PCA intensity scale. This has not often been done, however, and most of the measurements quoted in the literature refer simply to the untreated absorption measured by an actual riometer.

The dynamic range of the riometer extends from a lower limit of the order of 0.3 db to an upper limit in the neighborhood of 15-20 db. The lower limit is imposed by the usually unavoidable variability in the quiet-day signal recorded by the riometer, which amounts to a few tenths of a decibel over a period of a few days; short-period variations in absorption, such as frequently occur in association with electron precipitation in the auroral zones, can be measured down to values considerably lower than 0.3 db, and careful design and maintenance could undoubtedly increase the sensitivity to weak PCA events. The upper limit is more fundamental, being due to the fact that a good absorber is also a good radiator, so that as the ionospheric absorption increases the riometer begins to record thermal radiation from the absorbing ionosphere itself. In thermodynamic terms, the cosmic-noise signal received at the earth at a frequency of 30 MHz is equivalent to the black-body radiation that would be received from a source at a temperature of about $20,000°K$. The lower ionosphere has a temperature of about $200°K$, so that even when the absorption is total the riometer will still be recording the equivalent of $200°K$ black-body radiation, i.e., the recorded attenuation will reach a maximum of a factor of 100, or 20 db. In actual fact, this limit has been approached in only a few cases, and the practical limitation is not serious.

Although the bulk of the quantitative information on PCA events has come from the riometer technique, other ground-based radio techniques have also yielded important data. As described above, the first identification of the ionizing effects of solar protons was made by Bailey (1957, 1959) using recordings of VHF forward-scatter signals at high latitudes. VHF forward-scatter circuits operate at frequencies above the penetration frequency of the ionosphere, in contrast to conventional HF ionospheric circuits. They rely on the fact that the lower ionsphere always contains small-scale irregularities in electron concentration that are capable of scattering a small fraction of the VHF radio-wave energy incident on them from below, while the vast majority of the energy escapes through the ionosphere. Weak though they are, the scattered signals still provide a practical means of radio communication, and during the early 1950's a considerable amount of effort was devoted to develop-developing practical systems, especially for arctic regions, where conventional systems are highly susceptible to aurorally associated disruptions (Bailey et al., 1955). While relatively immune to these auroral effects, VHF forward-scatter signals are affected inmarked and characteristic ways by PCA events. Qualitatively, these effects represent a balance between an enhancement in signal strength brought about by an increase in the scattering efficiency of the irregularities (which are presumably themselves due to mesospheric turbulence) and an attenuation caused by the increased concentration of free electrons below the scattering level. As we shall see in the next section, there are marked differences in the PCA-produced D region between daytime and nighttime conditions, and these differences are reflected in the response of the forward-scatter signals. During daytime conditions the concentration of electrons below the scattering level is relatively high, even during moderate events, and absorption is the dominant effect, causing a large decrease in signal strength below the normal undistrubed level. At night, the number of free electrons at the lower levels is greatly reduced, and the enhancement in signal strength caused by the increased scattering at higher levels becomes the dominant feature.

VHF forward-scatter recordings provide an excellent indicator of the occurrence and morphology of PCA events, but quantitative interpretation of the effects have been hampered by the difficulty in separating the enhancement and attenuation effects, and by lack of adequate knowledge of the height of the scattering irregularities. There are grounds for believing this height to lie between 70 and 75 km by day and near 85 km by night, but there have been no direct measurements of the possible diurnal or seasonal variation. The irregularities certainly deserve more thorough investigation, as does the entire topic of mesospheric turbulence, but any discussion of these areas does not belong within our present scope.

The radio techniques discussed so far have employed frequencies above the ionospheric penetration frequency. Much qualitative information on PCA events has been obtained from conventional HF ionospheric sounding records, in which the radio-wave energy is reflected vertically from the upper regions of the ionosphere, passing twice through the absorbing D region. Since a fairly extensive network of ionosondes has been in existence for a number of years, such recordings have proven extremely valuable in extending our knowledge of the occurrence of PCA events backward in time, as mentioned above, and in examining the geographical distribution of selected events in some detail. However, the intensity of the absorption effects at ionosonde frequencies becomes very large even for moderate events, and only the presence of a "blackout" can be recorded. During very weak events, or in the early stages of an intense event, some quantitative information can be obtained in the form of an increase in the minimum frequency at which echoes are recorded (f_{min} in ionospheric sounding terminology). While this quantity has been used by several workers in connection with PCA, it has the major disadvantage of being strongly dependent on equipment parameters, and is thus unsuited to absolute measurement.

PCA events also have marked effects on the propagation of radio waves in the

LF (30-300 kHz) and VLF (3-30 kHz) frequency ranges. The normal mode of propagation of these waves is quite different from that of the higher frequencies, since their wavelength is comparable with the distance between the earth and the base of the ionosphere. They propagate essentially in a spherical-shell wave-guide bounded by the conducting surface of the earth and the lower ionosphere, and consequently the effects produced by PCA events are quite different in nature from the effects produced at the higher frequencies, where the simple ray analogy is an excellent approximation.

PCA effects on long-distance VLF circuits crossing the polar caps have been observed since the earliest days (Ortner *et al.*, 1960), and have been discussed extensively since (e.g. Potemra *et al.*, 1967; Westerlund *et al.*, 1969), but the complexity of the modes of propagation involved, coupled with the variability of conditions over the long paths, have hampered their usefulness as a quantitative technique. Both the amplitude and the phase of the received signal are affected by the existence of a PCA event, and these changes are sufficiently distinctive that they form an excellent qualitative indicator of the onset of an event, particularly at night, when the standard absorption techniques are relatively insensitive. The effects are also generally of much longer duration than the absorption effects, suggesting that the technique is sensitive to much lower fluxes of particles than the riometer. Much remains to be done, however, to place these qualitative effects on a sure quantitative basis.

As mentioned above, all of these ground-based techniques are limited in their usefulness by the fact that they integrate the ionospheric effects over a considerable range of altitudes. In the case of the riometer, the integration is throughout the entire ionosphere, while in other cases it is between the base of the ionosphere and the altitude of reflection or scattering, which is often itself not well known. Attempts have been made to overcome this difficulty, notably through the use of several riometers operating at a range of frequencies (Parthasarathy *et al.*, 1963) and more recently through the use of variable-frequency sounding at VLF (Helms and Swarm, 1969). These attempts, while revealing some of the features of the electron-density profile during PCA events, have not been completely successful as a substitute for direct probing of the region. Rocket measurements of the properties of the ionosphere during PCA events have been remarkably few in number until very recently, when a concerted rocket study of the event of 2 November 1969 was carried out at Fort Churchill. The results of this study, when fully analyzed, should greatly increase our knowledge of the structure and composition of the lower ionosphere during one of these events.

4 IONOSPHERIC ASPECTS OF PCA EVENTS

In this section we shall discuss the ionospheric aspects of PCA events, beginning with the ionization process itself, and treating the problem of the products of the ionization process, i.e., the steady-state concentrations of electrons and the various species of positive and negative ions, ending with a discussion of the empirical relationships that have been derived between the incoming proton flux and the ionospheric response.

4.1   The ionization process

As we shall see later, the altitude region of chief concern during PCA events lies below 100 km. In this region, the composition of the atmosphere differs little from that at sea level as far as the major constituents are concerned, and the use of the well-known range-energy relations for standard air is quite justified. Many of the minor constituents play an important role in the chemical reactions that determine the ambient electron and ion concentrations, but their presence can be neglected in calculating the energy-loss rate for energetic particles.

The range of protons and alpha-particles in standard air, and the rate of energy loss, can be found in many references (e.g., Bethe and Ashkin, 1953), and can be transformed into the corresponding quantities for the atmosphere by adopting some atmospheric model (e.g., CIRA, 1965). In Figure 1 we show the maximum penetration of the atmosphere, i.e., the stopping altitude for vertical incidence, for both protons and alpha-particles in the energy range from some hundreds of kev to a few Bev. Particles incident other than vertically, of course, will not penetrate so deeply, but this is a rather small effect since the concentration of energy loss toward the end of the path combines with the exponentially increasing density of the atmosphere to cause practically all of the energy loss to occur within the last few kilometers of altitude; increasing the angle of incidence thus has little effect on the stopping height until the angle becomes close to 90°. This effect is illustrated in Figure 2, which shows the stopping altitude for 3 proton energies as a function of angle of incidence. The angle of incidence must be more than 60° for the stopping altitude to be raised as much as 5 km, and must be near 80° for a 10 km increase.

The role of the geomagnetic field, which can be taken as vertically directed to a good approximation, is worth mentioning in this connection. Instead of travelling along straight lines, the particles spiral about the field as they enter the atmosphere. Provided the pitch-angle of the spiral remains constant, the spiralling does not change the energy-loss or range calculation, since only the azimuth of the path is affected, and the elementary segments of the spiral could in principle be reformed into a straight line by moving each of them horizontally. Due to the constancy of the magnetic moment of the particle, the pitch-angle of the spiral does not remain constant, tending to increase as the magnetic-field strength increases. This effect is negligible, however, since

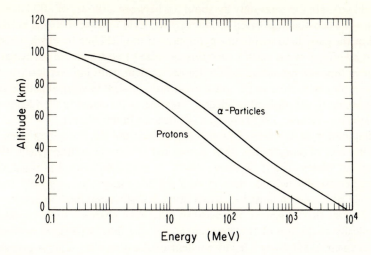

FIGURE 1 Altitude reached by protons and alpha-particles incident vertically on the atmosphere.

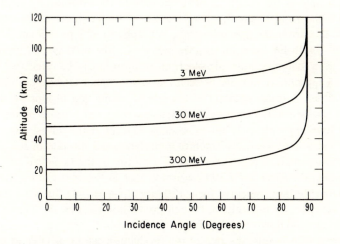

FIGURE 2 Altitude reached by protons with the initial energies shown incident on the atmosphere at different angles from the vertical.

the field strength increases only by about 3% between altitudes of 100 km and 30 km. The constraining action of the magnetic field restricts the total horizontal travel of the particle to within one gyroradius of the field line to which it is tied, so that all effects of the earth's curvature can also be neglected. In particular, the Chapman function cannot be used to transform slant paths into vertical paths, and even for angles close to $90°$ the correct ratio of slant to vertical path is given by the secant of the angle of incidence. Treatments that employ the Chapman function (e.g. Velinov, 1968) appear to be in error in this respect.

Computation of the rate of production of ions and electrons by a given incoming flux of particles has been carried out by several authors (Reid, 1961; Adams and Masley, 1965; Velinov, 1968). It is straightforward in principle, but mathematically tedious, since at each altitude integration over both energy and angle of incidence must be carried out. A slight simplification results from the fact that in general the particles have been observed to be incident isotropically at the top of the atmosphere, i.e., the flux of particles within a given element of solid angle is independent of the direction. This is probably more generally true for very high magnetic latitudes, well within the polar cap, but is apt to lead to erroneous results near the edge of the region affected by PCA, where anisotropic pitch-angle distributions have frequently been reported by satellites.

An example of the results of such calculations for isotropically incident protons is shown in Figure 3, where the rate of ion production is shown for an isotropic flux (over the upward-looking hemisphere) of 1 proton cm^{-2} sec^{-1} steradian^{-1} with the incident energies indicated. The total rate of production by any given incoming isotropic spectrum of protons can be calculated by adding an appropriate number of such curves for monoenergetic protons at each altitude. A rough estimate of the rate of production of ions by alpha-particles can be made by noting that alpha-particles and protons of the same velocity have approximately the same range (except at the lowest energies, when capture and loss of electrons becomes important), and the rate of ion production by an alpha-particle is four times that of a proton of the same velocity. Thus, for example, the curve for a 30-Mev proton flux in Figure 3 can be trans-formed roughly into the curve for a 120-Mev alpha-particle flux (the same particle velocity) by multiplying the ion production rates by 4. This procedure is valid down to energies of a few Mev.

In actual PCA events, the rate of ion production can be as large as several thousand per cm^3 sec at a typical altitude of 60 km; this is four or five orders of magnitude larger than the rate of production by galactic cosmic rays at this altitude. The magnitude of the D-region effects observed during PCA events is thus hardly surprising, even though the total energy input at the top of the atmosphere (less than 0.1 erg cm^{-2} sec^{-1}) is not large.

As mentioned above, appreciable fluxes of solar electrons are often present. The ionization produced by them, however is completely negligible in

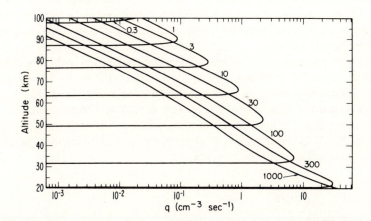

FIGURE 3   Rate of production (q) of ions and electrons by monoenergetic fluxes of protons with the initial energies shown. The protons are incident isotropically over the upper hemisphere, and the flux in each case is 1 proton cm^{-2} sec^{-1} steradian^{-1}.

comparison to that produced by the heavy particles, except possibly at the beginning of an event, since the electrons are often the first particles to arrive at the earth (Hakura, 1967).

4.2 Steady state ionization conditions

The radio-wave effects observed during PCA events are determined by the steady-state altitude distribution of free electrons, since the ions are too massive to be affected appreciably by radio frequencies higher than a few kHz. The distribution of electrons is in turn determined by a balance between the rate of production, discussed above, and the rate of loss, which will be discussed in this section.

A freshly produced electron reaches thermal equilibrium with its surroundings after about 10^4 collisions, i.e., in a few milliseconds in the altitude region we are considering. Most of these collisions will be with N_2 and O_2 molecules, and the attachment rates are low enough that the ultimate loss of an electron will occur long after it has 'thermalized'. At a temperature of the order of $200°K$ in the D region, an electron can be lost either through dissociative recombination with a positive ion, e.g.

$$e + O_2^+ \rightarrow O + O \tag{1}$$

or through attachment to a neutral species to form a negative ion. By far the most abundant neutral species in the D region are N_2 and O_2; N_2 does not form a stable negative ion, but O_2 does, requiring a three-body reaction of the type

$$e + O_2 + O_2 \rightarrow O_2^- + O_2 \tag{2}$$

at D-region temperatures. The ions themselves undergo a complicated series of chemical reactions with various minor neutral species, tending to become more complex and even to form relatively loosely bound clusters, especially with water vapor, as they progress along the reaction chain. Since both dissociative-recombination rates and detachment rates change as the ion species change, a detailed understanding of the steady-state electron distribution is not a simple matter, and would require a knowledge of the ion chemistry that is still beyond our grasp. We shall not attempt to review the subject of ion chemistry in the D region here (see, for example, Ferguson, 1971a), but will merely emphasize a few of its aspects that appear likely to have a direct bearing on the problem of electron densities in PCA events.

Considering positive ions first, the primary species produced by energetic-particle ionization are mostly N_2^+ and O_2^+ with smaller amounts of N^+ and O^+

due to dissociative ionization of the neutral molecules (Swider, 1969). Within a very short time these are chemically converted into mostly O_2^+ with a much smaller amount of NO^+. In the quiet D region, O_2^+ ions are rapidly changed into water-cluster ions of the type $H_3O^+.(H_2O)_n$ through a series of chemical reactions that begins with a clustering reaction with O_2 of the form

$$O_2^+ + O_2 + M \rightarrow O_4^+ + M \tag{3}$$

where M is any third body and O_4^+ is a cluster ion containing a relatively weak bond between an O_2^+ ion and an O_2 molecule. Following the formation of O_4^+ a rapid series of reactions leads to the formation of the water-cluster ions (Fehsenfeld and Ferguson, 1969). In the upper part of the D region, however, atomic oxygen is quite abundant, and the O_4^+ clusters can be destroyed through the reaction

$$O_4^+ + O \rightarrow O_2^+ + O_3 \tag{4}$$

(Ferguson, 1971b), instead of being converted to water clusters. The steep increase in O concentration with height in the upper D region, together with the increasing effectiveness of recombination as the electron production increases, leads to the disappearance of the water-cluster ions above a rather sharply marked altitude that is generally at about 82 km in the quiet daytime D region (Narcisi and Bailey, 1965).

During PCA events, it would be expected that the greatly increased concentration of electrons would make recombination a much more effective competitor with ion-molecule reactions, and that concentrations of O would increase due to both dissociation by proton impact and dissociative recombination of O_2^+. We would thus expect the 'top' of the water-cluster region to appear at lower levels during PCA events than during quiet conditions, and the limited evidence from rocket mass-spectrometer measurements (Narcisi et al., 1972a) has shown this to be the case. During a daytime flight into the intense PCA event of November 1969, NO^+ and O_2^+ were found to be the dominant species down to below 73 km; as mentioned above, they are generally dominant in the quiet D region only above about 82 km.

Since recombination coefficients for simple molecular ions are considerably smaller than those for water-cluster ions (Reid, 1970; Biondi et al., 1972), the region from 82 km to below 73 km contains more electrons during an intense PCA event than one would have expected on the basis of a simple comparison between quiet-time and PCA production rates. This illustrates the dangers inherent in attempting to extrapolate from quiet conditions to PCA conditions, and even from event to event, in an empirical way without considering the details of the chemistry involved.

In the case of negative ions, we have mentioned above that attachment to O_2 molecules can take place. A three-body reaction is required, however, with the result that negative ions are only formed rapidly below about 70 km (e.g., Reid, 1970). The negative-ion situation has a certain similarity to the positive-ion situation, in that once formed an O_2^- ion can react either with O_3

$$O_2^- + O_3 \rightarrow O_3^- + O_2 \tag{5}$$

leading ultimately to such complex and stable negative ions as NO_3^- and its hydrates (Ferguson, 1971a), or it can react with O in the associative-detachment reaction

$$O_2^- + O \rightarrow O_3 + e \tag{6}$$

in which the energy released in forming O_3 is sufficient to detach the electron. The ratio of the concentrations of O and O_3 is thus a very important factor in determing the steady-state negative-ion concentration and composition, and in addition the concentration of positive ions plays an important part, since mutual-neutralization reactions of the type

$$O_2^- + O_2^+ \rightarrow O_2 + O_2 \tag{7}$$

tend to destroy negative ions before they have time to form more complex species. Mutual neutralization thus plays a role analogous to that of dissociative recombination in the positive-ion situation. Electrons can be released from negative ions both through associative-detachment reactions like (6) above, and through photodetachment by sunlight during daytime. Very little is yet known about photodetachment cross-sections other than those of oxygen (O^-, O_2^-, and O_3^-).

Direct rocket measurements of negative-ion composition are considerably jnore difficult to make than those of positive-ion composition, and consequently the information available is sparse and somewhat conflicting. However, measurements were made during both daytime and nighttime conditions during the intense PCA event of November 1969 at Churchill (Narcisi et al., 1972b), and the results showed that large quantities of O_2^- existed between 72 and 94 km in daytime, and that the dominant ions in the same altitude region at night were apparently O^-. This particular result is surprising, and no satisfactory way of accounting for the formation of large quantities of O^- has yet been proposed. Much heavier negative ions, probably hydrates of NO_3^- and possibly CO_3^-, were also found. During quiet conditions, these heavy hydrates are generally the dominant species, so that PCA events are apparently characterized by having a less complex negative-ion composition than the quiet D region, a situation that is also true for positive ions.

As mentioned above, the quantity of chief direct concern in interpreting PCA effects is the concentration of free electrons, n_e . If we denote the concentration of the jth species of negative ion by n_j^- , we can define a quantity λ_j as the ratio of the concentrations of the jth negative ions and of electrons, i.e.,

$$\lambda_j = n_j^-/n_e \ . \tag{8}$$

The total ratio of negative-ion and electron concentrations at a given altitude is then defined as

$$\Lambda = \sum_j n_j^-/n_e = \sum_j \lambda_j \ . \tag{9}$$

Let us make the simplifying assumption that all mutual-neutralization reactions proceed at the same rate, regardless of the species involved (there is some experimental evidence that this may not be a bad assumption). Then we can write the following equation for the time dependence of the jth species of positive ion:

$$\frac{dn_j^+}{dt} = q_j + R_j - L_j n_j^+ - \alpha_j n_e n_j^+ - \alpha_i \Lambda n_e n_j^+ \tag{10}$$

where q_j is the rate of primary production of the positive ion by proton ionization, R_j is its rate of production from other positive-ion species by chemical reactions, $L_j n_j^+$ is its rate of disappearance to form other species by chemical reactions, α_j is its dissociative-recombination coefficient, and α_i is the assumed mutual-neutralization rate coefficient. In a steady state, we can set the left-hand side equal to zero, and we can add the equations for all the positive-ion species, obtaining

$$\sum_j q_j - n_e \sum_j \alpha_j n_j^+ - n_e \Lambda \alpha_i \sum_j n_j^+ = 0 \tag{11}$$

(since each term representing chemical production of one species must be accompanied by an equal term representing the disappearance of another species, the terms in R_j and L_j add to zero). Rearranging, we have

$$n_e = \frac{\sum_j q_j}{\sum_j \alpha_j n_j^+ + \Lambda \alpha_i \sum_j n_j^+} \ . \tag{12}$$

Now define a weighted mean recombination coefficient α' by

$$\alpha' = \frac{\sum\limits_{j} \alpha_j n_j^+}{\sum\limits_{j} n_j^+} \tag{13}$$

and we find

$$n_e = \frac{\sum\limits_{j} q_j}{(\alpha' + \Lambda \, \alpha_i) \sum\limits_{j} n_j^+} \tag{14}$$

Since the medium must be electrically neutral, we have

$$\sum\limits_{j} n_j^+ = (1 + \Lambda) \, n_e$$

and hence

$$n_e = \left\{ \frac{\sum\limits_{j} q_j}{(1 + \Lambda) \, (\alpha' + \Lambda \, \alpha_i)} \right\}^{1/2} \tag{15}$$

In the simple case in which there are no negative ions, and only one species of positive ion with recombination coefficient α, we can write

$$\frac{dn_e}{dt} = 0 = q - \alpha n_e^2 \tag{16}$$

leading to

$$n_e = \left(\frac{q}{\alpha} \right)^{1/2} \tag{17}$$

Eqs. (15) and (17) are formally similar, and this has led to the concept of an "effective recombination coefficient" or "effective loss coefficient" for a complicated multi-species situation. It is obvious from (15) that accurate determination of this quantity demands knowledge of the ion composition and of the recombination coefficients of the different species. An empirical approach has, however, been used by several authors (Adams and Masley, 1965; Potemra et al., 1969; Megill et al., 1971), using simultaneous measurements of the incoming particle flux and spectrum and of the steady-state electron concentration to derive an altitude profile for the loss coefficient.

Examples of a few such daytime profiles are shown in Figure 4, where major discrepancies are evident, though there seems to be general agreement on a value in the vicinity of a few times 10^{-7} cm^3 sec^{-1} in the upper D region. This value corresponds roughly with the dissociative-recombination coefficient for the dominant molecular-ion species (O_2^+ and NO^+), which is in accord with our knowledge of the ion chemistry, but in general the dangers of using an effective loss coefficient determined in this way are apparent. Unfortunately few direct measurements of the electron-density profile have been made during PCA events, so that in general the loss-coefficient profile can only be checked against some integrated parameter such as the 30-MHz absorption measured by a riometer. The same absorption can, however, be produced by a wide variety of electron-density profiles, and the fact that good agreement can be found between the measured absorption and the absorption calculated by use of these coefficients is no guarantee that the electron-density profile has been correctly predicted, and it would be dangerous to draw conclusions about the aeronomy of the D region on this basis.

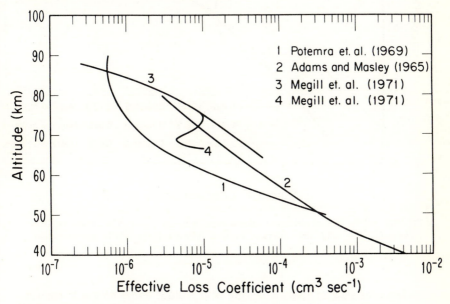

FIGURE 4   Examples of effective loss coefficients for the daytime lower ionosphere obtained during PCA events.

Few direct rocket measurements of the electron-density profile during PCA events have yet appeared in the literature, but examples of three such profiles, obtained by a combination of probe and radio techniques, are shown in Figure 5. Two of these were obtained during nighttime conditions, and show similar features--the electron density tends to a constant value in the upper D region, and there is a sharp 'ledge' below 80 km--at 76 km in one case, and at 71 km in the other. The normal quiet nighttime level of electron density in the D region is considerably less than the values shown, so that the nighttime profiles probably represent the electron densities that would be produced by the solar protons alone. This is also true of the daytime profile below 85 km, but the tendency for this profile to continue increasing upward beyond this level probably represents a merging with the normal daytime ionosphere.

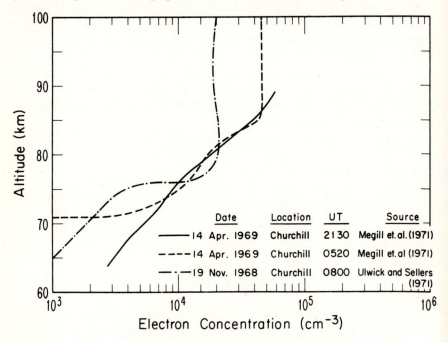

FIGURE 5   Profiles of electron concentration obtained during PCA events by rocket techniques. The solid curve refers to daytime conditions, the two broken curves to nighttime conditions.

The constancy of electron density with altitude above 85 km at night is worth a comment; since the recombination coefficient is probably not varying appreciably with height in this region, it implies that the rate of electron production is also relatively independent of height. Since the neutral-atmosphere density decreases roughly exponentially, the proton flux must be increasing in such a way as to compensate, i.e., the proton spectrum must be continuing to increase quite steeply with decreasing energy even at the energies responsible for ionization near 100 km, which are probably considerably below 1 Mev. Steep proton spectra in this energy range have frequently been observed by satellites (e.g., Lanzerotti, 1970). Alternatively, there may be some contamination by auroral-particle bombardment at these altitudes.

The steep ledge below 80 km at night is probably directly related to the importance of atomic oxygen in maintaining free electrons through associative detachment of O_2^- (Swider et al., 1971; Reid, 1969). A sharp gradient is expected to exist at night in the O concentration profile, since the rate of loss of O is strongly dependent on height. The 5-km difference in altitude of the ledge between the two profiles may be a consequence of seasonal variations in the atmosphere.

4.3   Twilight effects

Due to the importance of photochemical effects in the lower ionosphere, PCA phenomena change markedly in going from daytime to nighttime conditions. The transition occurs during the twilight period, and the twilight behavior of PCA has aroused a considerable amount of interest since the earliest days of the phenomenon.

Figure 6 shows an example of the variation in vertical absorption recorded by a 20-MHz riometer during sunset at the Antarctic station of Vostok on 4 September 1966. Throughout the period shown the absorption recorded at South Pole (where the solar zenith angle is essentially constant for the duration of a PCA event) showed very little variation, indicating that the particle flux was steady. The change in absorption at Vostok by a factor of about 5 in going from daytime to nighttime conditions is fairly typical of PCA events.

The simplest physical interpretation of the twilight variation bases it on photodetachment, which might be expected to maintain relatively high electron densities in the lower D region by day, but to make no contribution at night. In the early days of PCA observations, the only atmospheric negative-ion species whose properties were at all known was O_2^-, which suffered efficient photodetachment by sunlight in the visible region of the spectrum. Since visible sunlight is not greatly attenuated until the sun goes below the limb of the solid earth (or at least below the level of clouds in the lower atmosphere), one would expect the lower D region to respond to the passage of the shadow of the solid earth. The lower of the two scales at the top of Figure 6 shows the altitudes of solid-earth shadow corresponding to the solar zenith angles indicated, and yields the surprising result that approximately half of the sunset decay in electron density takes place while the shadow moves from ground level to about 30 km (taking proper account of attenuation by clouds and scattering in the lower atmosphere would add a few kilometers to this height range). At least half of the absorbing region would thus have to lie below 40 km for this explanation to be tenable, and this would be totally inconsistent with

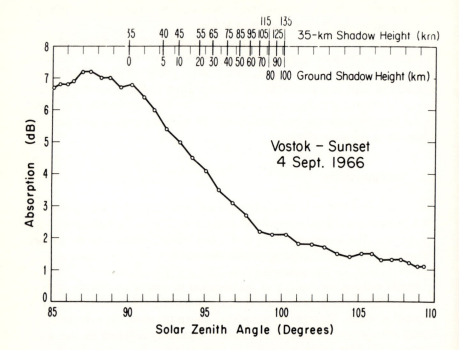

FIGURE 6   Sunset transition in polar-cap absorption recorded at a frequency of 20 MHz on 4 September 1966 at Vostok, Antarctica. The upper scales show the altitude above the station of the shadow of the earth and of a layer 35 km above the surface of the earth.

our knowledge of the properties of the lower ionosphere and of the incoming particle flux and spectrum. A much more realistic interpretation can be made if we assume that the radiation responsible for maintaining high daytime electron densities is strongly attenuated by the atmosphere below some altitude of the order of 35 km, when the shadow heights are as indicated in the upper scale at the top of Figure 6. Approximately half of the absorbing region now lies below 65 km, which is roughly in accord with what we know of the lower ionosphere. It forces us, however, to identify the radiation as lying in the ultraviolet, rather than the visible, region of the spectrum since radiation with wavelength less than about 2900 A is indeed heavily absorbed by the ozone lying below about 35 km.

These considerations led to the proposal that the day-night transition in PCA was caused by the change in photodetachment conditions, but that the photodetachment was taking place from a species of negative ion that was much more tightly bound than O_2^-, requiring ultraviolet photon energies. As discussed above, more recent direct measurements have shown that such tightly bound negative ions do exist in the lower D region (e.g., NO_3^-), but the role of photodetachment of these ions is still not clear. As mentioned above, associative detachment plays an important role in maintaining large electron densities, particularly through the reaction

$$O_2^- + O \rightarrow O_3 + e$$

After sunset, atomic oxygen disappears very rapidly below 75 km, leading to a rapid drop in electron density. The extent to which the twilight variation is due to a change in atomic-oxygen concentration, as opposed to direct photo-detachment, is still undecided, but it is probabley that both mechanisms are involved to some extent.

4.4   Radio-wave absorption

The basic mechanism responsible for the absorption of radio waves in the lower ionosphere is the occurrence of collisions between free electrons and neutral molecules and atoms. Collisions interrupt the organized motion of the electrons in the electric field of the radio wave, and transform the energy picked up by the oscillating electrons into thermal energy of the neutral particles.

Mathematically, absorption appears in the magneto-ionic theory of radio-wave propagation as the imaginary part of the complex refractive index of the ionosphere, the real part of which is responsible for deviating the radio waves. The original Appleton-Hartree development of magneto-ionic theory expresses the absorption coefficient, k, as

$$k = \frac{e^2}{2\epsilon_0 mc} \frac{1}{\mu} \frac{N\nu}{\nu^2 + (\omega \pm \omega_L)^2} \tag{18}$$

in *MKSA* units, where N is the electron number density, ν the electron collision frequency, ω the radio-wave angular frequency, ω_L the component of the gyrofrequency vector in the direction of propagation, and μ the (real) refractive index. e and m are, respectively, the charge and mass of the electron, c the speed of light, and ϵ_0 the permittivity of free space, which is expressed in *MKSA* units as 8.86×10^{-12} farads per meter. The expression is valid for the case of quasi-longitudinal propagation, which holds for most cases of interest. The positive and negative signs in the denominator of (18) refer respectively to the ordinary and extraordinary polarized components of the received wave. The absorption can be found from the relationship between the amplitudes of the incident and received waves:

$$I = I_0 \exp\left(-\int k \, ds\right) \tag{19}$$

where the integration is taken along the path of the wave, and is usually expressed in decibels (db), defined by

$$A \text{ (db)} = 20 \log_{10} (I_0/I) \tag{20}$$

Combining Eqs. (18), (19), and (20), and expressing the constant numerically, we find

$$A \text{ (db)} = 4.58 \times 10^{-5} \int \frac{1}{\mu} \frac{N\nu}{\nu^2 + (\omega \pm \omega_L)^2} \, ds \tag{21}$$

In the case of cosmic-noise observations, which are necessarily confined to frequencies appreciably above the F-region penetration frequency, the absorption is 'non-deviative', and μ can be taken as unity. Also at the commonly used operating frequencies of $20 - 50$ MHz, ω_L (whose maximum value is about $2\pi \times 1.6$ MHz) can safely be neglected in comparison with ω for most purposes.

The Appleton-Hartree development, on which the above relations are based, is unsatisfactory in that it neglects the variation of electron collision frequency with velocity. This variation is included in the more recent generalized magneto-ionic theory (Sen and Wyller, 1960), which replaces (18) above by

$$k = \frac{5e^2N}{4\epsilon_0 mc\omega} \alpha \, C_{5/2} (\alpha) \tag{22}$$

for the case of non-deviative absorption when $\omega \gg \omega_L$. Here α is the dimensionless ratio ω/ν_m, and $C_{5/2} (\alpha)$ is one of a class of integrals first evaluated in connection with semiconductor theroy, and tabulated by Dingle et al. (1957) and by Davies (1965). The quantity ν_m appearing here is the collision frequency associated with electrons of energy kT, i.e., with those electrons at the peak of the Maxwell-Boltzmann distribution. The altitude profile of ν_m has been discussed by several authors (e.g., Thrane and Piggot (1966)), but for many practical purposes it is quite well approximated by the relation

$$\nu_m = 8.40 \times 10^7 \, p \tag{23}$$

where p is the pressure in millimeters of mercury. Eqs. (19), (20), and (22) lead to the following expression for the absorption in decibels:

$$A \text{ (db)} = \frac{1.149 \times 10^{-4}}{\omega} \int N\alpha \, C_{5/2} \, (\alpha) \, ds \qquad (24)$$

in *MKSA* units. The quantity $\alpha \, C_{5/2} \, (\alpha)$ is shown as a function of altitude
for the CIRA 1965 model atmosphere in Figure 7 for a radio-wave frequency
of 30 MHz, the most common operating frequency for high-latitude riometers.
This quantity is a measure of effectiveness of a given electron density in
producing absorption, and maximizes near 45 km for 30 MHz. Since the
electron density itself during PCA events is almost certainly increasing rapidly
with increasing height at this altitude, the region that contributes most to
the absorption lies considerably higher than this level, probably in the vicinity
of 65 km during most daytime events.

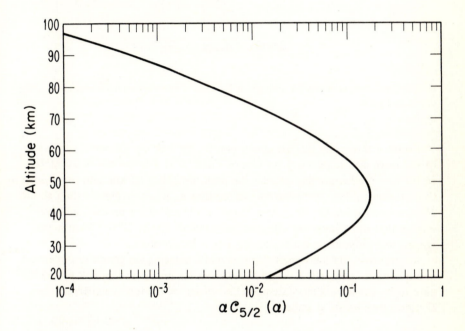

FIGURE 7   The quantity $\alpha \, C_{5/2} \, (\alpha)$ computed for a frequency of 30 MHz. This quantity
is a measure of the effectiveness of a given electron concentration in producing 30-MHz
radio-wave absorption

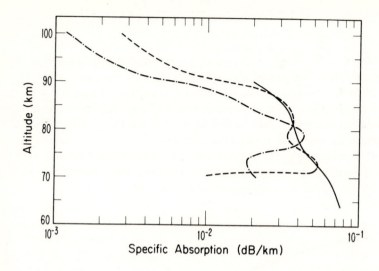

FIGURE 8   Computed specific absorption for each of the electron-concentration profiles
shown in Figure 5.

Figure 8 shows the specific absorption (in decibels per kilometer) for each
of the three electron-density profiles of Figure 6 at a frequency of 30 MHz.
In the case of the daytime profile, the peak lies below 65 km, and the absorption
thus responds primarily to the flux of protons with energy greater than about
10 Mev (see Figure 1). The steep 'ledge' in the nighttime profiles, however,
implies that the absorption will take place mainly in the 70 − 80 km height
ranges, thus responding mainly to the 1 − 5 Mev protons.

In summary, the ionospheric aspects of PCA events are generally understood
in a qualitative way, but a complete quantitative understanding will ultimately
have to be based on a more detailed knowledge of the ion chemistry of the
D region than exists at present.

5 MAGNETOSPHERIC ASPECTS OF PCA EVENTS

Apart from the regular diurnal variation discussed above, most of the
temporal and spatial variations of PCA are due to magnetospheric, rather than
ionospheric, effects. It is not possible to review these magnetospheric effects
without at the same time reviewing the extensive literature on magnetospheric
access of solar particles, which is beyond the scope of the present paper. In
what follows the discussion will be centered on deductions from PCA

observations themselves, and the interested reader is referred to existing review papers (e.g., Paulikas *et al.*, 1970; Bostrom, 1970; Lanzerotti, 1970, 1972) for summaries of the satellite work.

The particles responsible for PCA are of low rigidity, at least by cosmic-ray standards, and their access to the ionosphere is controlled by the magnetic-field configuration in the relatively weak fields of the outer magnetosphere. The simplest example of this control is the confinement of PCA to the polar caps-- the particles that ionize the *D* region heavily do not have enough energy to penetrate into the strong magnetic fields of the inner magnetosphere, and hence to reach low latitudes; particles that can reach these latitudes produce very little *D*-region ionization.

Quantitatively, however, the situation is more complicated. The classical Størmer theory of the motion of charged particles in a dipole field predicts that 100-Mev protons should have access to the earth only at geomagnetic latitudes higher than about 68°, and this is roughly in accord with the observation that PCA effects are normally seen only at latitudes higher than that of the auroral zones. More detailed examination, however, reveals marked departures from the predictions of Størmer theory. A station such as Thule, Greenland, with a geomagnetic latitude close to 90°, ought to have essentially zero cutoff for solar protons, while College, Alaska, at a geomagnetic latitude of about 65°, should lie almost outside the PCA region entirely, with a cut-off energy of about 120 Mev. In an examination of magnetic-storm effects on PCA, however, Leinbach *et al.* (1965) found that the ratio of PCA at College to that at Thule was about 0.8 under quiet conditions. Furthermore, observations at Farewell, Alaska, only about 3° south of College, showed that the absorption there was only about one-sixth of that at College. The observations thus showed that PCA events normally displayed a broad polar-cap 'plateau' of roughly 50° diameter in geomagnetic latitude, with a steep edge only a few degrees in width. Størmer theory, on the other hand, would predict a steady increase in PCA intensity from the edge, near 68° geomagnetic latitude, to the geomagnetic pole.

The solution to this problem became apparent after the discovery of the geomagnetic tail (Ness, 1965). The geomagnetic field lines from the polar caps are effectively stretched out into a long tail-like configuration on the anti-solar side of the magnetosphere by the action of the solar wind. The tail is apparently a sufficiently 'open' structure that it can be permeated by nearly all the protons responsible for PCA, accounting for the flat 'plateau' appearance normally observed across the polar cap. Calculations (Reid and Sauer, 1967a) based on a simple model for the magnetic field immediately inside the tail region, and assuming a completely accessible tail, yielded predicted proton cutoffs for the local-time period near midnight that were in good agreement both with ground-based observations of PCA and with direct satellite measurements of the protons, at least during fairly quiet geomagnetic conditions.

Detailed calculations of particle trajectories in model geomagnetic fields (Gall *et al.*, 1968) led to similar general conclusions, and showed that particles incident from the anti-solar direction can reach all longitudes by drifting around the earth in a pseudo-trapped mode on the closed field lines of the dayside magnetosphere.

The maps of Figure 9 show approximately the regions in the two polar caps that are affected by PCA phenomena, based on a combination of observation and semiquantitative theory. The inner closed curve is the boundary of the 'plateau' region that experiences essentially constant PCA, and corresponds roughly to an L-value of 5.5 (magnetic invariant latitude of $65°$). The outer curve marks the low-latitude boundary of the 'transition' region in which PCA effects climb from zero to reach their full level, and is roughly the $L = 4$ contour (magnetic invariant latitude of $60°$). In the southern hemisphere, the orientation of the geomagnetic axis is such that most of Antarctica lies well within the plateau region, while the transition region is located mainly over the oceans. In the northern hemisphere, on the other hand, the plateau region lies largely over the Arctic Ocean (except in northern Canada), while the transition region covers populated areas of Alaska, Canada, and Scandinavia. The geographical distribution of PCA has an important bearing on HF and VHF radiowave propagation characteristics that are beyond the scope of this paper.

5.1 Temporal variations

During geomagnetic storms, the area affected by PCA sometimes expands to latitudes considerably lower than the normal boundary, reaching as low as an L-value of about 3.5 on occasion. Leinbach *et al.* (1965) show several examples of these movements of the boundary, which are undoubtedly caused by storm-related changes in the topology of the outer magnetosphere, but are not yet fully understood quantitatively. The intense PCA event of May 24-28, 1967, afforded a good example of such a movement, and observations by a chain of riometers located in south-central Alaska are shown in Figure 10. For comparison, magnetograms from Honolulu and College are shown in the lower part of the diagram.

An intense PCA event had been in progress at the higher-latitude stations for about 24 hours prior to the period shown, but until the first of the two storm sudden commencements (SSC) the effective boundary of the PCA lay between L-values of 4.5 and 5.0. Unfortunately the first SSC, and the period between the two SSC's, occurred near local midnight, when the D region at the lower-latitude stations was in darkness (the normal sunset decrease in PCA intensity can be seen at the three high-latitude stations prior to the first sudden commencement). Sunrise occurred at about the time of the second SSC, and it is apparent that the lower-latitude stations thereafter record PCA at essentially

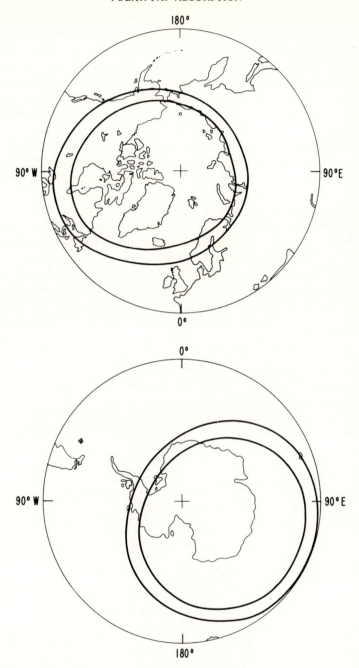

FIGURE 9   Maps showing the normal distribution of PCA effects. The regions inside the inner curves experience the full intensity, while regions outside the outer curves are normally unaffected, except during geomagnetic disturbances.

its full polar-cap intensity for a few hours. The absorption at these stations then decreases again to a minimum value near 2000 UT (1000 local time), then appears to increase once again. The broad decrease centered about the pre-noon hours is an example of the midday recovery phenomenon, to be discussed below, and in this example is superimposed on a general decrease in intensity. The period shown in the diagram is followed by a period of intense auroral absorption at the lower-latitude stations that obscures the later stages of the PCA event.

Comparing the riometer recordings with the magnetograms, there is an obvious general correspondence in activity, but the details of the connection are not easy to follow. The first SSC appears to have marked an important change in PCA characteristics at the higher-latitude stations, and was followed by the first appearance of PCA at L = 4.5, despite the lack of sunlight at the location. This SSC was not, however, accompanied by any noticeable geomagnetic activity. The second SSC was followed by intense substorm activity at College, and a large initial phase as shown by the increase in the H-component of the field at Honolulu. The rapid increase in PCA at the two lower latitude stations appears to be related to the sharp termination of the initial phase at Honolulu, which in turn is probably due to the build-up of a ring-current of trapped particles resulting from the substorm activity seen somewhat earlier at College.

The general relationship between depression in cutoffs and the build-up of a ring current has been recognized for many years (Winckler et al., 1961), but the quantitative details are not yet fully established. The net effect of the

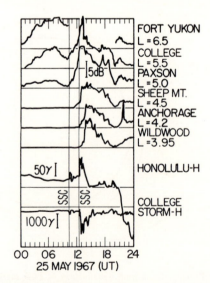

FIGURE 10   30-MHz absorption recorded by a chain of riometers in Alaska during the intense event of May 1967. The variation of the horizontal component of the geomagnetic field recorded at Honolulu and College is also shown.

ring current is to 'stretch' a magnetic field line originating at a given point on the night-side of the earth, so that it crosses the equatorial plane at a considerably greater distance from the earth than usual (Akasofu, 1963). The weaker field strength in this new location allows access to this field line by particles whose rigidity is too low to allow them access in the absence of the ring current.

Decreases in PCA intensity occurring during magnetic storms have been reported by Ortner et al. (1961) and by Reid (1970). These cases are the reverse of the more frequent increase at low latitudes discussed above, and their cause has not been determined. Figure 11 shows an example of one such decrease recorded at the magnetically conjugate stations of Great Whale River and Byrd (L = 7) at about the same time as the southward movement of the polar cap over Alaska shown in Figure 10. The local time at the two stations was about 0600 (in contrast to Alaska, where it was near midnight), and the sharp decrease in absorption occuring between the two SSC's was not accompanied by any significant effect at the higher-latitude station of South Pole, or by any change in proton flux as recorded by satellites. The decrease took place in the period between the two SSC's and comparison with the magnetograms of Figure 10 shows that there was no appreciable substorm or ring-current magnetic activity at the time. The absorption recovers to its original-level quite abruptly following the arrival of the second SSC.

Decreases of this kind were associated by Ortner et al. (1961) with the initial phase of magnetic storms. Qualitatively, one might expect that the compression of the lines of force associated with the initial phase would increase field strengths on the dayside of the magnetosphere, thus increasing cutoffs, but this interpretation impliciitly assumes that protons can enter the region of increased field directly from space. It now appears that the region of entry to the magnetosphere lies in the tail, and that the protons reach the dayside by drifting around in pseudo-trapped orbits. The interpretation of decreases in flux on the dayside is then far from clear.

5.2 Midday recoveries

A striking temporal variation in PCA is often observed at locations near the transition region at the edge of the polar cap. It takes the form of a broad decrease beginning in the morning hours. This is the 'midday recovery' phenomenon, an example of which can be seen in the riometer records shown in Figure 10. There the recovery appears to be most fully developed at Wildwood, presumably the station nearest the edge of the polar cap, but can be followed as far north as College, becoming steadily shorter in duration with increasing latitudes.

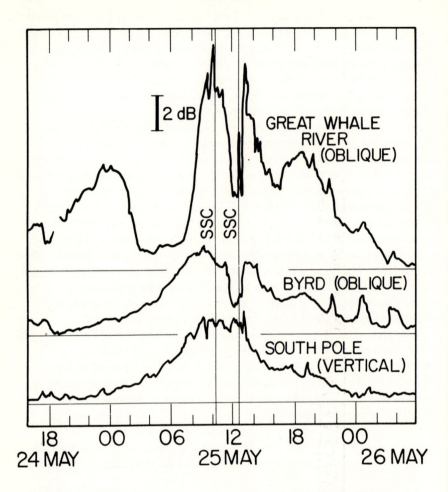

FIGURE 11   30-MHz absorption recorded at three stations during the PCA event of
May 1967. Great Whale River and Byrd are nearly conjugate magnetically, and the
riometer antennas are directed obliquely at an angle of 45° to the vertical; at Great
Whale River the antenna is aimed toward geomagnetic north, and at Byrd toward
geomagnetic south.

The properties of midday recoveries have been described in detail by
Leinbach (1967), who summarizes their characteristics as follows: (1)
midday recoveries reach their peak within the 0800-1500 local-time range, with
the majority in the 1000-1200 range; (2) they are most pronounced near the
edge of the polar cap, and are not observed deep inside the polar cap; (3) the
midday-recovery region remains at the edge of the polar cap as it expands
during a magnetic storm.

The midday recovery is undoubtedly caused by magnetospheric, rather than ionospheric, effects. Possible explanations that have been suggested include a simple increase in geomagnetic cutoff on the dayside of the magnetosphere, associated with the drift of pseudo-trapped solar protons in the geomagnetic field (Reid and Sauer, 1967a), and the development of an anisotropy in pitch-angle distribution, such that the flux of protons precipitating into the atmos-phere is reduced on the dayside, while the flux of protons mirroring above the atmosphere remains relatively unchanged (Leinbach, 1967). There is a certain amount of satellite evidence favoring the latter explanation, though the real cause may well be a mixture of the two mechanisms.

The midday recovery phenomenon still presents some mysterious features. however. If it were simply related to the drift of protons in pseudo-trapped orbits around to the dayside of the magnetosphere, one would expect it to be a regularly occurring feature of all PCA events. In fact, it is observed to occur only in some 20% of all PCA events (Leinbach, 1967), and then only to be pronounced during the first day of the event. These facts suggest some connection with a possible anisotropy of the proton flux in interplanetary space, since such anisotropies do not occur in every event, and are most pronounced in the early stages of an event. The physical connection between such an anisotropy in space and a corresponding anisotropy in fluxes on the dayside of the magnetosphere is not obvious, however.

5.3   Spatial variations in PCA

In addition to the temporal variations discussed above, PCA effects often display spatial variations that are also of magnetospheric origin. In particular, there is evidence that the two polar caps do not always show identical variations in absorption (Reid and Sauer, 1967b).

Figure 11 shows what may be an interesting case of this asymmetric behavior. The two stations of Great Whale River and Byrd are nearly conjugate magnetically, and would be expected to show identical absorption effects, assuming that the field line connecting them is closed. The major differences evident in Figure 11 are mainly due to the fact that Byrd is in continuous darkness, while Great Whale River is alternating between sunlit and dark conditions, the characteristic twilight variation being particularly pro-nounced between 0000 and 0300 UT (sunset) and between 0700 and 1000 UT (sunrise). In the short period between 0300 and 0700, however, both stations are in darkness yet the absorption increases steadily at Byrd while remaining nearly constant at Great Whale River. Recalling that the nighttime absorption is mainly due to protons with energies of a few Mev, this suggests that the low-energy proton flux had quite different properties over the two polar caps. The possibility of substantial differences in the chemical properties of the two

D regions, however, cannot be entirely rules out, though one would expect such differences to lead to differences in the absolute value of the absorption recorded, not to differences in the temporal behavior of the kind seen here.

Brief periods of asymmetry occurring deep inside the polar caps have been seen previously (Reid and Sauer, 1967b), and asymmetries in the particle flux have been recorded by satellites (Van Allen *et al.*, 1971). These were interpreted (Reid and Sauer, 1967b) in terms of a model in which the interplanetary and geomagnetic field lines are connected, and in which there is a pronounced anisotropy in proton flux in interplanetary space. In effect, one polar cap is connected to interplanetary field lines that can be traced back to the sun, while the other polar cap is connected to the other segment of the same field lines, going out beyond the earth. In the extreme case in which the entire flux of protons is directed outward from the sun, they would all impinge on one polar cap, and the other would be completely sheltered. In the more realistic case of a small anisotropy, the flux in one polar cap would be greater than that in the other.

6. SUMMARY

Summarizing, we have seen that the study of polar-cap absorption synthesizes a wide range of topics in the field of solar-terrestrial physics, from the mysterious and exotic plasma phenomena of the outer magnetosphere to the complexities of *D*-region ion chemistry. By the same token, it is inherently capable of providing new insights into the mechanisms operating in these diverse areas, though care has to be taken in interpreting PCA effects because of the intermingling of ionospheric and magnetospheric influence.

In the case of the *D* region, PCA events provide us with a strong source of ionization that is much more readily calculated quantitatively than are the normal undisturbed sources of ionization in the lower ionosphere. This, in principle, allows investigation of the electron loss processes and of the ion chemistry under fairly simple conditions. In the case of the magnetosphere, PCA effects can be used as 'tracers' to indicate the behavior of the field lines in the outer magnetosphere in response to changes in the solar wind and to geomagnetic disturbances. The full realization of this potential in quantitative terms remains a challenge for the future.

References

Adams, G.W. and A.J. Masley, (1965). *J. Atmos. Terrest. Phys.*, **27**, 289.
Akasofu, S.I., (1963). *Space Sci. Revs.*, **2**, 91.
Allan, A.H., D.D. Crombie, and W.A.Penton, (1957). *J. Atmos. Terrest. Phys.*, **10**, 110.
Anderson, K.A., (1958). *Phys. Rev. Letters*, **1**, 336.
Anderson, K.A. and R.P. Lin, (1966). *Phys. Rev. Letters*, **16**, 1121.
Bailey, D.K., (1957). *J. Geophys. Res.*, **62**, 431.
Bailey, D.K., (1959). *Proc. IRE*, **47**, 255.
Bailey, D.K., R. Bateman, and R.C. Kirby, (1955). *Proc. IRE*, **43**, 1181.
Belrose, J.S., M.H. Devenport, and K. Weekes, (1956). *J. Atmos. Terrest. Phys.*, **8**, 281.
Bethe, H.A. and J. Ashkin, (1953). In *Experimental Nuclear Physics*, Vol. 1, E. Segré (Ed).
  John Wiley & Sons, New York.
Biondi, M.A., M.T. Leu, and R. Johnsen, (1972). Paper presented at COSPAR
  Symposium on D- and E-Region Ion Chemistry, Urbana, Illinois, July 1971.
Bostrom, C.O., (1970). In *Intercorrelated Satellite Observations Related to Solar Events*,
  V. Manno and D.E. Page (Eds), D. Reidel Publishing Company, Dordrecht.
CIRA, (1965). *COSPAR International Reference Atmosphere 1965*, North-Holland
  Publishing Company, Amsterdam.
Collins, C., D.H. Jelly, and A.G. Matthews, (1961). *Canad. J. Phys.*, **39**, 35.
Davies, K., (1965). *Ionospheric Radio Propagation*, National Bureau of Standards
  Monograph 80.
Dingle, R.B., D. Arndt, and S.K. Roy, (1957). *Appl. Sci. Res.*, **6B**, 155.
Ellison, M.A. and J.H. Reid, (1956). *J. Atmos. Terrest. Phys.*, **8**, 291
Fehsenfeld, F.C. and E.E. Ferguson, (1969). *J. Geophys. Res.*, **74**, 2217
Ferguson, E.E., (1971a). *Revs. Geophys. and Space Phys.*, **9**, 997.
Ferguson, E.E., (1971b). In *Mesospheric Models and Related Experiments*, G.Fiocco (Ed),
  D. Reidel Publishing Company, Dordrecht.
Fichtel, C.E. and D.E. Guss, (1961). *Phys. Rev. Letters*, **6**, 495.
Freier, P.S., E.P. Ney and J.R. Winckler, (1959). *J. Geophys. Res.*, **64**, 685.
Gall, R., J. Jimenez and L. Camacho, (1968). *J. Geophys. Res.*, **73**, 1593.
Hakura, Y., (1967). *J. Geophys. Res.*, **72**, 1461.
Hakura, Y., Y. Takenoshita and T. Otsuki, (1958). *Rep. Ionosph. Res. in Japan*, **12**, 459.
Helms, W.J. and H.M. Swarm, (1969). *J. Geophys. Res.*, **74**, 6341.
Hultqvist, B. and J. Ortner, (1959). *Nature*, **183**, 1179.
Lanzerotti, L.J., (1970). In *Intercorrelated Satellite Observations Related to Solar Events*,
  V. Manno and D.E. Page (Eds), D. Reidel Publishing Company, Dordrecht.
Lanzerotti, L.J., (1972). *Revs. of Geophys. and Space Phys.*, **10**, 379.
Leinbach, H., (1967). *J. Geophys. Res.*, **72**, 5473.
Leinbach, H. and G.C. Reid, (1959). *Phys. Rev. Letters*, **2**, 61.
Leinbach, H., D. Venkatesan and R. Parthasarathy, (1965). *Planet. Space Sci.*, **13**, 1075.
Little, G.G. and H. Leinbach, (1958). *Proc. IRE*, **46**, 334.
Little, C.G. and H. Leinbach, (1959). *Proc. IRE*, **47**, 315.
Megill, L.R., G.W. Adams, J.C. Haslett and E.C. Whipple, (1971). *J. Geophys. Res.*,
  76, 4587.
Meyer, P. and R. Vogt, (1962). *Phys. Rev. Letters*, **8**, 387.
Narcisi, R.S. and A.D. Bailey, (1965). *J. Geophys. Res.*, **70**, 3687.
Narcisi, R.S., C.R. Philbrick, D.M. Thomas, A.D. Bailey, L.E. Wlodyka, R.A. Wlodyka,
  D.Baker, G. Federico and M.E. Gardner, (1972a). Paper presented at COSPAR
  Symposium on November 1969 Solar Particle Event, Boston College, June 1971.
Narcisi, R.S., C. Sherman, C.R. Philbrick, D.M. Thomas, A.D. Bailey, L.E. Wlodyka,
  R.A. Wlodyka, D. Baker and G. Federico, (1972b). Paper presented at COSPAR
  Symposium on November 1969 Solar Particle Event, Boston College, June 1971.
Ness. N.F., (1965). *J. Geophys. Res.*, **70**, 2989.
Ortner, J., A. Egeland and B. Hultqvist, (1960). *IRE Trans. on Ant. and Prop.* **AP-8**, 621.
Ortner, J., H. Leinbach and M. Sugiura, (1961). *Ark. f. Geophys.*, **3**, 429.

Parthasarathy, R., G.M. Lerfald and C.G. Little, (1963). *J. Geophys. Res.*, **68**, 3581.

Paulikas, G.A., J.B. Blake, and A.L. Vampola, (1970). In *Intercorrelated Satellite Observations Related to Solar Events*, V. Manno and D.E. Page (Eds), D. Reidel Publishing Company, Dordrecht.

Pierce, J.A., (1956). *J. Geophys. Res.*, **61**, 475.

Potemra, T.A., A.J. Zmuda, C.R. Haave and B.W. Shaw, (1967). *J. Geophys. Res.*, **72**, 6077.

Potemra, T.A., A.J. Zmuda, C.R. Haave and B.W. Shaw, (1969). *J. Geophys. Res.*, **74**, 6444.

Reid, G.C., (1961). *J. Geophys. Res.* **66**, 4071.

Reid, G.C., (1969). *Planet. Space Sci.*, **17**, 731.

Reid, G.C., (1970). In *Intercorrelated Satellite Observations Related to Solar Events*, V. Manno and D.E. Page (Eds), D. Reidel Publishing Company, Dordrecht.

Reid, G.C., and C. Collins, (1959). *J. Atmos. Terrest. Phys.*, **14**, 63.

Reid, G.C. and H. Leinbach, (1959). *J. Geophys. Res.*, **64**, 1801.

Reid, G.C. and H.H. Sauer, (1967a). *J. Geophys. Res.*, **72**, 197.

Reid, G.C. and H.H. Sauer, (1967b). *J. Geophys. Res.*, **72**, 4383.

Rothwell, P. and C. McIlwain, (1959). *Nature*, **184**, 138.

Sen, H.K. and A.A. Wyller, (1960). *J. Geophys. Res.*, **65**, 3931.

Swider, W., (1969). *Revs. Geophys. and Space Phys.*, **7**, 573.

Swider, W., R.S. Narcisi, T.J. Keneshea and J.C. Ulwick, (1971). *J. Geophys. Res.*, **76**, 4691.

Thrane, E.V. and W.R. Piggott, (1966). *J. Atmos. Terrest. Phys.*, **28**, 721.

Ulwick, J.C. and B. Sellers, (1971). In *Space Research XI*, K.Y. Kondratyev, M.J. Rycroft and C. Sagan (Eds), Akademic-Verlag, Berlin.

Van Allen, J.A. and S.M. Krimigis, (1965). *J. Geophys. Res.*, **70**, 5737.

Van Allen, J.A., J.F. Fennell and N.F. Ness, (1971). *J. Geophys. Res.*, **76**, 4262.

Velinov, P., (1968). *J. Atmos. Terrest. Phys.*, **30**, 1891.

Westerlund, S., F.H. Reder and C. Abom, (1969). *Planet. Space Sci.*, **17**, 1329.

Winckler, J.R., P.D. Bhavsar and L. Peterson, (1961). *J. Geophys. Res.*, **66**, 995.

The Physical Mechanisms of the Inner Van Allen Belt

MARTIN WALT
Lockheed Missiles and Space Co., Inc. Palo Alto, Ca. 94304
and
THOMAS A. FARLEY
Institute of Geophysics and Planetary Physics, University of California, Los Angeles, Ca. 90024

I. INTRODUCTION

Since 1958 measurements with particle and magnetic field detectors on space-craft have revealed many of the details of the near-earth space environment. A simplified schematic diagram of the basic structural configuration is shown in Figure 1.

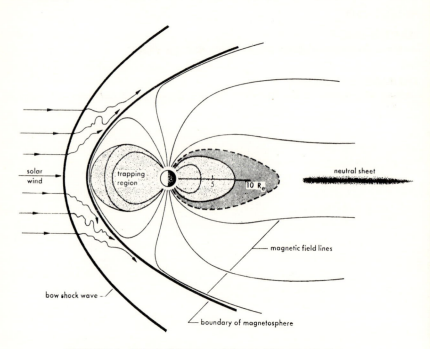

FIGURE 1   Principal features of the earth's magnetosphere.

The solar wind is a rarefied neutral plasma consisting chiefly of protons, helium nuclei and electrons. The wind originates in the solar atmosphere and streams toward the earth at an average speed of about 400 km sec^{-1}. The geomagnetic field represents a fixed obstacle, and the solar wind is diverted around it, confining the geomagnetic field to a cavity. The cavity is known as the magnetosphere, and the surface between the magnetosphere and the solar wind is known as the magnetopause.

Since the wind impinges at supersonic speed, a standing shock surface is formed on the sunward side of the earth. Within the shock the plasma is partially thermalized, allowing it to flow around the subsolar portion of the magnetopause at subsonic speed. As it flows along the flanks of the magneto-sphere it speeds up again and drags the polar connecting field lines far downstream, creating the magnetotail. Between the northern and southern lobes of the tail lies a sheet of hot plasma. A so-called neutral sheet of low magnetic field lies within the plasma sheet, separating the tail lobes. While the structural features pictured in Figure 1 are supported by much experimental evidence, there still remains considerable controversy over the details of the processes leading to the observed configuration.

Deep within the magnetosphere, in the regions where the field does not drastically depart from that of a dipole, lies the Van Allen radiation belt. The belt consists of energetic charged particles—electrons, protons, alphas, and a very small fraction of heavier ions—spiraling from north to south and back along the field lines in quasi-periodic orbits. The outer zone, somewhat arbitrarily defined as that portion of the radiation belt lying beyond 2·5 earth-radii on the geomagnetic equator, is subject to extreme fluctuations in intensity associated with major geomagnetic storms which occur several times each year, and to lesser fluctuations associated with magnetospheric substorms, which usually occur several times each day. The inner zone, with which this paper is concerned, shows fewer and smaller variations. The largest changes which have been observed were the result of a nuclear bomb detonation, not to natural phenomena.

The proper identification of the physical mechanisms responsible for the existence and maintenance of the earth's radiation belt has proved frustratingly difficult. An optimistic argument can be constructed to demonstrate that the problem is basically a simple one. Consider some particular location within the radiation belt and some single particle species. Of all the source mechanisms that one can imagine, the odds are that one is so much more important than all the rest that the others may be ignored. If there is a second source whose strength is within an order of magnitude of the first, it is an unlikely coincidence; and surely there will not be a third. Similarly, the odds are that there will be only one important loss mechanism. This argument has been found by investigation to be substantially correct. Why then is our understanding of the radiation belt so meager? Why are we so often limited to simple qualitative descriptions?

There are two important complications. First, the strength of a source or loss mechanism may vary by orders of magnitude between two locations only a few thousand (sometimes only a few hundred) kilometers apart. Second, the two locations are usually connected by a physical mechanism which transports particles from one of the locations to the other. Furthermore, the transport rates are also strong functions of position, and there may be simultaneous transport in energy in addition to spatial transport. Therefore, a complete description of the particle intensity and its time variation at some location usually depends on a detailed knowledge of several mechanisms over a large region of space. Progress in understanding the important processes has therefore been slow.

The conceptual division of the radiation belt into an inner zone and an outer zone is based in part on the nonspecific response of the early particle detectors. These detectors responded to high-energy protons at low L values and to high-energy electrons at high L values, with an apparent minimum between. (Throughout this paper L refers to the geomagnetic parameter originated by McIlwain (1961) to indicate a position in the earth's magnetic field. The L value of a magnetic field line is approximately equal to the dimensionless ratio of the distance from the dipole to the equatorial crossing of the line in question divided by the radius of the earth.) Subsequent investigations have shown that a minimum often does exist in the electron intensity between $L = 2$ and $L = 3$; it is sometimes called the electron "slot". The proton distribution has no "slot", but the average proton energy decreases continuously with increasing L. The term "inner zone" as used in this paper refers to the region below $L \approx 2 \cdot 5$. The physical mechanisms to be discussed are those which appear to be important for electrons above about 20 keV and protons above a few MeV. This choice, while arbitrary, reflects the emphasis of published research and the rather poorly known intensity and behavior of those particles having energies between these values and the much lower thermal energy of the residual atmosphere.

The inner radiation zone is characterized by its stability. The time scale for such changes as do occur are on the order of weeks, months or years. The geomagnetic field in this region is relatively strong and little distorted by magnetospheric current systems. Time dependent perturbations of the field are therefore very weak. These conditions have made possible a more rapid advance in the theoretical description of the inner zone than has been possible for the more dynamic outer zone. The injection of electrons by nuclear explosions in the inner zone has also proved to be a valuable source of information from which many insights into the trapped electron behavior have been obtained.

These advantages have brought the study of the inner zone radiation from simple qualitative descriptions to a fairly sophisticated, but still incomplete, theory in which the effect of a substantial number of physical mechanisms on the trapped particles can be described quantitatively. It is the purpose of this paper to explain how each process affects the particle distribution function and to show how it is incorporated into an equation containing terms representing the other known mechanisms.

This paper is neither a review nor a presentation of new research results. It rests in large measure on the published work of others, often rewritten here with the benefit of hindsight to simplify the presentation, to emphasize what is currently thought to be important, and to standardize notation and terminology. The brief historical reviews are limited to one specific aspect of inner zone research—the attempt to describe quantitatively the effect of various mechanisms on the distribution function of energetic electrons and protons in the inner radiation zone.

The reader is expected to be familiar with the basic aspects of the adiabatic theory of charged particle motion in a magnetic field. For convenience, a brief summary of the knowledge required is given in the following section.

II. TRAPPED PARTICLE MOTION IN THE EARTH'S MAGNETIC FIELD

The motion of an energetic charged particle in a static magnetic field B is described by

$$m\frac{d\mathbf{w}}{dt} = q(\mathbf{w} \times \mathbf{B}) \tag{2-1}$$

where m, q, and $\mathbf{w}$ are the mass, charge and velocity of the particle. If $\mathbf{B}$ is uniform the solution to Eq. (2-1) is a helix; the velocity parallel to $\mathbf{B}$ is unaffected by the field, and the motion perpendicular to $\mathbf{B}$ is a circle with radius of gyration

$$\text{radius} = \frac{mw_\perp}{qB} \tag{2-2}$$

In a nonuniform electric and magnetic field, if the gyration radius of the particle is much less than the scale size of the fields and if the electric field magnitude is small, the particle trajectory will be very nearly a helix. However, the small deviations from helical motion are crucial and lead directly to the ability of a dipole-like field to trap charged particles. Under these conditions the trajectory can be approximated as a circular motion about a moving guiding center, and for most purposes only the location of the guiding center is of interest. The equations of motion of the guiding center can be decomposed into components perpendicular and parallel to $\mathbf{B}$ (Alfvén and Fälthammar, 1963)

$$\mathbf{v}_\perp = c\frac{\mathbf{E} \times \mathbf{B}}{B^2} + \frac{mc}{qB^3}\left(\frac{w_\perp^2}{2} + w_\parallel^2\right)\mathbf{B} \times \nabla B \tag{2-3}$$

$$\frac{dv_\parallel}{dt} = -\frac{w_\perp^2}{2B}\frac{\partial B}{\partial s} + qE_\parallel \tag{2-4}$$

where $v_\perp$ and $v_{\shortparallel}$ are the components of the guiding center velocity perpendicular and parallel to **B** and $\partial B/\partial s$ is the derivative of the magnetic field with respect to distance along a magnetic field line. From Eqs. (2-3) and (2-4) the general characteristics of particle motion are immediately apparent. Neglecting the electric field terms it is clear from Eq. (2-4) that the guiding center will be repelled from a region of increasing B. In the geomagnetic field this force leads to the reflection of particles as they move along field lines toward higher latitudes. From Eq. (2-3) we see that both the perpendicular and parallel velocity components contribute to $\mathbf{v}_\perp$, the guiding center drift. In a pure dipole field $\mathbf{B} \times \nabla B$ will be in the azimuthal direction. In the geomagnetic field $\mathbf{B} \times \nabla B$ will be largely azimuthal also, but the magnetic irregularities cause some deviation from purely azimuthal drift. Thus, the overall motion of a charged particle in the geomagnetic field will be a superposition of a gyration about a field line, a bounce motion between the northern and southern hemispheres, and an eastward or westward drift in longitude dependent on the sign of q. Positive particles move westward while for negative particles $\mathbf{v}_\perp$ is eastwards.

Except for special geometries it is not possible to integrate Eqs. (2-3) and (2-4) to derive the long-term motion of trapped particles. However, these guiding center equations are used to calculate instantaneous particle velocities and to derive the effects of perturbing fields on trapped particles. To understand the long term motion one must resort to a perturbation theory describing the motion in an almost uniform, almost stationary field. This theory is developed in detail elsewhere (Northrop, 1963; Northrop and Teller, 1960) and only the results will be given here. As described above the trajectory of a trapped particle in the geomagnetic field can be conveniently decomposed into three almost periodic motions, a gyration about magnetic field lines, a bounce along field lines between northern and southern hemispheres, and a drift in longitude. (As will be seen later the longitudinal drift path closes on itself so that the overall trajectory of the guiding center is a closed magnetic shell whose elements are field lines.) Each of these motions has associated with it an adiabatic invariant, a quantity which is conserved throughout the particle trajectory. The adiabatic invariants are not true constants of motion; each is constant only under the conditions that (a) time changes in the fields are small during the period associated with that motion and (b) spatial variations are small between successive cycles. Table I gives the definitions of the adiabatic invariants and the approximate periods required for each component of the motion. The periods given here are typical of particles in the inner radiation belt but depend on particle type, energy, equatorial pitch angle, and location. The important point to note is the large ratio between the periods associated with each invariant, a fact which makes practical the decomposition of motion into gyration, bounce, and drift.

TABLE I

Motion	Adiabatic invariant	Typical periods for 1 MeV protons	1 MeV electrons
Gyration about field line	$\mu = p_\perp^2/2Bm_0$	2×10^{-2} sec	10^{-4} sec
Bounce between hemispheres	$J = \oint p_\parallel \; ds$	3·6 sec	0·2 sec
Longitudinal drift	$\Phi = \oint \mathbf{B} \cdot d\mathbf{A}$	0·5 h	0·5h

In the definitions of the invariants $p_\perp$ and $p_\parallel$ are the momentum components perpendicular and parallel to the magnetic field, and m_0 is the particle rest mass. The integral for J is taken along the magnetic field over a complete bounce, and the integral for Φ is over the area enclosed by the drift path of the particle.

Because the drift path of a geomagnetically trapped particle surrounds the earth and the magnetic dipole, the enclosed magnetic flux is equal to the northward directed flux of the field minus the southward return flux passing through the dipole position. This net flux is also equal to the northward flux outside the drift shell, and it is generally more convenient to calculate Φ by integrating B over the equatorial plane outside the particle drift shell.

The quantities μ, J and Φ are frequently called the first, second, and third adiabatic invariants. Note that μ differs from the classical magnetic moment for relativistic particles due to the fact that the rest mass, rather than the total mass, appears in the denominator. The conservation of the first invariant, μ, describes the change in pitch angle, or $p_\perp$, accompanying changes in B as the particle moves along the field line. As the particle moves toward either pole, B increases causing $p_\perp$ to increase. If $\mathbf{E} = 0$ the total momentum p cannot change, so an increase in $p_\perp$ implies that the pitch angle has increased. When $p_\perp = |\mathbf{p}|$ the particle will be moving at right angles to B and will be reflected. The value of B at which the particle mirrors is obtained immediately as

$$B_m = \frac{p^2}{2m_0\mu} \qquad (2\text{-}5)$$

The second adiabatic invariant J specifies the trace of a particle during its longitudinal drift. At each longitude, the particle will mirror at a field value B_m, and it will be located on that field line for which the integral J is equal to the required value. The constancy of J and μ requires that the particle guiding center returns to the initial field line after drifting completely around the earth, thus closing the magnetic shell.

The third adiabatic invariant is useful primarily for describing the inward or outward motion of the drift shell during slow changes in the geomagnetic field. If the overall magnetic field is slowly increasing (decreasing) the constancy of Φ requires that the particle drift shell move outward (inward).

The time periods shown in Table I for the gyration, bounce, and longitudinal drift indicate the approximate time intervals over which changes in B or E must occur to alter the values of an adiabatic invariant. For example, μ for electrons can be altered by whistler waves with frequencies above 10 kHz, J can be affected by micropulsations near 1 Hz, and Φ can be changed by field variation with frequencies greater than $\sim 2/h$.

III. TREATMENT OF NON-ADIABATIC EFFECTS

While the adiabatic invariant theory provides a zeroth order description of trapped particle motion, experimental data show that the adiabatic invariants are not conserved rigorously. Fluctuations in the magnetic field, time-dependent electric fields, and collision processes are present. All of these can alter the adiabatic invariants of a trapped particle, giving the radiation belt its dynamic character as well as determining its basic structure. The theoretical description of the radiation belts will be developed using the adiabatic theory as the basic particle motion but will include deviations from such motion as perturbations. The cumulative effect of the frequent perturbations is large and must be calculated in some detail in order to evaluate the overall effects. In this section the mathematical foundation which has proven useful in radiation belt problems is developed; in following sections these general principles will be applied to obtain a quantitative description of radiation belt populations.

A rather large number of physical mechanisms have been shown to affect the geomagnetically trapped electrons and protons of the inner radiation zone. In many situations several mechanisms are operating simultaneously, and the most practical way to describe their influence is to compute their effects on the distribution function of the particles. The distribution function F is the number of trapped particles per unit increment of each of the six independent variables which are required to describe the position and velocity of a particle. Physical mechanisms affect the distribution function by adding particles to it, removing them from it, or transporting particles from one region of the six-dimensional space to another. A large and important class of transport processes can be described by the particle continuity equation, which is a fundamental statement of the conservation of particles:

$$\frac{\partial F}{\partial t} + \nabla \cdot \mathbf{j} = 0 \qquad (3\text{-}1)$$

The quantity $\mathbf{j}$ is the directional current of particles in the particular coordinate system selected, and $\nabla \cdot$ is the six-dimensional divergence operator. Not all transport processes can be conveniently described by particle currents. For example, if a nuclear collision occurs, the particle will disappear at one point in coordinate space and simultaneously appear at another (at least on the level of detail considered here). Such processes require a more general form of transport equation than Eq. (3-1). However, all transport processes to be discussed in this paper can be described by Eq. (3-1) with suitable definition of the particle current.

When written in the six phase space coordinates p_i, q_i with time derivatives $\dot{p}_i$ and $\dot{q}_i$, Eq. (3-1) becomes

$$\frac{\partial F}{\partial t} + \sum_i \frac{\partial}{\partial q_i} \dot{q}_i F + \sum_i \frac{\partial}{\partial p_i} \dot{p}_i F = 0 \qquad (3\text{-}2)$$

The current components $\dot{q}_i F$ and $\dot{p}_i F$ depend on the velocities $\dot{q}_i$ and the momentum changes $\dot{p}_i$ (which are, of course, proportional to the forces).

Even in the absence of forces, $\dot{q}_i$ will generally not be zero, and transport of particles will occur. The $\dot{q}_i$ are then entirely determined by the initial conditions, and the resulting transport terms are called streaming terms. Streaming terms can often be eliminated by an appropriate choice of coordinate system. This choice is quite important and is discussed later in this chapter; the development which follows is carried out in phase space coordinates in order to establish clearly the generality of the results.

When forces act on the particles, they determine the $\dot{p}_i$, and the $\dot{q}_i$ are obtained by integrating the equations of motion. Insertion of these values into Eq. (3-2) establishes the appropriate transport equation. In addition to transport processes, there may be other processes which add new particles to the distribution (such as the addition of protons by neutron decay) or remove existing particles (such as the catastrophic loss of protons by collisions with the nuclei of atmospheric atoms). Such processes may be included in Eq. (3-2) by adding an appropriate source or loss term for each process.

There are some processes in which $\dot{q}_i$ and $\dot{p}_i$ are fluctuating functions of time too laborious or too little known to compute and use directly in Eq. (3-2). If the changes in q_i and p_i occur in a large number of uncorrelated steps, each of which is very small in magnitude compared to the range of values under consideration, it is sometimes possible to obtain approximations to the required transport terms using average characteristics of the perturbing forces. As might be expected, the resulting distribution function will be valid only for time scales which are long compared with the interval between the uncorrelated steps.

The basic technique for describing this type of process is the Fokker–Planck equation, which may be derived in the following way. Let x_1, $x_2, \ldots x_6$ represent the six phase space coordinates so that a position in phase space is given by the six dimensional vector $\mathbf{x}$. The distribution function $F(\mathbf{x}, t)d^6x$ gives the number of particles at time t in the phase space volume $d^6x = dq_1dq_2dq_3dp_1dp_2dp_3$. Let $\Psi(\mathbf{x}-\Delta\mathbf{x}, \Delta\mathbf{x}, \Delta t)$ be the probability that in time interval Δt a particle at $\mathbf{x}-\Delta\mathbf{x}$ will suffer an increment $\Delta\mathbf{x}$ and be transferred into unit d^6x at $\mathbf{x}$. The function Ψ is determined by whatever forces produce changes in the particle distribution and is assumed to be independent of time and of the particle distribution.

The interval Δt must be long compared to the time interval between the elementary perturbations producing the overall changes in $\mathbf{x}$. It must also be long enough to include a representative sample of the individual perturbations. On the other hand Δt must be short compared with time required for significant changes in $F(\mathbf{x}, t)$ to occur. The distribution function at time $t+\Delta t$ will be

$$F(\mathbf{x}, t+\Delta t) = \int d(\Delta\mathbf{x})F(\mathbf{x}-\Delta\mathbf{x}, t)\Psi(\mathbf{x}-\Delta\mathbf{x}, \Delta\mathbf{x}, \Delta t) \tag{3-3}$$

If the left hand side is expanded in a Taylor series about t and the right-hand side in a Taylor series about $\mathbf{x}$, the equation becomes

$$F(x, t)+\frac{\partial F}{\partial t}\Delta t+\ldots = \int d(\Delta\mathbf{x})\left\{ F(\mathbf{x})\Psi(\mathbf{x}, \Delta\mathbf{x}, \Delta t) - \left(\frac{\partial}{\partial x_i}\Psi F \right)\Delta x_i \right.$$
$$\left. +\frac{1}{2}\left(\frac{\partial^2}{\partial x_i\partial x_j}\Psi F \right)(\Delta x_i\Delta x_j)+\ldots \right\} \tag{3-4}$$

Throughout this section if an index occurs twice in a single term, the term is to be summed over that index. The first term on the right of Eq. (3-4) is immediately seen to be equal to $F(\mathbf{x}, t)$. In the other terms the sequence of integration and differentiation can be changed since F is not a function of $\Delta\mathbf{x}$, and $\Delta\mathbf{x}$ is not a function of $\mathbf{x}$. Hence

$$\frac{\partial F}{\partial t}\Delta t+\ldots = -\frac{\partial}{\partial x_i}\int d(\Delta\mathbf{x})(\Psi F)(\Delta x_i)+\frac{1}{2}\frac{\partial^2}{\partial x_i\partial x_j}\int d(\Delta\mathbf{x})(\Psi F)(\Delta x_i\Delta x_j)+\ldots \tag{3-5}$$

If the average time rate of change of a quantity is denoted by angular brackets, then

$$\frac{\partial F}{\partial t}+\ldots = -\frac{\partial}{\partial x_i}\langle\Delta x_i\rangle F+\frac{1}{2}\frac{\partial^2}{\partial x_i\partial x_j}\langle\Delta x_i\Delta x_j\rangle F+\ldots \tag{3-6}$$

where

$$\langle \Delta x_i \rangle = \frac{1}{\Delta t} \int (\Delta x_i) \Psi(\mathbf{x}, \Delta \mathbf{x}, \Delta t) d(\Delta \mathbf{x})$$

$$\langle \Delta x_i \Delta x_j \rangle = \frac{1}{\Delta t} \int \Delta x_i \Delta x_j \Psi(\mathbf{x}, \Delta \mathbf{x}, \Delta t) d(\Delta \mathbf{x}) \qquad (3\text{-}7)$$

In almost all applications of the Fokker–Planck equation the terms containing derivatives of higher than second order can be neglected. The validity of this truncation depends on the transition probability $\Psi(\mathbf{x}, \Delta \mathbf{x}, \Delta t)$. If the integrals in Eq. (3-7) depend on Δt to higher than first order, the term can be neglected as was done in the case of the term $\frac{1}{2}(\partial^2 F/\partial t^2)(\Delta t)^2$ on the left-hand side of Eq. (3-4). Also, even if the integral varies as Δt but the value of the integral is much smaller than other terms of first order in Δt, the term can be neglected. Frequently, the cross terms in Eq. (3-6) vanish since a specific process may affect only one coordinate, in which case $\langle \Delta x_i \Delta x_j \rangle_{i \neq j} = 0$.

The usual streaming terms of the transport equation are included in Eq. (3-6) and the prescription for $\langle \Delta x_i \rangle$. If the forces $\mathscr{F}$ acting on the particles are constant over the time period Δt, $\Delta \mathbf{q} = \mathbf{p}\Delta t/m$ and $\Delta \mathbf{p} = \mathscr{F}\Delta t$. The transition probability becomes

$$\Psi(\mathbf{x}, \Delta \mathbf{x}, \Delta t) = \delta(\Delta \mathbf{q} - \mathbf{p}\Delta t/m)\, \delta\,(\Delta \mathbf{p} - \mathscr{F}\Delta t) \qquad (3\text{-}8)$$

and the integration for the $\langle \Delta x_i \rangle$ follows directly to give

$$\langle \Delta \mathbf{q} \rangle = \mathbf{p}/m = \mathbf{w}$$

$$\langle \Delta \mathbf{p} \rangle = \mathscr{F} \qquad (3\text{-}9)$$

The first derivative terms in Eq. (3-6) then become

$$\frac{\partial}{\partial x_i}\langle \Delta x_i \rangle F = \nabla_q \cdot (\mathbf{w}F) + \frac{1}{m}\nabla_w \cdot (\mathscr{F}F) \qquad (3\text{-}10)$$

which are immediately recognized as streaming terms in configuration and velocity space.

Equation (3-6) is a generalized Fokker–Planck equation suitable for describing a large class of problems. The coefficients $\langle \Delta x_i \rangle$ and $\langle \Delta x_i \Delta x_j \rangle$ are frequently called the first and second Fokker–Planck coefficients. The equation is linear in F because the coefficients are independent of F, a feature which greatly simplifies solution in practical cases. The linearity follows directly from the assumption that Ψ was independent of the particle distribution. This assumption is valid only for noninteracting particles; hence, the Fokker–Planck equation in this form cannot be used for situations involving collisions of the trapped particles with each other.

The Fokker–Planck equation is an alternative method of describing the time dependence of the distribution function in which one has sacrificed accuracy (by ignoring higher order terms) and time resolution (because the solution will be valid only for time scales long compared with the interval between individual perturbations). In return for this loss, only the time-averaged characteristics of the perturbing influences need be specified in order to obtain a solution. A great virtue of the Fokker–Planck equation is that it provides a simple prescription for deriving the terms in the continuity equation; one need only calculate $\langle \Delta x_i \rangle$ and $\langle \Delta x_i \Delta x_j \rangle$ to express the time dependence of F.

Suppose, however, that the perturbing forces are completely known as a function of the coordinates and time. Then in principle Eq. (3-2) could be used to describe changes in the distribution function. If the motion of the particles under the action of the forces can be derived from Hamilton's canonical equations (as, for example, the motions due to fluctuating electric and magnetic fields), then for each component

$$\dot{q}_i = \frac{\partial H(p, q)}{\partial p_i}, \qquad \dot{p}_i = -\frac{\partial H(p, q)}{\partial q_i} \tag{3-11}$$

where $H(p, q)$ is the Hamiltonian. When Eq. (3-11) is combined with Eq. (3-2), the Liouville equation is obtained

$$\frac{\partial F}{\partial t} + \dot{q}_i \frac{\partial F}{\partial q_i} + \dot{p}_i \frac{\partial F}{\partial p_i} = 0 = \frac{dF}{dt} \tag{3-12}$$

From Eq. (3-12) it is apparent that a distribution which is uniform in phase space $(\partial F/\partial q_i = \partial F/\partial p_i = 0)$ will remain constant in time, i.e. $\partial F/\partial t = 0$ irrespective of the applied forces. The left hand side of Eq. (3-12) is the total derivative of F along a dynamical path. Because the total derivative $dF/dt = 0$ and the number of particles described by Eq. (3-12) remains constant, the phase space volume element containing a group of particles remains the same size throughout the particle motion.

If the Liouville equation demonstrates that a distribution function without gradients in phase space must remain constant in time, then so must the Fokker–Planck equation since it is an alternative method (subject to the limitations indicated above) to describe the same situation. Therefore, for a uniform distribution F_0,

$$\frac{\partial F_0}{\partial t} = 0 = -\frac{\partial}{\partial x_i}\langle \Delta x_i \rangle F_0 + \frac{1}{2}\frac{\partial^2}{\partial x_i \partial x_j}\langle \Delta x_i \Delta x_j \rangle F_0 \qquad (3\text{-}13)$$

The terms on the right-hand side can be expressed as the divergence of a vector current $\mathbf{j}$ so that Eq. (3-13) may be rewritten as

$$\nabla \cdot \mathbf{j} = 0 \qquad (3\text{-}14)$$

where

$$j_i = \langle \Delta x_i \rangle F_0 - \frac{1}{2}\frac{\partial}{\partial x_j}\langle \Delta x_i \Delta x_j \rangle F_0$$

$$= F_0\left(\langle \Delta x_i \rangle - \frac{1}{2}\frac{\partial}{\partial x_j}\langle \Delta x_i \Delta x_j \rangle\right) \qquad (3\text{-}15)$$

For a uniform distribution function the current $\mathbf{j}$ can be shown to be everywhere zero by the following argument due to Haerendel (1968). Since $\nabla \cdot \mathbf{j} = 0$, the flow lines of the current must either pass through the phase space volume under consideration or close on themselves. If a rigid reflector (which does not invalidate Hamilton's equations) is inserted across the flow lines in configuration space, any systematic flow will be blocked producing an excess of particles upstream of the barrier and a depletion of particles downstream. This change would contradict Eq. (3–13), which requires that the distribution function remain constant. The assumption that there is a finite current is therefore shown to be false, and $\mathbf{j}$ must vanish everywhere. This argument does not exclude the possibility of a net flow in momentum space keeping all q_i constant. However, if Hamilton's Eqs. (3-11) apply, a systematic change in the momentum of a particle is not possible without a corresponding change in position. Thus, for a uniform distribution of particles being perturbed by forces which can be represented by a Hamiltonian the net current is zero and

$$\langle \Delta x_i \rangle = \frac{1}{2}\frac{\partial}{\partial x_j}\langle \Delta x_i \Delta x_j \rangle \qquad (3\text{-}16)$$

Both sides of Eq. (3-16) depend only on the perturbing forces (as yet unspecified) and not on the distribution function. Consequently, the equation represents a fundamental relation between the coordinate change of particles caused by arbitrary perturbing forces. Changing F from a uniform distribution to any other will have no effect on the coordinate changes of individual particles caused by perturbing forces, and thus this relation holds for an arbitrary distribution function.

If now $\langle \Delta x_i \rangle$ from Eq. (3-16) is substituted into Eq. (3-6) a diffusion equation requiring only the second Fokker–Planck coefficients results

$$\frac{\partial F}{\partial t} = \frac{\partial}{\partial x_i}\left[\frac{\langle \Delta x_i \Delta x_j \rangle}{2}\frac{\partial}{\partial x_j}F\right] \tag{3-17}$$

with current components given by

$$j_i = -\frac{\langle \Delta x_i \Delta x_j \rangle}{2}\frac{\partial}{\partial x_j}F \tag{3-18}$$

The 36 quantities $\langle \Delta x_i \Delta x_j \rangle/2$ are the diffusion coefficients and are hereafter labeled D_{ij}. From the symmetry of the definitions there are only 21 independent coefficients. In practice diffusion in more than one dimension is extremely difficult to treat numerically. Fortunately, for many radiation belt situations it is possible by the choice of variables to restrict diffusion to a single dimension and the summation in Eq. (3-17) is not required.

Equation (3-18) indicates that the particle currents are in a direction opposite to the gradient of the distribution function. That is, particles flow away from a region where the distribution function has a maximum with respect to any coordinate (unless of course the diffusion coefficient vanishes). Consequently, diffusion tends to reduce local particle concentrations. If diffusion is occurring in all coordinates and no other processes are present, the distribution function will evolve towards a distribution which is uniform in phase space. This conclusion is valid for all systems described by Eq. (3-17) (and also Eq. 3-6) irrespective of the physical nature of the forces or their variation in space or time.

Finally, the general procedure for describing the time evolution of the distribution function in the presence of several mechanisms is to write a continuity equation of the form of Eq. (3-2) for all processes for which $\dot{q}_i$ and $\dot{p}_i$ can be fully specified, adding Fokker–Planck terms (Eq. 3-6) or diffusion terms (Eq. 3-17) for processes being described by these methods and then adding source or loss terms to represent the rate at which particles are added to or removed from the existing distribution.

Choice of Coordinates

Phase space coordinates are almost never a convenient choice with which to describe the distribution of geomagnetically trapped particles. Let y_1, $y_2, \ldots y_6$ be the six coordinates of a particle in a desired coordinate system, and let $Y(\mathbf{y}, t)\mathrm{d}^6 y$ represent the number of particles contained in the volume element $\mathrm{d}^6 y$ (i.e. Y is the distribution function in these coordinates). The equations previously derived can be transformed to the new (not necessarily orthogonal) coordinate system by standard methods.

The continuity equation becomes

$$\frac{\partial Y}{\partial t} + \frac{\partial}{\partial y_i}(\dot{y}_i Y) = 0 \qquad (3\text{-}19)$$

The Fokker–Planck equation is

$$\frac{\partial Y}{\partial t} = -\frac{\partial}{\partial y_i}\langle \Delta y_i \rangle Y + \frac{1}{2}\frac{\partial^2}{\partial y_i \partial y_j}\langle \Delta y_i \Delta y_j \rangle Y \qquad (3\text{-}20)$$

and the diffusion equation is

$$\frac{\partial Y}{\partial t} = \frac{\partial}{\partial y_i}\left[\frac{\langle \Delta y_i \Delta y_j \rangle \mathscr{J}}{2}\frac{\partial}{\partial y_j}\frac{Y}{\mathscr{J}}\right] \qquad (3\text{-}21)$$

The relation between $F(\mathbf{x})$ and $Y(\mathbf{y})$ where the x_i are phase space variables is

$$Y(\mathbf{y}, t) = F(\mathbf{x}, t)\mathscr{J}(\mathbf{x}; \mathbf{y}) \qquad (3\text{-}22)$$

where the Jacobian $\mathscr{J}$ is the ratio of the volume element $\mathrm{d}^6 x$ in phase space to the volume element $\mathrm{d}^6 y$ in the selected coordinates.

The most important coordinate system for geomagnetically trapped particles is the one in which the six variables are the three adiabatic invariants μ, J, Φ and their associated phase variables $\varphi_1, \varphi_2,$ and φ_3. The construction of the Jacobian relating a volume element in phase space to one in the space defined by adiabatic invariants and their phase variables is not readily accomplished. Fortunately, the basic theory which proves the existence of adiabatic invariants can be used to determine $\mathscr{J}$. According to the formal theory (Kruskal, 1962) the equations of motion (3-11) can be transformed by a series of canonical transformations to a Hamiltonian formulation in which the new variables are the adiabatic invariants and their conjugate angle variables. A property of canonical transformations is that phase space volumes are invariant under such transformations (see for example Goldstein (1950), page 250). Thus, the Jacobian for the transformation from phase space to invariant space is a constant, and Eqs. (3-19 to 3-21) have exactly the same form in invariant space as they do in phase space. Furthermore, the Jacobian $\mathcal{J}$ required for any arbitrary coordinates $y_1, \ldots y_6$ is proportional to the Jacobian for the transformation from invariant space to the y variables, and transformations from phase space coordinates need never be performed in practice. A corollary to this theorem is that a particle distribution which is uniform in phase space will be uniform in adiabatic invariant space.

For a large majority of radiation belt problems the three angle variables conjugate to the adiabatic invariants are of little interest. For example, magnetic and electric perturbations will destroy an initially uniform distribution momentarily, but the subsequent motion of the particles will gradually restore uniformity in the angle variables. The Fokker–Planck coefficients (which are long term rates of change) containing the changes in angle variables will consequently vanish. A uniform distribution among the angle variables is by definition one in which the particle flux observed by a stationary observer will be time independent. For the angle variables corresponding to the first, second and third invariants of particles trapped in a dipole field this distribution will have particles equally distributed in gyration phase, in position along their spiral bounce paths, and in longitude. For these problems a three-dimensional distribution function may be defined by integrating over phase variables

$$Z'(\mu, J, \Phi) = \int Z(\mu, J, \Phi; \varphi_1, \varphi_2, \varphi_3) d\varphi_1 d\varphi_2 d\varphi_3 \qquad (3\text{-}23)$$

The general equation for treating multidimensional diffusion in the three adiabatic invariants is therefore

$$\frac{\partial Z'}{\partial t} = \frac{\partial}{\partial z_i} \left[\frac{\langle \Delta z_i \Delta z_j \rangle}{2} \frac{\partial Z'}{\partial z_j} \right] \qquad (3\text{-}24)$$

where the z_1 are the three adiabatic invariants. For any other set of variables y_i and distribution function $Y(y_1, y_2, y_3, t)$ related to the adiabatic invariants Eq. (3-24) becomes

$$\frac{\partial Y}{\partial t} = \frac{\partial}{\partial y_i} \frac{\langle \Delta y_i \Delta y_j \rangle \mathcal{J}}{2} \frac{\partial}{\partial y_j} \frac{Y}{\mathcal{J}} \qquad (3\text{-}25)$$

where the Jacobian $\mathscr{J}$ relates the phase space volume in invariant space to the variable y_i chosen

$$\mathscr{J}(\mu, J, \Phi; y_1, y_2, y_3) = \frac{\partial(\mu, J, \Phi)}{\partial(y_1, y_2, y_3)} \tag{3-26}$$

Most of the radiation belt diffusion calculations have considered processes in which only one adiabatic invariant is changed. Hence, the summation in Eq. (3-25) is seldom required, and the partial differential equation to be solved is second order in only one of the variables. An approximate method of treating simultaneous diffusion in pitch angle and L value was derived by Walt (1970), but existing data are not precise enough to require this treatment.

The choice of coordinates in which to formulate the equation for the distribution function is dictated largely by the desire to simplify the equation and to find applicable boundary conditions. If a mechanism preserves two independent variables while causing transport in a third, the equation will obviously be simplest if the two conserved variables are chosen as coordinates. Some common substitutions for μ, J, and Φ are kinetic energy E for μ, cosine of the equatorial pitch angle for J, and McIlwain parameter L for Φ. Roederer (1970) has an extensive discussion of possible coordinate systems and the associated Jacobians.

When suitable coordinates have been selected, theoretical treatment of a physical mechanism is often simpler. But to relate the theoretical result to experimental measurement, it is usually necessary to express the distribution function in terms of the commonly measured quantity, the particle flux or, as it is sometimes known, the directional intensity. Particle flux, for which the usual symbol is j, is defined as the number of particles per unit area, per unit energy, per unit solid angle, per unit time passing through a plane area oriented perpendicular to the direction of arrival of the particles. In general, j is a function of particle energy, pitch angle, and L.

The relation between F, the phase space distribution function, and j, the particle flux is established by calculating the number of particles of energy between E and $E+dE$ which in time dt pass through an elemental area $dA = dxdy$. The particles' velocity vectors lie within solid angle $d\Omega$ which is at angle θ to the z-axis. From the definition of j the number of such particles is equal to

$$dN = j \, dE \, dt \, d\mathbf{A} \cdot d\Omega$$

$$= j \, \frac{p dp}{m} \, dt \, dxdy \cos \theta \, d\Omega \tag{3-27}$$

Using the phase space density F the quantity dN is equal to

$$\mathrm{d}N = F\,\mathrm{d}x\mathrm{d}y\mathrm{d}z\,\mathrm{d}p_x\mathrm{d}p_y\mathrm{d}p_z \tag{3-28}$$

The volume element containing the particles which will pass through dxdy in dt is dxdyd$z = $\mathrm{d}x\,\mathrm{d}y\,\cos\theta\,p\mathrm{d}t/m$, and the momentum space containing these particles is $\mathrm{d}p_x\mathrm{d}p_y\mathrm{d}p_z = p^2\,\mathrm{d}\Omega\mathrm{d}p$ giving

$$\mathrm{d}N = F\,\mathrm{d}x\mathrm{d}y\,\cos\theta\,\frac{p}{m}\,\mathrm{d}tp^2\,\mathrm{d}\Omega\,\mathrm{d}p \tag{3-29}$$

By equating (3-27) and (3-29) and cancelling the common factors, the result is

$$F = j/p^2. \tag{3-30}$$

Since the Liouville theorem requires F to be constant along a dynamica particle path, (3-30) indicates that j/p^2 will also be constant. This result is quite useful in a variety of radiation belt problems.

It is sometimes desirable to compare the relative importance of two or more mechanisms producing particle transport in the same coordinate at some particular location in coordinate space. When the terms have the form of Eq. (3-19), it is a simple matter to compare the corresponding values of $\dot{y}_i$, the particle transport velocities. In the common one-dimensional case inspection of the form of the diffusion term in Eq. (3-25) reveals that diffusion is formally equivalent to a transport process in which all particles at a given location move with a common velocity given by

$$\frac{1}{Y}\,D_{ii}\mathscr{J}\,\frac{\partial}{\partial y_i}\left(\frac{Y}{\mathscr{J}}\right)$$

This quantity may reasonably be called the diffusion velocity, and it should properly be compared with the other transport velocities. Unlike other transport velocities, the diffusion velocity depends upon the gradient of the distribution function and cannot be estimated until the distribution function is known.   It is sometimes assumed that the gradient $\partial/\partial y_i\,Y/\mathscr{J}$ will be of the order of $Y/\mathscr{J}\Delta S_i$ where ΔS_i is a characteristic scale size for y_i, so that the diffusion velocity becomes $D_{ii}/\Delta S_i$. In most radiation belt problems the actual gradient varies so greatly over the range of y_i that this estimate is of limited usefulness.

It is also possible to define a characteristic time for a physical process by dividing the characteristic scale size by the corresponding transport or diffusion velocity. Again, the velocity often varies so much over the range of y_i that comparison of characteristic times is useful only for order-of-magnitude estimates.

IV. EFFECTS OF VARIOUS PHYSICAL PROCESSES ON TRAPPED PARTICLE DISTRIBUTION FUNCTIONS

A. Collisions of trapped electrons with atmospheric constituents

The details of the electron collision problem were worked out by MacDonald and Walt (1961); the development here follows their work closely. The coordinates (in addition to time) which they selected for the description are the cosine of the equatorial pitch angle x, the kinetic energy E, and the McIlwain parameter L. Since collisions do not change the L value of a particle, the equations considered here are understood to apply to a single L shell, and the L dependence of the distribution function will be suppressed.

The Fokker–Planck equation for the time evolution of the distribution function f due to collisions is

$$\frac{\partial f(x, E, L, t)}{\partial t} = -\frac{\partial}{\partial x}f\langle\Delta x\rangle - \frac{\partial}{\partial E}f\langle\Delta E\rangle + \frac{1}{2}\frac{\partial^2}{\partial x^2}f\langle(\Delta x^2)\rangle$$

$$+\frac{\partial^2}{\partial x\partial E}f\langle\Delta x\Delta E\rangle + \frac{1}{2}\frac{\partial^2}{\partial E^2}f\langle(\Delta E)^2\rangle + \text{higher order terms.} \quad (4\text{-}1)$$

There is no *a priori* reason for excluding higher order terms; however when the evaluation is made, terms of higher order than second order will be found to be negligible for this process.

Terms 1 and 3 are not independent, and the first and second Fokker–Planck coefficients in these terms are related according to the argument in Section III. The following discussion applicable to this particular mechanism allows further insight into the general proof.

Suppose that the electrons are subjected to a process that scatters them but produces no energy losses. Then the Fokker–Planck equation contains only terms 1 and 3. Such a process tends to produce an isotropic distribution of particle velocities—any concentration of velocity vectors within a particular solid angle is eliminated by an increased rate of scattering out of that solid angle. An isotropic distribution will be time independent in the presence of scattering centers. Furthermore, once a distribution is produced which is isotropic anywhere within a flux tube, it will be isotropic everywhere, according to the Liouville equation. The general relation between the distribution function in the variables x, E, t and the directional flux $j(x, E, t)$ is given by

$$f(x, E, t) = 2\pi x \tau_b j(x, E, t) \qquad (4\text{-}2)$$

where τ_b is the full bounce period of the particle.

The symmetry argument above indicates that j is independent of time whenever it is isotropic; the Fokker–Planck equation must yield the same result. Consequently

$$\frac{\partial f(\mu, E, t)}{\partial t} = 0 = -\frac{\partial}{\partial x}\langle\Delta x\rangle 2\pi x \tau_b j + \frac{1}{2}\frac{\partial^2}{\partial x^2}\langle(\Delta x)^2\rangle 2\pi x \tau_b j \qquad (4\text{-}3)$$

Since j is independent of x, it may be divided out yielding the relation

$$\frac{\partial}{\partial x}\langle\Delta x\rangle x\tau_b = \frac{1}{2}\frac{\partial^2}{\partial x^2}\langle(\Delta x)^2\rangle x\tau_b \qquad (4\text{-}4)$$

Now $\langle\Delta x\rangle$ and $\langle(\Delta x)^2\rangle$ are derived from details of the scattering cross sections not yet specified. The cross sections are not affected by the distribution function; therefore the relation (Eq. 4-4) is a general one which is valid for all distribution functions.

Note that this argument, like the general argument of Section III, depends on the existence of an equilibrium distribution. For the scattering process the existence of such a distribution has been deduced from a symmetry argument rather than from the Liouville equation. The fact that the two derivations lead to the same result is not accidental. For the scattering situation considered above, the scattering centers are infinitely massive and are not displaced by the collision process. Hence, the trajectories of the scattered electrons can be described by Hamilton's equations, and the Liouville equation applied to the electron distribution. The case of infinitely heavy scattering centers is therefore a special case of the general theory derived in Section III.

When terms 1 and 3 of Eq. (4-1) are combined using Eq. (4-4), the result is

$$\frac{\partial f}{\partial t} = \frac{\partial}{\partial x}\left[\frac{\langle(\Delta x)^2\rangle}{2}xS\frac{\partial}{\partial x}\left(\frac{f}{xS}\right)\right] - \frac{\partial}{\partial E}f\langle\Delta E\rangle + \frac{\partial^2}{\partial x\partial E}f\langle\Delta x\Delta E\rangle$$

$$+ \frac{1}{2}\frac{\partial^2}{\partial E^2}f\langle(\Delta E)^2\rangle \quad (4\text{-}5)$$

where $S = v\tau_b$ is twice the spiral distance between particle mirror points. The diffusion form of the pitch angle scattering term in Eq. (4-5) was derived by MacDonald and Walt (1961) without utilizing the formal theory of Section III. However, the above derivation is much simpler. It will be shown in the calculations below that with reasonable assumptions the last two terms in Eq. (4-5) can be discarded as small in relation to the first two. The effect of collisions of electrons with the ambient atmosphere thus results in a diffusion in pitch angle with a diffusion coefficient $D = \langle (\Delta x)^2 \rangle / 2$ and a simultaneous *transport* in energy at a rate $\langle \Delta E \rangle$.

The $\langle \Delta x \rangle$ and $\langle \Delta E \rangle$ which appear in Eq. (4-5) are the time rates of change of the variables x and E due to collisions. The angular brackets in Eq. (4-5) indicate that these rates and products of rates are averaged over a time period Δt long compared with the interval between collisions and also long compared with any periodicities present in these rates. It is an assumption of the Fokker–Planck method that the time Δt which will meet these requirements is still short in comparison to the time for the x and E coordinates of the particle to change significantly. Any bracketed quantity h (i.e. $(\Delta x)^2$, ΔE, etc.) is given by

$$\langle h \rangle = \frac{1}{\Delta t} \int_{\Delta t} dt \int \int d(\Delta x) d(\Delta E) h \Psi(\Delta x, x; \Delta E, E) \qquad (4-6)$$

where $\Psi(\Delta x, x; \Delta E, E)$ is the probability per unit time for the coordinates x and E to have increments Δx and ΔE as a result of collisions. The inner integrals, which will be called the collision sum, are taken over all possible increments, while the outer integral produces the appropriate time average.

The integral which represents the collision sum is more conveniently evaluated by a change of variable from Δx, the change in the cosine of the equatorial pitch angle, to $\Delta \alpha$, the change in the cosine of the local pitch angle at which the scattering event occurs. The dominance of small angle collisions allows a number of approximations which simplify the derivation. The change Δx produced by a change $\Delta \alpha$ is to second order in $\Delta \alpha$,

$$\Delta x = \Delta \alpha \frac{dx}{d\alpha} + \frac{1}{2} (\Delta \alpha)^2 \frac{d^2 x}{d\alpha^2} \qquad (4-7)$$

from which is obtained to second order in $\Delta \alpha$

$$(\Delta x)^2 = (\Delta \alpha)^2 \left(\frac{dx}{d\alpha} \right)^2 \qquad (4-8)$$

The average quantities required for Eq. (4-5) are

$$\langle(\Delta x)^2\rangle = \frac{1}{\Delta t}\int_{\Delta t} dt \int\int d(\Delta\alpha)d(\Delta E)(\Delta\alpha)^2\left(\frac{dx}{d\alpha}\right)^2 \Psi(\Delta\alpha, \alpha; \Delta E, E)$$

$$\langle\Delta E\rangle = \frac{1}{\Delta t}\int_{\Delta t} dt \int\int d(\Delta\alpha)d(\Delta E)\Delta E\Psi(\Delta\alpha, \alpha; \Delta E, E)$$

$$\langle\Delta x\Delta E\rangle = \frac{1}{\Delta t}\int_{\Delta t} dt \int\int d(\Delta\alpha)d(\Delta E)\Delta E\Delta\alpha\frac{dx}{d\alpha}\Psi(\Delta\alpha, \alpha; \Delta E, E)$$

$$\langle(\Delta E)^2\rangle = \frac{1}{\Delta t}\int_{\Delta t} dt \int\int d(\Delta\alpha)d(\Delta E)(\Delta E)^2\Psi(\Delta\alpha, \alpha; \Delta E, E) \qquad (4\text{-}9)$$

where $\Psi(\Delta\alpha, \alpha; \Delta E, E)$ is the probability per unit time for the cosine of the local pitch angle α to change by an amount $\Delta\alpha$ in a collision in which the kinetic energy changes by an amount ΔE.

1. *Angular deflections*

Significant angular deflections of energetic electrons occur from elastic Coulomb collisions with the nuclei of neutral atoms and with ambient free electrons and protons. Details of a scattering event in laboratory coordinates are illustrated in Figure 2. An electron with initial velocity **w** is scattered through an angle θ_s, acquiring a final velocity **w'**. The azimuthal angle ψ describes the orientation of the scattering plane with respect to the plane defined by the initial velocity and the magnetic field **B** directed along the z-axis.

It is more convenient to perform the collision sums in center of mass coordinates and then transform to laboratory coordinates to obtain the required changes in pitch angle cosine $\Delta\alpha$ and energy ΔE. The center of mass scattering angle will be called η. The angle ψ describes the orientation of the scattering plane in both coordinate systems, since the scattering planes in the two systems coincide.

An elastic scattering event producing a change in the cosine of the local pitch angle $\Delta\alpha$ and a change in kinetic energy ΔE may be equivalently described as a scattering through angle η into the solid angle $\sin \eta\, d\eta\, d\psi$. The probability per unit time of such a scattering event is $nw\sigma(\eta)$, where n is the number of scatterers per unit volume, and $\sigma(\eta)$ is the differential scattering cross section. If the sum over all collisions is denoted by the use of the curly brackets, the calculation of collision sums of any quantity $\tilde{h}$ is accomplished by using the equation

$$\{\tilde{h}(\Delta\alpha, \Delta E)\} = \int d(\Delta\alpha) \int d(\Delta E) \tilde{h} \Psi(\Delta\alpha, \alpha; \Delta E, E)$$

$$= \int_0^{2\pi} d\psi \int_0^{\pi} d\eta \sin \eta n w \sigma(\eta) \tilde{h}(\eta, \psi) \qquad (4\text{-}10)$$

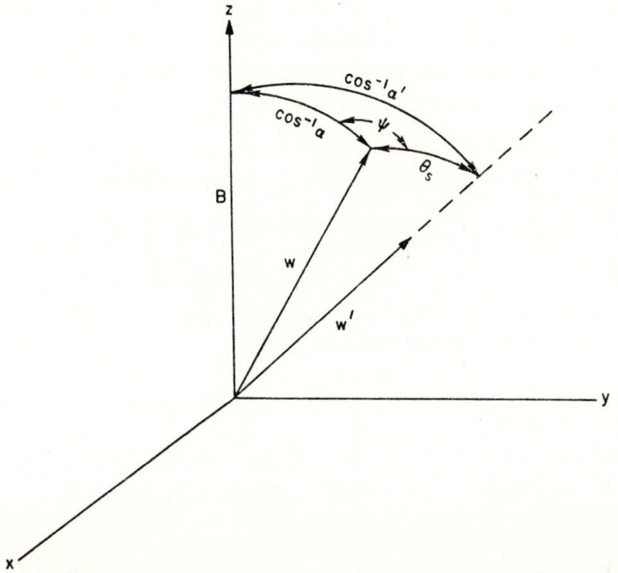

FIGURE 2 Laboratory coordinate system for the scattering of an electron with initial velocity w and local pitch angle cosine α through a scattering angle θ_s. For scattering by° protons and by nuclei of neutral atoms, θ_s is identical to η, the center of mass scattering angle. For scattering by electrons, θ_s must be obtained from η by a relativistic transformation. Note that ψ describes the orientation of the scattering plane in both cases, since the scattering plane in laboratory coordinates coincides with the scattering plane in center of mass coordinates.

The cross sections for the scattering of electrons by a heavy point charge (proton or nucleus of a neutral atom) and for the scattering of electrons by other electrons are well established both theoretically and experimentally. When they are inserted into Eq. (4-10) for $\tilde{h} = \Delta\alpha$, $(\Delta\alpha)^2$, and ΔE, the largest term in the integral over η is found to vary as $\ln 1/\eta$ for small η, and the integral diverges at the lower end point, $\eta = 0$. This result is a consequence of the long range nature of the Coulomb force; the electron suffers a deflection in an arbitrarily distant collision. In the case of atmospheric scattering, charges of the opposite sign will screen the point scatterer at sufficiently large distances, establishing a limit to the distance at which a collision occurs and therefore a minimum angle of deflection η_m. The appropriate lower limit of the integral over η in Eq. (4-10) is therefore η_m and not zero. While the integral is now finite at the lower end point, it is some ten to twenty times larger than at the upper end point, and evaluation of the integral at the upper end point may be neglected. Since only small angle scattering is important, the integrand in Eq. (4-10) may be expanded in a power series in η in which only the leading term (lowest power of η) is retained.

The impact parameter l in a collision is defined as the distance of closest approach between the electron and the scatterer if there were no deflection. For small angle deflections the minimum scattering angle η_m of an electron is related to the maximum impact parameter l_{max} for which scattering occurs by

$$\eta_m = \frac{Ze^2}{\frac{1}{2}m_r w^2 l_{max}} \quad \text{when } Ze^2/\hbar w \gg 1 \qquad (4\text{-}11a)$$

$$= \frac{\hbar}{m_r w l_{max}} \quad \text{when } Ze^2/\hbar w \ll 1 \qquad (4\text{-}11b)$$

The quantity m_r, called the reduced mass, is defined by

$$m_r = \frac{m m_s}{m + m_s} \qquad (4\text{-}12)$$

where m is the mass of the electron and m_s is the mass of the scattering particle. Equation (4-11a) represents the limit in which classical orbit theory for a Coulomb field determines the deflection angle as a function of the impact parameter. Equation (4-11b), which is appropriate for the electron energies of interest here, represents the limit in which the minimum angle of deflection at a given impact parameter is determined by the uncertainty principle, which may be written

$$\Delta\eta\Delta K \approx \hbar. \qquad (4\text{-}13)$$

K, the angular momentum, cannot have an uncertainty greater than its total value $m_r w l_{max}$, and therefore the uncertainty in η cannot be less than $\hbar/m_r w l_{max}$. Angular deflections of less than this value are thus not physically meaningful. The maximum impact parameter l_{max} is determined by the previously mentioned screening effects. For scattering by free electrons and protons, screening is effective beyond the Debye length. The Debye length of an ionized medium is a characteristic distance beyond which the electrostatic field of a charge is effectively shielded by displaced charges in the medium. The Debye length is given by

$$\lambda_D = \left(\frac{kT}{4\pi n_e e^2} \right)^{1/2} \tag{4-14}$$

where k is the Boltzmann constant, T the kinetic temperature, and n_e the number density of ambient free electrons. From Eq. (4-11b)

$$\eta_m = \eta_D = \frac{\hbar}{m_r w \lambda_D} \tag{4-15}$$

For scattering by the nuclei of neutral atoms, the screening is accomplished by the orbital electrons. A statistical model of the distribution of atomic electrons within an atom yields a characteristic length from which it may be computed that

$$\eta_m = \eta_a = \frac{2 \cdot 10 Z^{1/3} (1-\beta^2)^{\frac{1}{2}}}{137\beta} \tag{4-16}$$

for collisions between electrons and the nuclei of neutral atoms (Mott and Massey, 1950).

The differential cross section for the scattering of electrons by protons and by the nuclei of neutral atoms is

$$\sigma(\eta) = \frac{Z^2 e^4}{4m_0^2 c^4} \frac{1-\beta^2}{\beta^4} \frac{1}{\sin^4(\eta/2)} \tag{4-17}$$

where m_0 is the electron rest mass, Z is the atomic number of the scattering nucleus and $\beta = w/c$.

From the geometry of Figure 2 the cosine of the pitch angle after collision is found to be

$$\alpha' = \alpha \cos \theta_s + (1-\alpha^2)^{\frac{1}{2}} \sin \theta_s \cos \psi \tag{4-18}$$

Since the center of mass frame of reference and the laboratory frame are coincident for scattering by a heavy point charge, $\theta_s \doteq \eta$. The change in α produced by a particular collision is

$$\Delta\alpha = \alpha' - \alpha = -2\alpha \sin^2(\eta/2) + (1-\alpha^2)^{\frac{1}{2}} \sin \eta \cos \psi \tag{4-19}$$

The cross section for the scattering of relativistic electrons by free electrons is given by the Moller (1932) scattering formula in center of mass coordinates

$$\sigma(\eta) = \frac{2e^4}{m_0^2 c^4} \frac{(1-\beta^2)+(1-\beta^2)^{\frac{1}{2}}}{\beta^4} \left\{ \frac{4}{\sin^4\eta} - \frac{3}{\sin^2\eta} \right.$$

$$\left. + \frac{[1-(1-\beta^2)^{\frac{1}{2}}]^2}{4} \left[1+\frac{4}{\sin^2\eta} \right] \right\} \quad (4\text{-}20)$$

For these electron-electron collisions, the scattering angle θ_s in lab coordinates is not identical to the scattering angle η in center of mass coordinates but must be obtained from it by a relativistic coordinate transformation. The necessary transformation for this two-body relativistic elastic collision may be found, for example, in Morrison (1953).

$$\tan \theta_s = \frac{(1-\beta'^2)^{\frac{1}{2}}\sin \eta}{1+\cos \eta} \quad (4\text{-}21)$$

where β' is the ratio of the center of mass speed of the electron to the speed of light. β' is given in terms of β by Morrison as

$$\beta' = \beta[1+(1-\beta^2)^{\frac{1}{2}}]^{-1} \quad (4\text{-}22)$$

Equation (4-18), with the help of Eq. (4-21), becomes

$$\alpha' = [\alpha \cos \eta/2 + (1-\alpha^2)^{\frac{1}{2}} \sin \eta/2(1-\beta'^2)^{\frac{1}{2}} \cos \psi]$$

$$[1-\beta'^2 \sin^2\eta/2]^{-\frac{1}{2}} \quad (4\text{-}23)$$

If Eq. (4-22) is used to eliminate β' then $\Delta\alpha = \alpha' - \alpha$ becomes, keeping terms to order η^2,

$$\Delta\alpha = \eta(1-\alpha^2)^{\frac{1}{2}}(m_0 c^2)^{\frac{1}{2}}(2E+4m_0 c^2)^{-\frac{1}{2}} \cos \psi - \eta^2 \alpha m_0 c^2 (4E+8m_0 c^2)^{-1} \quad (4\text{-}24)$$

Use has been made of the relation

$$\beta = E^{\frac{1}{2}}(E+2m_0 c^2)^{\frac{1}{2}}(E+m_0 c^2)^{-1} \quad (4\text{-}25)$$

The sum $\{(\Delta\alpha)^2\}$ can now be calculated by inserting the square of Eq. (4-19) and Eq. (4-24) into Eq. (4-10), and performing the indicated integration. The integral over η is evaluated at the appropriate value of η_m as the lower limit, and the upper limit is neglected. The result is

$$\{(\Delta\alpha)^2\} = \frac{4\pi e^4 c(E+m_0c^2)}{E^{3/2}(E+2m_0c^2)^{3/2}}[n_e ln\eta_D^{-1} + n_p ln\eta_D^{-1}$$

$$+ \sum_i Z_i^2 n_i ln\eta_{ai}^{-1}][1-\alpha^2] \quad (4\text{-}26)$$

In these equations n_e, n_p and n_i denote the number density of free electrons, free protons, and neutral atoms of the ith species, respectively.

2. *Energy losses*

Significant energy losses of the incident electron will arise principally from inelastic collisions with atoms and from elastic Coulomb collisions with free electrons. Electron-nucleus collisions result in negligible energy loss of the electron because of the large mass of the scattering nucleus. The collision sums for the terms describing the change in energy, viz., $\{\Delta E\}$ and $\{(\Delta E)^2\}$ can be calculated in a manner similar to that employed for $\{(\Delta\alpha)^2\}$ once ΔE is expressed in terms of the scattering angles.

The sum $\{\Delta E\}$ for inelastic collisions with the bound electrons of neutral atoms is a well-known result. It is given by

$$\{\Delta E\} = -\frac{4\pi e^4}{m_0 c\beta} \sum_i Z_i n_i \left[ln\frac{E(E/m_0c^2+2)^{\frac{1}{2}}}{I_i} - \tfrac{1}{2}\beta^2 \right] \quad (4\text{-}27)$$

where I_i is the geometric mean ionization potential for the atom of the ith species and $Z_i n_i$ is the number of atomic electrons of the ith atomic species per unit volume.

The calculation of the time rate of energy loss for high energy electrons colliding with free electrons can be found from the expression (Morrison, 1953) for the energy lost in a single collision of scattering angle η.

$$\Delta E = -E \sin^2\eta/2 \quad (4\text{-}28)$$

The quantity $\{\Delta E\}$ for these collisions is obtained by inserting Eq. (4-28) into Eq. (4-10). When the results for both bound and free electron collisions are added,

$$\{\Delta E\} = \frac{-4\pi e^4}{m_0 c} \frac{(E+m_0c^2)}{E^{\frac{1}{2}}(E+2m_0c^2)^{\frac{1}{2}}}$$

$$\left[n_e ln\eta_D^{-1} + \sum_i Z_i n_i \left(ln\frac{E(E/m_0c^2+2)^{\frac{1}{2}}}{I_i} - \frac{1}{2}\frac{E(E+2m_0c^2)}{(E+m_0c^2)^2} \right) \right] \quad (4\text{-}29)$$

The sums $\{\Delta E^2\}$ and $\{\Delta E \Delta x\}$ contain no logarithmic terms and will be dropped from the Fokker–Planck equation. The approximation of neglecting $\{(\Delta E)^2\}$ amounts to neglecting the effect of energy straggling in spreading an initial distribution. For particles which have lost an appreciable fraction of their initial energy, the approximation is not good and the consequences should be considered. However, if the energy spectrum of the particles under consideration varies smoothly with energy, and if there are additional sources of low energy electrons other than those which are slowed down from higher energies, straggling will not have an appreciable effect on the resulting distribution function.

3. *Time averages*

The time average in Eq. (4-9) must be taken over a time period sufficiently long to eliminate both the statistical and periodic variations in the atmospheric density encountered by the electron. The periodic variations are due to the bounce motion and to the longitudinal drift motion. An average over a time period long in comparison to the drift period will be the same as an average over a single drift period, and the latter will be computed. Thus it is necessary to integrate Eq. (4-9) over a complete drift path of the particle, inserting at each point of the orbit the densities of the atmospheric species encountered at that point. Under the approximation that the change in E or x is small during a complete drift period, it is permissible to average over longitude first, in effect constructing a longitudinally averaged atmosphere. For a given value of x it is necessary to know the atmospheric density experienced at each value of α averaged over longitude. Since the local pitch angle of the particle is determined by the magnetic field, one needs an average atmosphere for each B value of the magnetic shell under consideration. These averages are found by tracing paths of constant B, L (which will not be paths of constant altitude) through the geomagnetic field and finding the average density of each constituent along the path. At each point in the path it is necessary to weight the density at that point with a factor inversely proportional to the longitudinal drift velocity at that point. With these average atmospheric densities used in Eqs. (4-26) and (4-29) the remaining time integration involves only the bounce motion. Therefore

$$\langle h \rangle = \frac{1}{\Delta t} \int_{\Delta t} dt\{h\} = \oint \frac{ds}{v\alpha}\{h\} \Big/ \oint \frac{ds}{v\alpha} \tag{4-30}$$

where $\oint ds/v\alpha$ is one full bounce period.

4. *Equation for the distribution function*

From Eqs. (4-9), (4-26), and (4-30) the pitch angle diffusion coefficient, expressed as a function of E and x, is

$$\frac{\langle(\Delta x)^2\rangle}{2} = 2\pi e^4 c E^{-3/2}(E+2m_0c^2)^{-3/2}(E+m_0c^2)\mathscr{L}(x, E)(1-x^2) \qquad (4\text{-}31)$$

while the transport rate in the energy coordinate is

$$\langle\Delta E\rangle = -\frac{4\pi e^4}{m_0c}E^{-\frac{1}{2}}(E+2m_0c^2)^{-\frac{1}{2}}(E+m_0c^2)\mathfrak{M}(x, E) \qquad (4\text{-}32)$$

with

$$\mathscr{L}(x, E) = \oint\frac{ds\;\alpha^2(1-x^2)}{v\alpha\;x^2(1-\alpha^2)}\left[\bar{n}_p ln\eta_D^{-1}+\bar{n}_e ln\eta_D^{-1}+\Sigma Z_i^2\bar{n}_i ln\eta_{ai}^{-1}\right]\bigg/\oint\frac{ds}{v\alpha}$$

$$\mathfrak{M}(x, E) = \bar{n}_e ln\eta_D^{-1}+\Sigma Z_i\bar{n}_i\left(ln\frac{E(E/m_0c^2+2)^{\frac{1}{2}}}{I_i}-\frac{1}{2}\frac{E(E+2m_0c^2)}{(E+m_0c^2)^2}\right)$$

$$\bar{n}_i = \oint\frac{ds}{v\alpha}\bar{n}_i\bigg/\oint\frac{ds}{v\alpha} \qquad (4\text{-}33)$$

In performing the average over the particle bounce period the weak dependence of $ln\;\eta_D^{-1}$ on n_e has been noted, and η_D has been evaluated in the equatorial plane where $n_e = n_0$ on the field line under consideration.

Insertion of these coefficients into Eq. (4-5) gives the final equation describing the distribution function of magnetically confined electrons in a scattering atmosphere:

$$\frac{\partial f}{\partial t} = \frac{4\pi e^4}{m_0c}\frac{\partial}{\partial E}\left[\frac{(E+m_0c^2)}{E^{\frac{1}{2}}(E+2m_0c^2)^{\frac{1}{2}}}\mathfrak{M}(x, E)f\right]$$

$$+\frac{2\pi e^4 c(E+m_0c^2)}{E^{3/2}(E+2m_0c^2)^{3/2}}\frac{\partial}{\partial x}\left[\mathscr{L}(x, E)S(1-x^2)x\frac{\partial}{\partial x}\left(\frac{f}{xS}\right)\right] \qquad (4\text{-}34)$$

B. Albedo neutrons

Protons from the decay of albedo (earth-escaping) neutrons produced by cosmic rays in the earth's atmosphere are an important source of the high-energy protons trapped in the inner zone. This source of trapped particles was the first to be identified (Singer, 1958; Vernov et al., 1959; Hess, 1959), but the difficulty in making accurate measurements of the differential neutron intensity above 10 Mev in space has delayed a complete evaluation of the role played by this source. Neutron flux measurements in this energy range have recently been made by Heidbreder *et al.* (1970), Eyles *et al.* (1971), and Preszler *et al.* (1972).

Albedo neutrons result when high energy galactic cosmic rays (primarily protons) penetrate the earth's atmosphere and produce secondary neutrons in collisions with atmospheric constituents. The first collision initiates a cascade in which the secondary nucleons can produce further collisions and additional secondary nucleons. There is also an important non-nucleonic secondary component consisting of pions, muons, electrons, and gamma rays.

In the nuclear collisions neutron production occurs by one of two physical processes. Fast neutrons whose energies range from 1 Mev to 1 Gev or more are produced in spallation reactions. They are preferentially emitted in the direction of the incident proton, especially so at the higher energies. Somewhat slower neutrons are emitted by deexcitation of nuclei excited by the nucleon collision in what is known as an evaporation process. Evaporation neutrons have an energy distribution which is roughly Maxwellian, peaked at about 1 Mev. They are emitted isotropically in the center of mass system. Production of both fast and evaporation neutrons takes place at all altitudes in the earth's atmosphere.

Both the fast neutrons and the evaporation neutrons are slowed and scattered by further collisions until they are ultimately captured by atmospheric nuclei or escape from the top of the earth's atmosphere. At energies below a few Mev the scattering of neutrons by nitrogen and oxygen nuclei is nearly isotropic, and after a few collisions the neutron flux in the atmosphere becomes distributed almost uniformly over all angles. It is therefore possible to use diffusion theory to estimate the escape flux, although some more rigorous form of transport theory is needed to derive neutron flux values within a few mean free paths of the upper boundary of the atmosphere. On the other hand the neutrons which escape with very high energies (>10 Mev) are produced near the top of the atmosphere, have scattering cross sections which are sharply peaked in the forward direction and generally have not suffered enough collisions for the flux to become isotropic. As a result it is necessary to use Monte Carlo methods to predict their angular distributions and leakage rates.

Calculations performed by Hess *et al.* (1961) indicated that the leakage flux of neutrons below about 10 Mev is proportional to $\cos^2\zeta$ where ζ is the local zenith angle. A simple qualitative argument was used to show that at the opposite extreme very high energy (~ 1 Gev) neutrons must be emitted at zenith angles near 90°. Such high energy neutrons are emitted within a narrow cone of angles about the incident cosmic ray velocity vector, and if they are escaping with such high energies, they must have suffered few or no collisions. Consequently, they must have been produced by collisions of cosmic rays travelling nearly tangential to the top of the atmosphere, and their velocity vectors will lie at zenith angles only slightly less than 90°. Hess *et al.* made rough estimates for the angular distribution at intermediate energies.

Most of the neutrons which leak from the top of the atmosphere have lost nearly all of their energy in atmospheric collisions, and the differential energy spectrum of escaping neutrons has a peak near 0·1 ev. Those whose energies are below 0·66 ev, the energy corresponding to escape velocity, are gravitationally trapped, falling back into the atmosphere after executing ballistic trajectories. High energy neutrons experience negligible gravitational effects and travel outward in straight lines.

The neutron is a radioactive particle, undergoing beta decay with a half-period of about 12 minutes. If it decays after its escape from the atmosphere and before it departs from the vicinity of the earth, both the proton and electron which are produced are candidates for geomagnetic trapping.

The neutron decay beta spectrum has an end point at 782 kev. Because of the large abundance of low energy neutrons in the leakage flux, it is reasonable to assume that the birth spectrum of electrons in space is virtually identical to the neutron decay electron spectrum and that the production in space is isotropic. There is a small high energy electron component produced by energetic neutrons from which the decay electron is emitted in the forward direction, so that there is no fixed electron energy cutoff at 782 kev.

Those electrons emitted in forward directions have greater energies in the earth-fixed frame while those emitted in backward directions have lesser energies. At a neutron energy of 100 Mev, for example, a 782 kev electron emitted in the direction of the neutron velocity will have an energy of about 1·5 Mev in the earth frame. A complete discussion of the resulting electron energy spectrum has been given by Nakada (1963).

In contrast with the electrons produced by albedo neutron decay, the protons retain very nearly the kinetic eneigy of the decaying neutrons and maintain their direction of motion to an excellent approximation. The production of energetic protons therefore depends entirely on the decay of neutrons at the same energy.

At 10 Mev a neutron will travel approximately 0·3 astronomical units during one mean life. Consequently there is no significant depletion of the neutron flux by radioactive decay in the vicinity of the earth at this and higher energies.

Even though the differential energy spectrum of albedo neutrons above 10 Mev has only recently been measured, previous experiments have established several other characteristics of the neutron intensity which must be taken into account in a complete treatment of proton injection from this source. As high energy charged cosmic rays approach the earth, their trajectories are deflected by the geomagnetic field. Even though the radii of curvature of their paths are very large, the field acts to exclude them from regions in which the magnetic dipole has a large horizontal component, i.e., the geomagnetic equator. For example, protons of energy less than about 15 Gev cannot be vertically incident on the top of the earth's atmosphere at the geomagnetic equator regardless of the direction from which they come. On the other hand there is no vertical energy cutoff at the poles of a dipole field. As a result cosmic ray bombardment is much less intense (but of higher average energy) at the equator than at the poles. This introduces an important latitude dependence in the albedo neutron intensity whose magnitude is well-established by experiments which respond to the total neutron spectrum (Lockwood and Friling, 1968). The ratio of polar intensity to equatorial intensity is about 10 to 1. Due to the very few measurements of the differential energy spectrum of albedo neutrons above 10 Mev, the latitude variation of these neutrons is not well known. Presumably it should be less than that of the total neutron flux.

A time variation in the neutron flux arises from the variation in galactic cosmic ray intensity caused by the eleven-year cycle of solar activity. During the active portion of the cycle, the sun emits more plasma, which is embedded in a higher average solar magnetic field. The heliosphere (region of the solar system dominated by solar rather than interstellar magnetic fields) expands. Galactic cosmic rays, particularly at lower energies ($\gtrsim 1$ Gev) suffer greater deflections in the region of the solar system, and the number reaching the earth is reduced. The consequent solar cycle variation of the 10 Mev albedo neutron flux has been estimated by Lingenfelter (1963) to be about 20 percent at the geomagnetic equator.

1. Neutron leakage rates

The neutron albedo flux j_n is defined as the number of neutrons $\sec^{-1}$ ster^{-1} which will pass through one square centimeter oriented perpendicular to the direction of arrival of the neutrons. This is the flux measured by a directional detector in space which views some small portion of the top of the atmosphere. It is important to realize that the measured value of j_n is independent of the distance of the detector from the earth, provided the angular view of the instrument is adjusted to include only that portion of the top of the atmosphere from which the flux j_n is emerging.

The neutron leakage rate is the number of neutrons per second passing through a horizontally-oriented square centimeter at the top of the atmosphere. This rate has occasionally been called leakage flux (e.g. Lingenfelter, 1963) but this name is misleading and should not be used. In terms of the neutron flux j_n (neutrons $\mathrm{cm}^{-2}\sec^{-1}\mathrm{ster}^{-1}$), the leakage rate s_n is given by

$$s_n = \int_\Omega j_n \cos \zeta d\Omega \qquad (4\text{-}35)$$

where ζ is the zenith angle and the solid angle Ω is the upper half-hemisphere.

There are several simple functions of zenith angle which have been used as approximate representations of $j_n(\zeta)$ in different neutron energy ranges. If j_0 is a constant flux independent of ζ, then $j_n = j_0$ represents isotropic emission, $j_n = j_0 \sec \zeta$ is a distribution peaked at the horizontal, and $j_n = j_0$ for $\zeta \geq \zeta_m$, $j_n = 0$ for $\zeta < \zeta_m$ is a distribution with no neutrons within a cone of half angle ζ_m about the vertical direction. In each case, of course, j_0 is chosen to fit experimental data or theoretical calculations.

From Eq. (4-35), the leakage rates s_n for these three representations of $j_n(\zeta)$ are, respectively, πj_0, $2\pi j_0$, and $\pi j_0 \cos^2 \zeta_m$.

The isotropic distribution has generally been assumed to apply to lower energy ($\gtrsim 50$ Mev) neutrons, even though the estimates of Hess, et al. (1961) indicate a peak at the zenith for such neutrons. The latter two distributions are thought to be more appropriate at the higher energies. The $\sec \zeta$ distribution has the awkward property of an infinite flux at $\zeta = 90°$ (local horizontal) but is still useful because s_n is finite.

The distribution characterized by a minimum zenith angle ζ_m was used by Lenchek and Singer (1962). They considered the results of nuclear star production experiments carried out with nuclear emulsions and made a rough approximation for ζ_m as

$$\cos \zeta_m = 300 \, p^{-1} \qquad (4\text{-}36)$$

where p is the neutron momentum in units of Mev/c. At an energy of 100 Mev, for example, $p \approx 450$ Mev/c and $\zeta_m \approx 50°$.

2. Proton injection coefficients

A geomagnetically trapped proton having a particular equatorial pitch angle moves along a helical path of increasing pitch angle as it moves from the equator to its mirror point. Each differential element of its path may be described by a vector $d\mathbf{r}$ whose sense is defined to be the same as the velocity vector of the proton. If $d\mathbf{r}$, extended backward, intersects the top of the earth's atmosphere, it is possible for a neutron originating at that point to decay while traversing the element $d\mathbf{r}$ and thus add a proton to the population traversing that particular element $d\mathbf{r}$. If the extension of $d\mathbf{r}$ does not intersect the earth's atmosphere, injection is not possible in that element.

Let n_p be the number of protons per unit volume per unit solid angle at $d\mathbf{r}$ having velocity vectors of magnitude w in the direction $d\mathbf{r}$. The time rate of change of n_p is equal to the number of neutrons similarly located and directed multiplied by the probability per unit time that a neutron will decay and produce a proton:

$$\frac{dn_p}{dt} = \frac{j_n}{w}\frac{1}{\gamma\tau_n} \tag{4-37}$$

In this expression $\gamma\tau_n$ is the neutron mean life, with $\gamma = (1-\beta^2)^{-\frac{1}{2}}$ and $\tau_n = 1013$ seconds. To find the average rate at which protons are being produced in this particular helical path, this quantity must be averaged over all elements dr composing the path.

$$\left.\frac{dn_p}{dt}\right|_{avg} = \frac{1}{\gamma\tau_n w} \oint j_n d\mathbf{r}/\oint d\mathbf{r} \tag{4-38}$$

Because the proton retains the neutron velocity to an excellent approximation, the rate of increase of the proton flux due to albedo neutron decay may be written

$$\frac{dj_p}{dt} = \frac{1}{\gamma\tau_n} \oint j_n d\mathbf{r}/\oint d\mathbf{r} \tag{4-39}$$

The ratio of the two integrals in Eq. (4-39) has units of flux. It has become customary to normalize this ratio by dividing it by a convenient but arbitrary neutron flux characterizing the problem. The resulting normalized ratio is a dimensionless quantity, usually less than unity, called the injection coefficient. In general it is a function of L, the proton equatorial pitch angle, and the assumed angular and latitudinal variation of j_n. Different workers have used different normalization for the injection coefficient, a fact which must be considered when comparisons are made. For example, Dragt et al. (1966) normalized the ratio to a neutron flux j_n which, if uniform over the earth and isotropic over the upper hemisphere, would produce one-half of the actual total neutron leakage rate of the whole earth. Their injection coefficient was designated χ, so that

$$\frac{dj_p}{dt} = \frac{\chi \bar{j}_n}{\gamma\tau_n} \tag{4-40}$$

The calculation of the injection coefficient will first be illustrated in the simple case of equatorial trapping. For protons of 90° pitch angle the orbits are approximately circles, and the injection coefficient may be readily computed analytically. The geometry is illustrated in Figure 3. In one gyroperiod,

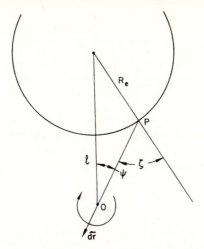

FIGURE 3   Equatorial plane view showing geometry of injection along dr at point 0 by neutrons leaving the earth's atmosphere at P with zenith angle ζ.

the rotation of dr generates a flat disk. The figure shows the element dr in a position in which a neutron from point P at zenith angle ζ can decay along dr. The equatorial flux j_n is first assumed to be isotropic and independent of longitude. Using the normalization of Dragt et al. (1966),

$$\chi = \frac{1}{j_n} \oint j_o \mathrm{dr}/\oint \mathrm{dr} \tag{4-41}$$

$$= \frac{2}{2\pi j_n} \int_0^{\text{arcsin}(R_e/L)} j_o \mathrm{d}\psi \tag{4-42}$$

$$= \frac{j_o}{j_n} \left[\frac{1}{\pi} \arcsin (R_e/L) \right] \tag{4-43}$$

$$= \frac{s_n}{2\pi j_n} \left[\frac{2}{\pi} \arcsin (R_e/L) \right] \tag{4-44}$$

If the neutron flux is not isotropic, but is given by $j_0 \sec \zeta$ then

$$\chi = \frac{2}{2\pi j_n} \int_0^{\text{arcsin } R_e/L} j_o \sec \zeta \mathrm{d}\zeta \tag{4-45}$$

$$= \frac{j_o}{j_n}\left[\frac{R_e}{\pi L}\int_0^{\pi/2}(1-\frac{R_e^2}{L^2}\sin^2\zeta)^{-\frac{1}{2}}d\zeta\right] \qquad (4\text{-}46)$$

$$= \frac{s_n}{2\pi j_n}\left[\frac{1}{\pi}\int_0^{\pi/2}\left(\frac{L^2}{R_e^2}-\sin^2\zeta\right)^{-\frac{1}{2}}d\zeta\right] \qquad (4\text{-}47)$$

For the same leakage rate s_n at the geomagnetic equator, the two assumptions concerning the angular distribution of the neutrons lead to injection coefficients which have different factors enclosed in the brackets of Eqs. (4-44) and (4-47). The two bracketed factors are compared in Figure 4. The comparison

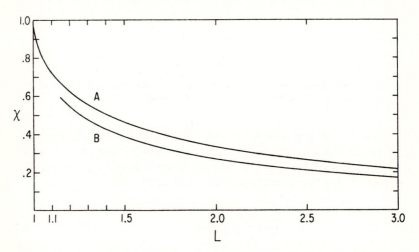

FIGURE 4   A comparison of the two geometry factors in the injection coefficients for equatorially trapped protons. Curve A, for an isotropic neutron flux, is a plot of $2\pi^{-1}$ arcsin (R_e/L). Curve B, for a sec ζ angular dependence of the neutron flux is a plot of

$$\pi^{-1}\int_0^{\pi/2}[(L/R_e)^2-\sin^2\zeta]^{-\frac{1}{2}}d\zeta.$$

illustrates the fact that the injection coefficients do not have a strong dependence on the angular distribution of the albedo neutrons.

The more general case of trapping at arbitrary equatorial pitch angle has been described by Lenchek and Singer (1962). Figure 5 illustrates the injection of protons at an arbitrary point 0 along the helical trajectory of a geomagnetically trapped proton. In one gyroperiod the element $d\mathbf{r}$ generates a cone whose half-angle is Θ, the local particle pitch angle. Another cone, called the δ cone, is defined by the solid angle subtended by the earth from point 0. Whenever the Θ cone intersects the δ cone, injection can occur for that fraction η of the gyroperiod in which $d\mathbf{r}$ is within the δ cone. η is evidently a function of L, Θ, and λ, the magnetic latitude. For the case of

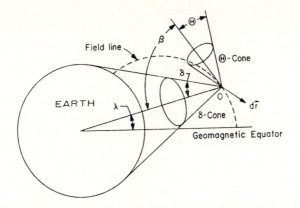

FIGURE 5   Meridian view showing the Θ and δ cones for the injection of protons at point 0 with local pitch angle Θ. In the case illustrated the extension of dr is never within the δ cone, so injection does not occur. After Lenchek and Singer (1962).

a uniform isotropic neutron flux j_0 the injection coefficient χ with the normalization of Dragt, *et al.* (1966) is

$$\chi = \frac{j_o}{j_n}\oint\frac{\eta(L, \Theta, \lambda)ds}{\cos\Theta}\bigg/\oint\frac{ds}{\cos\Theta} = \frac{j_o}{j_n}\bar{\eta} \qquad (4\text{-}48)$$

where the integration is carried out between particle mirror points. The element ds is the length element along the dipole field line.

To perform the integration, it is necessary to express the integrand as a function of L and a convenient variable of integration, say λ. η may be evaluated from the geometry illustrated in Figures 5 and 6. The angle between the axes of the Θ and δ cones is labeled β in Figure 5. Figure 6 shows how β is related to λ by making use of the dipole field line equation,

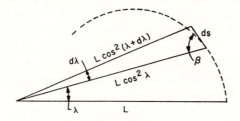

FIGURE 6   Geometry for the determination of the relation $\tan\beta = (\tfrac{1}{2})\cot\lambda$.

$r = L \cos^2 \lambda$. By using trigonometric relations in the differential triangle and by expanding the trigonometric functions of the differential angles in a power series, keeping lowest order non-vanishing terms only, it is found that

$$\tan \beta = \tfrac{1}{2} \cot \lambda \tag{4-49}$$

If now a sphere is imagined centered at 0 in Figure 5, the intersection of the Θ and δ cones and their axes with this sphere establish the vertices of a spherical triangle from which may be obtained the relation

$$\cos (\eta \pi) = \frac{\cos \delta - \cos \Theta \cos \beta}{\sin \Theta \sin \beta} \tag{4-50}$$

Note that the equatorial case corresponds to $\Theta = \beta = 90°$, $\cos \delta = R_e/L$, so that

$$\eta = \frac{1}{\pi} \arccos (R_e/L) \tag{4-51}$$

reproducing the result derived earlier for this case.

The local pitch angle Θ may be eliminated by using the first adiabatic invariant. The function $\cos \delta$ is readily expressed in terms of λ using the dipole field line equation. The variation of B along a dipole field line and the relation between ds and $d\lambda$ are well known. With these changes of variable, and with the help of Eqs. (4-49) and (4-50) the integrand in Eq. (4-48) may be expressed as a function of λ. Since the sense of $d\mathbf{r}$ reverses at the mirror point, the integrand is not an even function of latitude and the integral must be taken from the mirror latitude in one hemisphere to the mirror latitude in the other. Lenchek and Singer (1962), who presented the results of numerically integrating Eq. (4-48) for various L and α_e accidentally overlooked this fact, leading to some error, particularly at high latitudes.

In the analysis of Lenchek and Singer (1962) the presumed departure of the neutron flux from isotropy at high energies is most easily taken into account by using the disc or pancake distribution for the higher energies. If albedo neutrons with zenith angles less than ζ_m are not present in the angular distribution at the top of the atmosphere then there will exist a δ' cone within and coaxial with the δ cone within which no neutron velocity vectors are directed. The situation is illustrated in Figure 7. When the element $d\mathbf{r}$ dips within the δ' cone no proton injection takes place; all injection takes place when $d\mathbf{r}$ is within the δ cone and simultaneously outside the δ' cone. From the Figure it is evident that

$$\sin \delta' = R_e \sin \zeta_m/r \tag{4-52}$$

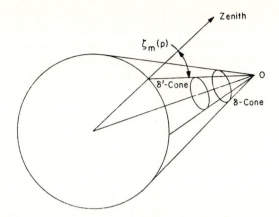

FIGURE 7 Diagram showing how the restriction of neutrons to zenith angles greater than ζ_m creates the δ' cone within which no neutron velocity vectors will be present at point 0. Injection occurs only when the neutron velocity vector is within the conical shell between the δ and δ' cones. After Lenchek and Singer (1962).

The injection coefficient must be reduced by the fraction of the proton orbit in which $d\mathbf{r}$ is within δ'. Thus, the injection coefficient for this case is simply the injection coefficient for the δ cone minus that calculated for the δ' cone. If j_0 is the uniform neutron flux, isotropic for $\zeta \geq \zeta_m$ and zero for $\zeta < \zeta_m$ then

$$\chi = \frac{j_o}{j_n}\left[\bar{\eta}(\delta) - \bar{\eta}(\delta')\right] \tag{4-53}$$

Lenchek and Singer (1962), using the estimate of ζ_m given in Eq. (4-36), made numerical calculations of the difference $[\eta(\delta) - \eta(\delta')]$ as a function both of equatorial pitch angle and energy.

The above examples illustrate the limit of success of mathematical analysis of injection coefficients which has thus far been achieved. There is yet one more feature of the albedo neutron distribution which cannot readily be taken into account by these techniques—the latitude variation. As the extension of $d\mathbf{r}$ sweeps across the earth during a single gyroperiod it cuts a wide arc across various latitudes, and the position of this arc varies with each point along the trapped proton path. The latitude variation of the neutron albedo flux, which is of the order of 10 to 1 between the pole and the equator, is much too large to be ignored in these coefficients.

Accordingly, Dragt et al. (1966) adopted a sampling approach to the calculation of injection coefficients by imagining the launching of many neutrons from the top of the atmosphere and calculating where and with what local pitch angle each neutron decayed.

Suppose that the top surface of the earth's atmosphere is divided into equal increments of area ΔA_k and, at each ΔA_k, into equal increments of zenith angle $\Delta \zeta_i$ and azimuthal angle $\Delta \psi_j$. Let one neutron be launched within each increment. Each neutron event will be given a weight ω_{ijk} designating the contribution of neutrons from that increment to the neutron leakage rate of the earth. Thus,

$$\omega_{ijk} = j_{nijk} \Delta \Omega_{ij} \cdot \Delta \mathbf{A}_k \qquad (4\ 54)$$

$$= j_{nijk} \sin \zeta_i \cos \zeta_i \Delta \zeta_i \Delta \psi_j \Delta A_k \qquad (4\ 55)$$

where j_{nijk} is the actual flux, measured or assumed, which is emitted from ΔA_k within the solid angle determined by $\Delta \zeta_i$ and $\Delta \psi_j$.

Since the increments $\Delta \zeta_i$, $\Delta \psi_j$ and ΔA_k were chosen to be equal, the subscripts on the increments may be dropped. The total number of neutrons launched is

$$N_n = \frac{\pi}{2\Delta \zeta} \cdot \frac{2\pi}{\Delta \psi} \cdot \frac{A}{\Delta A} = \frac{\pi^2 A}{\Delta \zeta \Delta \psi \Delta A} \qquad (4\text{-}56)$$

so that the weight given each neutron becomes

$$\omega_{ijk} = \frac{\pi^2 A}{N_n} j_{nijk} \cos \zeta_i \sin \zeta_i \qquad (4\text{-}57)$$

The weight is zero for neutrons at $\zeta = 0$ because the upward flux is contained within a vanishingly small solid angle. The weight is also zero for horizontal neutrons ($\zeta = 90°$) because this flux is emitted from a vanishingly small area, i.e. ΔA seen edge-on. It is evident that

$$\sum_{ijk} \omega_{ijk}$$

is the total neutron leakage rate of the earth.

Since the probability of neutron decay is the same in each path interval, each neutron launched may be allowed to decay many times at equal arbitrary intervals $\Delta \ell$ of its path. This is equivalent to assigning a probability of decay per unit time of $w/\Delta \ell$ instead of its true probability per unit time of $1/\gamma \tau_n$, a fact which is later taken into account.

When a neutron decays the L value and equatorial pitch angle of the resulting proton are determined, and the weight factor for that neutron is placed in a bin of size $\Delta L \, \Delta \Theta_e$ which contains the point L, Θ_e. The choice of the optimum size of ΔA, $\Delta\zeta$, $\Delta\psi$ and $\Delta\ell$ which will minimize the number of neutron decays required to determine the proton production rate in each bin is not at all obvious. However, it should be a sufficient condition to choose them in such a way that there are many neutron decays in each $\Delta L\,\Delta\Theta_e$ bin. Dragt et al. (1966) launched 16,000 neutrons and typically chose $\Delta\ell = \cdot01\ R$, which proved satisfactory for $\Delta L = 0\cdot1\ R_e$ and $\Delta\Theta_e = 4\cdot5°$. The proton production rate dN_p/dt in each bin is simply the sum of the weights in the bin (designated $\overset{b}{\Sigma}$) multiplied by the ratio $\Delta\ell/w\gamma\tau_n$

$$\frac{dN_p(\Delta L,\, \Delta\Theta)}{dt} = \frac{\Delta l}{w\gamma t_n}\,\frac{\pi^2 A}{N}\,\overset{b}{\sum} j_{nijk}\cos\zeta_i \sin\zeta_i \qquad (4\text{-}58)$$

To obtain the rate of change of proton flux dj/dt it is necessary first to compute the proton production rate dn_p/dt per unit volume per unit solid angle from the bin production rate. Since dj/dt will be uniform throughout the $\Delta L\,\Delta\Theta_e$ bin by Liouville's theorem, n_p will also be uniform, and any convenient subvolume within the dipole shell of thickness ΔL may be used for the calculation. The simplest volume to use is an annular ring extending from L to $L+\Delta L$ and having the thickness Δz. The quantity dn_p/dt is computed by dividing dN_p/dt by the solid angle increment $2\pi \sin\Theta_e\Delta\Theta_e$, dividing by the ring volume $2\pi L\Delta L\Delta z$, and then multiplying the result by the fraction of the time which a proton of equatorial pitch angle Θ_e at L spends within Δz thereby contributing to the density within the ring. This fraction is given by $\Delta z/\cos\Theta_e S$ where S is the spiral distance between mirror points. Thus

$$\frac{dn_p}{dt} = \frac{\Delta l}{w\gamma\tau_n}\,\frac{\pi^2 A}{N}\,\frac{\overset{b}{\sum} j_{nijk}\sin\zeta_i\cos\zeta_i}{2\pi L\Delta L\Delta z 2\pi \sin\Theta_e\Delta\Theta_e}\,\frac{\Delta z}{\cos\Theta_e S} \qquad (4\text{-}59)$$

From Eq. (4-40) the injection coefficient may now be written

$$\chi(L,\,\Theta_e) = \frac{\Delta l}{j_n}\,\frac{A}{2N}\,\frac{\overset{b}{\sum} j_{nijk}\sin\zeta_i\cos\zeta_i}{\sin 2\Theta_e SL\Delta L\Delta\Theta_e} \qquad (4\text{-}60)$$

The method of Dragt et al. (1966) is a very powerful technique since it can include completely arbitrary variations of j_n with direction and location at the top of the atmosphere. Furthermore, its use is not restricted to a strictly dipole field. An accurate field representation can be used to determine L and Θ_e for the protons as they are born if the calculation to be made justifies that much accuracy in the injection coefficients.

3. *Electron production by albedo neutrons*

In contrast to the protons, the electrons from albedo neutron decay acquire their kinetic energy from the mass energy available in the neutron decay rather than from the kinetic energy of the neutron. Consequently most of the electrons originate in the decay of the low energy ($\sim$0·1 ev.) neutrons which constitute the most abundant fraction of the albedo neutron energy spectrum. Distortion of the energy spectrum and angular distribution due to the center of mass motion of the system is not important, and the electrons are, to an excellent approximation, emitted isotropically with an energy distribution identical to that of neutrons decaying at rest.

Estimates of the spatial distribution of the density of decaying neutrons, including those gravitationally trapped, have been made by Hess *et al.* (1961). This density falls rather rapidly with distance from the earth since these low energy neutrons are rapidly depleted by radioactive decay as they move outward. The density is also lower near the equator because of latitude variation in albedo neutron production. Hess *et al.* (1961) estimate the decay density n at $L = 1\cdot5$ on the geomagnetic equator to be about 2×10^{-12} neutron decays cm^{-3}sec^{-1}.

The rate of increase of electron number density may be calculated in a manner similar to that described for protons. Let n_e be the number of electrons of all energies per unit volume per unit solid angle at $d\mathbf{r}$ on the spiral path having velocity vectors in the direction $d\mathbf{r}$. The rate of change of n_e due to the decay of n neutrons per unit volume per unit time is

$$\frac{dn_e}{dt} = \frac{n}{4\pi} \qquad (4\text{-}61)$$

The production rate must be averaged over all elements composing the path to obtain

$$\left.\frac{dn_e}{dt}\right|_{avg} = \frac{1}{2\pi} \oint n \; d\mathbf{r}/\oint d\mathbf{r} \qquad (4\text{-}62)$$

where the integrals are taken between mirror points.

The neutron decay differential electron energy spectrum, normalized to one electron over the whole energy range, is given in electrons per kev by

$$\frac{dn_e}{dE} = [1\cdot2 \times 10^{-3}/m_0c^2][E/m_0c^2(E/m_0c^2+2)]^{\frac{1}{2}}$$

$$[E/m_0c^2+1][1\cdot53-E/m_0c^2]^2 \quad (4\text{-}63)$$

and the rate of change of electron flux j at energy E is given by

$$\frac{dj}{dt} = \frac{dn_e}{dt}\bigg|_{avg} \cdot \frac{dn_e}{dE}w = \frac{dn_e}{dE}\frac{w}{2\pi}\oint n \; dr/\oint dr \qquad (4\text{-}64)$$

C. Solar neutrons

According to Lingenfelter *et al.* (1965a, b) substantial numbers of neutrons should be produced by nuclear reactions occurring among the high-energy particles accelerated in a solar flare. These neutrons will move outward from the sun, eventually reaching the position of the earth's orbit. Some of them will decay and leave a proton in a geomagnetically trapped orbit in the same way as earth albedo neutrons.

The detection of energetic solar neutrons is even more difficult than the detection of energetic albedo neutrons since solar flares occur at unpredictable intervals, and the neutrons are likely to be accompanied by large fluxes of high energy protons ejected from the sun by the solar flare. At any rate, they have not yet been unambiguously detected. Their contribution to the trapped protons of the inner radiation belt therefore remains speculative.

The relations necessary to predict the change in the proton distribution function resulting from solar neutrons have been worked out in detail by Claflin and White (1970). Their technique is similar to the statistical calculation of injection coefficients for albedo neutron decay made by Dragt, *et al.* (1966).

The situation is illustrated in Figure 8. The sun is sufficiently distant that the neutrons constitute a parallel beam whose intensity is designated j_n^u measured in neutrons $cm^{-2}sec^{-1}$ passing through a plane perpendicular to

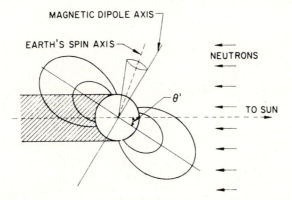

FIGURE 8   Geometry for the injection of solar neutrons. No neutron decays take place in the shadow of the earth.

the earth-sun line. The quantity $j_n^u/2\pi$ neutrons cm^{-2} sec^{-1} ster^{-1} provides a convenient directional intensity for normalization of the injection coefficient (see albedo neutron injection discussion) so that

$$\frac{dj_p}{dt} = \frac{\chi j_n^u}{2\pi\gamma\tau_n} \tag{4-65}$$

The injection coefficient χ as a function of L, Θ and θ' is determined by launching neutrons from the plane surface of Figure 8 and allowing them to decay at equal intervals along their flight paths. When a decay occurs the proton pitch angle is determined, and an event is added to the appropriate $\Delta L\Delta\Theta$ bin. Injection coefficients are calculated from the bin populations as for albedo neutrons.

The injection coefficients depend on the tilt angle θ', and whether the coefficients should be time-averaged over a year (a full time-period for θ') depends on whether injection from a particular solar production event or injection averaged over many events is being considered. Claflin and White present examples of both averaged and unaveraged injection coefficients, one of which is shown in Figure 9.

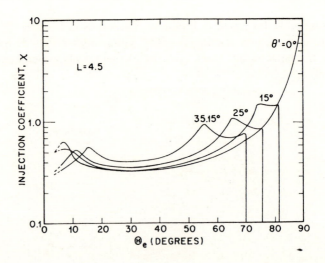

FIGURE 9   Injection coefficients for solar neutrons at $L = 4.5$. The angle θ' is measured between the earth-sun line and the magnetic equatorial plane, and Θ_e is the equatorial pitch angle (see Figure 7). After Claflin and White (1970).

D. Nuclear reactions

Protons whose energies are more than several hundred Mev have a significant probability of an inelastic collision with the nuclei of the atmospheric constituents. Such collisions cause the incident protons to lose a large fraction of their incident energies in a single event, while producing a variety of secondary particles of lesser kinetic energy. The loss of particles from the distribution due to inelastic nuclear collisions is given by

$$\frac{\partial f}{\partial t} = f \sum_i n_i \sigma_i w \qquad (4\text{-}66)$$

where n_i and σ_i are the number of nuclei cm^{-3} and nonelastic cross section for the ith atmospheric constituent, respectively, and w is the proton speed.

As an example, consider protons trapped at the geomagnetic equator at $L = 1\cdot25$. According to the atmospheric model used by Farley and Walt (1971), the number density n_i of the various atmospheric constituents at this location is: H, $4\cdot3 \times 10^3$ atoms cm^{-3}; He, $8\cdot0 \times 10^4$ atoms cm^{-3}, 0, $3\cdot8 \times 10^2$ atoms cm^{-3}, and others negligible. The mean time τ_{col} for a proton to suffer a nonelastic collision can be computed by summing the probabilities for the various constituents. Thus,

$$\frac{1}{\tau_{col}} = \sum_i \sigma_i n_i w \qquad (4\text{-}67)$$

The cross sections σ_i for H, He, and 0 are approximately 22, 150, and 350 mb (Morrison, 1953) respectively, for a 750 Mev proton whose velocity w is $2\cdot5 \times 10^{10}$ cm sec^{-1}, and therefore τ_{col} is approximately 104 years.

An estimate of the importance of nonelastic nuclear collisions at this energy may be made by comparing this result with the time required for a 750 Mev proton to lose one-half its energy by ionization in helium (the major component) at this density. Range-energy curves for helium indicate a path length in helium of 148 gm cm^{-2} will be required for this energy loss, and this will require a time τ_E, given by

$$\tau_E = \frac{148 \text{ gm cm}^{-2}}{n_{He} w m} = 353 \text{ years} \qquad (4\text{-}68)$$

Clearly, nonelastic nuclear collisions must be taken into account in this energy range.

Cross-sections for elastic nuclear scattering are comparable to those for nonelastic scattering at these energies, but this scattering is strongly peaked in the forward direction with little energy loss by the incident proton. It therefore has much less effect on the proton distribution function.

Any complete treatment should also consider the energy and angular distribution of secondary particles from ineleastic nuclear collisions, some of which are protons which are added to the distribution. Detailed calculations on these effects have not been made, primarily because of the paucity of data on trapped protons in the very high energy range, particularly with regard to their angular distribution.

E. Secular variation of geomagnetic field

The existence of the geomagnetic field is an obvious prerequisite for the existence of a radiation belt. The field is known to have important long term variations. According to paleomagnetic evidence the dipole field has undergone many polarity reversals, the last of which took place about 850,000 years ago. The average rate of change of the dipole field over the last 135 years in which detailed observations have been made is about five percent per century, or about 16 gamma per year. Therefore over this period the average rate of change of the dipole moment $\mathcal{M}$ is given by

$$\frac{1}{\mathcal{M}} \frac{d\mathcal{M}}{dt} = a = -5 \times 10^{-4} \, \text{yr}^{-1} \qquad (4\text{-}69)$$

From this average a characteristic time of 2000 years may be inferred. Since the geomagnetic field not only provides the magnetic trap for charged particles but also exerts major control over such important source and transport mechanisms as the supply of albedo neutrons and the magnitude of the diffusion coefficients, gross changes in the radiation belt certainly take place on a 2000 year time scale. The very nearly complete absence of detailed information on the field and its variations over this time period precludes any accurate reconstruction of the history of the radiation belt. However the trapping times of some of the higher-energy protons appear to be of the order of 200 years or more, and they have been affected by the field changes which have taken place. These effects must be taken into account in order to reconcile the present belt with any theoretical model of its existence.

The long-term decrease of the dipole field gives rise to an azimuthal electric field which simultaneously accelerates the trapped protons to higher energies, and convects them inward toward the earth. These effects were independently recognized by Schulz and Paulikas (1972) and by Heckman and Lindstrom (1972). The effects were incorporated quantitatively into a model of the proton belt by Farley, et al. (1972), from which the following description of the effects has been adapated.

It is the order of magnitude separation between the 200 year proton trapping time and the 2000 year characteristic time for the field change which permits the model to be treated as a quasi-stationary approximation to the actual time-dependent radiation belt. The continuous transport and acceleration of the protons by the secular field variation are included, while the variations of those quantities expected to change on a 2000 year time scale—boundary conditions, diffusion coefficients, and albedo neutron source—are ignored.

The secular field changes are sufficiently slow that the three adiabatic invariants μ, J, Φ of the trapped protons are conserved in the process. If $f_1 d\mu dJ d\Phi$ is the number of particles in the volume $d\mu dJ d\Phi$, then the distribution function f_1 is time independent under the secular variation of the field

$$\frac{\partial f_1(\mu, J, \Phi)}{\partial t} = 0 \qquad (4\text{-}70)$$

While the equation is admirably simple when expressed in these coordinates, it will not be so when other source, transport, and loss terms are added. Consider, for example, the energy loss due to interaction between the protons and atmospheric constituents. These losses depend fundamentally on the proton's L shell (which determines what average atmospheric density it will encounter) and its energy or momentum (which determines its rate of ionization energy loss). The coordinates μ, J, Φ do not have a time-independent relation to L and p since by definition

$$L = \frac{2\pi \mathscr{M}}{\Phi} = \frac{2\pi \mathscr{M}_0 e^{at}}{\Phi} = L_0 e^{at} \qquad (4\text{-}71)$$

where $\mathscr{M}_0$ is the dipole moment of the earth at time $t = 0$. Also

$$\mu = \frac{p_\perp^2}{2mB(L, t)} = \frac{p_\perp^2}{2me^{at}B(L, t = 0)} \qquad (4\text{-}72)$$

The averaged atmospheric density and ionization energy loss rate will thus have explicit time dependence in μ, J, Φ coordinates and no time-stationary solutions can exist in these coordinates. A similar problem exists with respect to the albedo neutron source term since it will not be time-independent in μ, J, Φ coordinates. If, however, a time-dependent coordinate transformation of Eq. (4-70) can be made to coordinates in which the source, transport, and loss terms are time-independent, then time-stationary solutions can be sought. Suitable new coordinates are L, defined by Eq. (4-71) and μ' where

$$\bar{\mu}' = \frac{p_\perp^2}{2mB(L,\, t = 0)} = \mu e^{at} \tag{4-73}$$

The transformation from momentum and position coordinates to J is not time dependent because

$$\left.\frac{dJ}{dt}\right|_{\text{constant } \mathbf{p,\, r}} = \frac{d}{dt}\left\{2p\oint\left[1 - \frac{B(s)e^{at}}{B_m e^{at}}\right]^{\frac{1}{2}}ds\right\} = 0 \tag{4-74}$$

with the integral taken between mirror points. Thus J need not be replaced. The Jacobian for the coordinate transformation is $2\pi\mathcal{M}_0/L^2$ so that

$$f_1(\mu,\, J,\, \Phi;\, t) = (L^2/2\pi\mathcal{M}_0)f_2(\mu',\, J,\, L,\, t) \tag{4-75}$$

where $f_2(\mu',\, J, L; t)\, d\mu'dJdL$ is the number of particles in the volume $d\mu'dJdL$. Equation (4-63), expressed in the new coordinates, becomes

$$\frac{\partial f_2(\mu',\, J,\, L;\, t)}{\partial t} = -\frac{\partial}{\partial L}\left(\frac{dL}{dt}f_2\right) - \frac{\partial}{\partial\mu'}\left(\frac{d\mu'}{dt}f_2\right)$$

$$= -\frac{\partial}{\partial L}(aLf_2) - \frac{\partial}{\partial\mu'}(a\mu'f_2) \tag{4-76}$$

The appearance of the two new terms has a familiar analogy in the study of gravitational forces. When Newton's law, formulated in an inertial system, is transformed by a time-dependent coordinate transformation to the rotating frame of the earth, two new forces appear. They are called the centrifugal force and the Coriolis force. The new terms which appear above are not forces in this case, but new transport terms.

The first term is the divergence of a particle current directed toward smaller L. The individual particles have a velocity aL. This is the drift velocity or convection rate caused by the azimuthal electric field which is produced, according to Maxwell's Laws, by the changing magnetic field.

The second term is the divergence of a particle current directed toward smaller μ'. Since μ' is an unfamiliar coordinate, it is useful to compute the rate of change of momentum of the protons:

$$\frac{dp_\perp^2}{dt} = \frac{d}{dt}(\mu' 2mB(L, t = 0)) = \frac{2m\mathcal{M}_0}{L^3}\frac{d\mu'}{dt} + 2m\mu'\frac{d}{dt}\frac{\mathcal{M}_0}{L^3}$$

$$= \frac{2m\mathcal{M}_0}{L^3}a\mu' + 2m\mu'\mathcal{M}_0\left(\frac{-3}{L^4}\right)aL$$

$$= ap_\perp^2 - 3ap_\perp^2 = -2ap_\perp^2 \tag{4-77}$$

Since a is a negative quantity, the fractional change of $p_\perp^2$ per unit time is positive and is equal to twice the fractional change per unit time of the dipole moment, or about ten percent per century. For non-relativistic particles $p_\perp^2$ is proportional to the perpendicular kinetic energy. This change in kinetic energy is due to the work done on the particle by the azimuthal electric field. There is no change in $p_\parallel$.

Farley, *et al.* (1972) noted that if t in Eq. (4-73) is chosen to be zero at present, then μ' and μ are identical now but diverge from each other with the 2000 year characteristic time of the secular variation. Consequently, they approximated Eq. (4-76) by substituting μ for μ', obtaining finally

$$\frac{\partial f_2(\mu, J, L; t)}{\partial t} = \frac{\partial}{\partial L}(aLf_2) - \frac{\partial}{\partial \mu}(a\mu f_2) \tag{4-78}$$

Similar reasoning may be applied to the selection of a boundary condition necessary for a time-stationary solution of Eq. (4-78). If it can be assumed that $f_1(\mu', J, \Phi; t)$ is constant in time for some value of Φ then it may be assumed that $f_2(\mu, J, L; t)$ is also constant at the corresponding value of L since f_1 and f_2 diverge from each other with a characteristic time of 2000 years.

Farley, *et al.* (1972) argued that changes in the non-dipole portion of the earth's field can be ignored even though they are not small. Such changes serve only to produce continuous deformations of the L shells, and the associated electric fields must serve only to transport and accelerate the particles in space in such a way that they remain on the same L shell, preserving all of the adiabatic invariants.

F. Radial diffusion

If the magnetospheric electric and magnetic fields are perturbed during a time interval which is long compared to a particle bounce period but shorter than or comparable to the longitudinal drift period, the third adiabatic invariant of a trapped particle will be changed. Since the bounce periods of energetic particles are on the order of seconds while the drift periods are some 10's of minutes, perturbations over a wide range of frequencies can produce changes in the third invariant. Examples of the types of disturbances which will affect the third invariant are sudden commencements, sudden impulses, and the expansion phase of magnetic substorms.

If the first and second adiabatic invariants are conserved and the third varies, the particle orbit will move across L shells. However, the constancy of the first and second invariant will constrain the variation of the particle mirror point and energy. The motion of the mirror point of a trapped particle during changes in the third invariant is given in Figure 10, the

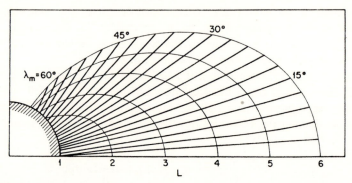

FIGURE 10   Drift paths of the mirroring points of trapped particles diffusing in L with constant first and second adiabatic invariants. Narrow lines are magnetic field lines for several L values.

heavy lines showing the position of the mirror point as a function of L. Note that particles which initially mirror at the equator remain on the equatorial plane ($J = 0$). The mirroring points of off-equatorial particles move to slightly lower latitudes as the particle is transported to lower L values, although this change in mirroring latitude is not large. The constancy of the first invariant fixes the particle energy at each L shell, the energy increas- as the particle moves inward.

The physical basis for third invariant diffusion can be illustrated by considering the effects of an idealized magnetic perturbation on equatorially trapped particles whose guiding centers are initially located on a narrow ring. The sequence of Figures 11a, b, and c indicates the evolution of the radial distribution when subjected to a rapid, asymmetric magnetic field compression followed by a slow relaxation of the field to the original configuration. The field magnitude is shown schematically below the illustrations which give the particle distribution at three stages of the magnetic perturbation. The motion of the trapped particles under a magnetic perturbation depends on both the magnetic field and on the induced electric field. The

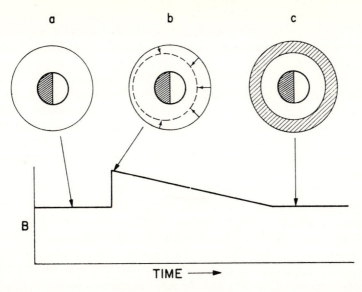

FIGURE 11   Response of a shell of trapped particles to a sudden commencement type magnetic disturbance. In the initial compression particles move along arrows to the dashed ring in *b*. After the recovery period the particles are located in the shaded band of figure *c*.

induced electric field in turn depends on the conducting properties of the medium, i.e., the boundary conditions. If one assumes that the cold plasma in the magnetosphere is a good electrical conductor in the direction parallel to the magnetic field lines and that the earth is enclosed by a perfectly conducting surface which fixes the feet of the field lines, then the induced electric field is completely specified. Birmingham and Jones (1968) have shown that when such conditions apply, the field lines can be considered to move with a velocity related to the induced electric field by

$$\mathbf{v}_\perp = (\mathbf{E} \times \mathbf{B})/B^2 \tag{4-79}$$

Field line motion has also been discussed by Stern (1966) and Vasyliunas (1972). The instantaneous position of an individual magnetic field line is obtained by tracing its position through space beginning with the fixed location at the conducting surface. The change in its position with time gives the field line velocity from which the induced electric field can be computed by Eq. (4-79). The velocity $\mathbf{v}_\perp = (\mathbf{E} \times \mathbf{B}/B^2$ is also the velocity of cold plasma motion, and thus this method of defining field line motion produces correspondence between field line motion and cold plasma motion.

This correspondence is called the frozen field condition. Energetic trapped particles will also experience this electric field drift although it will be superimposed on the normal drifts caused by curvature and gradient of the geomagnetic field.

The 'frozen field' assumption is certainly open to question, particularly in the outer zone where the plasma density can be very low. Even in the inner zone, the insulating atmosphere at low altitude may uncouple the magnetospheric field from the interior field and result in a different induced electric field from that determined here. However, in view of current limited knowledge, the frozen field assumption is probably the most reasonable way to proceed.

In the example of Figure 11 the rapid asymmetric compression carries the trapped particles and the cold plasma inward until at the end of the compression, the particles occupy positions shown by the dashed line in Figure 11b. The third invariant is not conserved during this compression although the first and second invariants remain constant. Since the compression is assumed to be stronger on the right hand (sunward) side, the particles on that side are moved further into the magnetosphere and accelerated more than those on the left (night) side. During the slow restoration of the field, a process which takes place over many drift periods, the particles maintain all three invariants, and when the field is restored will occupy the shaded band in Figure 11c. An individual particle may be moved inward or outward depending on its longitude at the time of the rapid compression. Note that the compression introduces a longitudinal asymmetry in the particle distribution. However, because particles with slightly different energies, mirroring points, or L values will have different drift velocities, these azimuthal inhomogeneities will gradually disappear after many drift periods. This homogenizing is needed to allow the phase averaging referred to in Section III.

While the magnetic disturbances produced by sudden commencements have a time history similar to the above model, other types of perturbations will have different time variations. However, from the above example it can be recognized that in order to cause radial diffusion the disturbance must have two characteristics. These characteristics are a) some part of the variation must be rapid enough to alter the third adiabatic invariant and b) the disturbance must have an azimuthal asymmetry.

Electric potential fields can also lead to radial diffusion of trapped particles, and the recognition that large scale convection and therefore large scale electric fields are present in the magnetosphere makes this mechanism a likely candidate for driving radial motion. Diffusion by electric potential fields is similar to the magnetic case in that fluctuations on a time scale shorter than or comparable to the drift period are required, and an azimuthal asymmetry must be present.

The process of radial diffusion is of great importance to radiation belt physics since it is a mechanism for transporting particles across magnetic shells. In this way various regions of the magnetosphere can communicate with each other, transferring particles either inward toward the earth or outward depending on the location of sources and sinks of trapped particles and on the character of the diffusing forces. The influence of radial diffusion on a distribution of trapped particles can be calculated using the formalism developed in Section III. In that section the Fokker–Planck equation was reduced to a diffusion form, which for diffusion in the third invariant only becomes

$$\frac{\partial f(\mu, J, \Phi)}{\partial t} = \frac{\partial}{\partial \Phi}\left[D_{\Phi\Phi}\frac{\partial f(\mu, J, \Phi)}{\partial \Phi} \right] \qquad (4\text{-}80)$$

For most purposes Φ, the third adiabatic invariant, is an inconvenient quantity with which to work, and results are more easily visualized using L as the radial coordinate. The transformation of the independent variables of Eq. (4-80) to μ, J, L is accomplished using Eq. (3-25), and since $\mathscr{J} \propto L^{-2}$

$$\frac{\partial f(\mu, J, L)}{\partial t} = \frac{\partial}{\partial L}\left[\frac{D_{LL}}{L^2}\frac{\partial}{\partial L}(L^2 f) \right] \qquad (4\text{-}81)$$

where $D_{LL} = \langle(\Delta L)^2\rangle/2$. The methods of computing D_{LL} for various field perturbations will be described in the remainder of this section. The calculations will be done assuming the unperturbed field is a dipole, since only rarely are the higher order terms in the static geomagnetic field important. For the dipole case the radial coordinate $r = LR_e$ when r is in the equatorial plane, and the determination of $\langle(\Delta L)^2\rangle$ is equivalent to finding $\langle(\Delta r)^2\rangle$.

1. *Radial diffusion by magnetic field fluctuations*

The radial diffusion caused by large scale fluctuations of the geomagnetic field has been considered by Davis and Chang (1962), Nakada and Mead (1965), Tverskoy (1965), Fälthammar (1968) and others. The general approach followed by these authors is to construct an idealized model of the disturbance field and to calculate the radial displacement of a trapped particle experiencing such a disturbance. By averaging the square of the displacement over many such disturbances occurring at random times, the diffusion coefficient as a function of the statistical properties of the disturbance is obtained.

In this section the method of calculating the radial diffusion coefficient for magnetic fluctuations will be illustrated. The calculation will be done in detail for the relatively simple case of equatorially trapped particles being diffused by magnetic field perturbations which are symmetric above and below the equatorial plane. These restrictions greatly simplify the derivation while retaining the major physical and mathematical arguments. The more general case of off-equator trapping will be described briefly and the results presented.

The drift velocity of the guiding center of an equatorially trapped particle is perpendicular to the local magnetic field and is given by

$$\mathbf{v}_\perp = \frac{-\mathbf{B}}{eB^2} \times (e\mathbf{E} - \mu \nabla B) \tag{4-82}$$

where $\mathbf{E}$ and $\mathbf{B}$ are the electric and magnetic fields and μ and e are the magnetic moment and electric charge of the particle. Even in the absence of electric potential fields it is necessary to include the electric field term in Eq. (4-82) since a changing magnetic field will be accompanied by an induced electric field. The magnetic field will be composed of the normal dipole field $\mathbf{B}_0$ and a disturbance field $\mathbf{b}(t)$.

$$\mathbf{B} = \mathbf{B}_0 + \mathbf{b}(t) \tag{4-83}$$

where it will be assumed that $|\mathbf{b}| \ll |\mathbf{B}_0|$ at all positions.

In the following calculations the magnetic disturbances are presumed to result from changes in the solar wind characteristics. These changes alter the flow of current on the magnetospheric boundary and thereby produce magnetic field alterations within the cavity. If these currents are on the boundary, the field perturbations within the cavity will be equal to the negative gradient of a potential function and can be expressed by a spherical harmonic expansion

$$\mathbf{b} = -\nabla V = -\nabla R_e \sum_{n=1}^{\infty} \left(\frac{r}{R_e}\right)^n \sum_{m=0}^{n} (g_n^m \cos m\phi + h_n^m \sin m\phi) P_n^m(\cos \theta) \tag{4-84}$$

where r, θ and φ are spherical polar coordinates.

The P_n^m are the Schmidt normalized associated Legendre functions and R_e is the radius of the earth. For small perturbations near the earth the leading terms are the most important so that only $n \leq 2$ will be needed. Additional simplification results from setting the solar wind perpendicular to the dipole axis and labeling the meridian containing the sun $\varphi = 0$. With this symmetry, $h_n^m = 0$ for all n, m and $g_n^m = 0$ when $n + m$ is even. The only surviving terms are g_1^0 and g_2^1. Hence the time dependent disturbance field is

$$\mathbf{b}(t) = -\nabla V = -\nabla\left[g_1^0 r \cos\theta + \sqrt{3}\frac{r^2}{R_e}g_2^1 \sin 2\theta \cos\phi \right]$$

$$= [-S(t)\cos\theta - A(t)r\sin 2\theta \cos\phi]\hat{\mathbf{e}}_r$$

$$+ [S(t)\sin\theta - A(t)r\cos 2\theta \cos\phi]\hat{\mathbf{e}}_\theta$$

$$+ A(t)r\cos\theta \sin\phi\,\hat{\mathbf{e}}_\phi \qquad (4\text{-}85)$$

where $\hat{\mathbf{e}}_r$, $\hat{\mathbf{e}}_\theta$ and $\hat{\mathbf{e}}_\phi$ are unit vectors in the r, θ and φ directions. The characteristics of the field changes are described by the time dependent coefficients $S(t) = g_0^1$, and $A(t) = 2\sqrt{3}\,g_2^1/R_e$. The notation $S(t)$ and $A(t)$ indicates which terms are symmetric and which are asymmetric with respect to longitude. The values of $A(t)$ and $S(t)$ must be obtained from a model of the geomagnetic field distortion produced by a change in solar wind characteristics. Since it is possible to carry through the calculation with arbitrary $A(t)$ and $S(t)$, these functions will be retained so that the result will have greater generality.

The induction electric field to be used in Eq. (4-82) cannot be derived from the magnetic field variations, since according to Maxwell's equations the field variations determine only curl E. Boundary conditions must be specified to obtain E, and these depend on physical conditions of the medium in which the magnetic field changes occur. As discussed above it has generally been assumed that enough low energy or thermal plasma is available to cause the electrical conductivity parallel to the field lines to be infinite. It is also assumed that the field lines extend downward to a perfectly conducting ionosphere or spherical shell enclosing the dipole. Under these conditions the electrical field $E(t)$ can be obtained from the cold plasma velocity $\mathbf{v}_p$ by the relation

$$\mathbf{E} = -\mathbf{v}_p \times \mathbf{B} \qquad (4\text{-}86)$$

The instantaneous values of $\mathbf{v}_p$ are obtained by tracing the equatorial crossing point of a field line whose feet are fixed at some lower position, which for convenience is taken near the origin.

The equations in polar coordinates for the disturbed magnetic field lines are given in differential form by

$$\frac{dr}{B_r} = \frac{r d\theta}{B_\theta}$$

$$\frac{r\sin\theta\,d\phi}{B_\phi} = \frac{r d\theta}{B_\theta} \qquad (4\text{-}87)$$

with $\mathbf{B} = \mathbf{B}_0 + \mathbf{b}$.

The Eqs. (4-87) are to be integrated from the origin to the equatorial plane. The integration cannot be carried out directly since B_r, B_θ and B_ϕ are complicated functions of the integration variables. However, by replacing r and φ in B_r, B_θ and B_ϕ by their unperturbed dipole values

$$r = R_0 \sin^2 \theta$$

$$\varphi = \varphi_0 \qquad\qquad (4\text{-}88)$$

the integration can be performed. The integration is equivalent to tracing a field line from the origin to the equator, thus obtaining the radius and longitude of the equatorial crossing. The approximation (Eq. 4-88) replaces the field $\mathbf{B}_0 + \mathbf{b}$ at each position along the integration path by the value of the distorted field at the position of the equivalent undistorted field line. For $|\mathbf{b}| \ll |\mathbf{B}_0|$ the error introduced by this approximation is small. With the assumption that $S(t)/B \ll 1$ and $A(t)r/B \ll 1$ the resulting two equations can then be integrated from $\theta = 0$ to $\pi/2$ to give r and φ in terms of the perturbation quantities $A(t)$ and $S(t)$ and the constants R_0 and φ_0. These constants represent the equatorial crossing radius and longitude of the undistorted field line. For a given R_0, φ_0 the equatorial crossing position (r, φ) of a distorted field line can be determined as a function of S and A. Changes in r and φ as a function of $A(t)$ and $S(t)$ can then be interpreted as a motion of the field line and its associated plasma, and Eq. (4-86) can be used to evaluate the induced electric field at that position. In the equatorial plane the electric field thus obtained is

$$\mathbf{E} = \frac{1}{7}r^2\frac{dA}{dt} \sin \phi \, \hat{\mathbf{e}}_r + r\left(\frac{1}{2}\frac{dS}{dt} + \frac{8}{21}r\frac{dA}{dt} \cos \phi\right)\hat{\mathbf{e}}_\phi \qquad (4\text{-}89)$$

Note that in order to calculate the equatorial values of the induced electric field it is necessary to know the position of the disturbed magnetic field line along its entire length.

With the magnetic and electric field symmetry assumed ($b_r = b_\phi = E_\theta = 0$ in the equatorial plane) the radial component of the drift velocity of equatorial particles given by Eq. (4-82) reduces to

$$\frac{dr}{dt} = \left(-\frac{E_\phi}{B} + \frac{\mu}{eBr}\frac{\partial b_\theta}{\partial \phi}\right) \qquad (4\text{-}90)$$

where B can be approximated by the dipole value B_0.

After substituting Eqs. (4-85) and (4-89) for the field values, equation (4-90) gives

$$\frac{dr}{dt} = -r\left(\frac{1}{2B_0}\frac{dS}{dt} + \frac{8}{21}\frac{r}{B_0}\frac{dA}{dt}\cos\phi\right) - \frac{\mu}{eB_0}(A\sin\phi) \qquad (4\text{-}91)$$

The time dependence on the right hand side is contained in the disturbance parameters $A(t)$ and $S(t)$ and the particle coordinates r and φ. The usual policy of perturbation theory is to set the particle positions equal to their unperturbed values. Hence, on the right hand side, $r = r_0$, and $\varphi = \Omega_0 t + \eta$ where Ω_0 is the angular drift velocity of the particle and η is a phase constant giving the longitude at $t = 0$. The magnetic moment μ in equation (4-91) can be replaced by its value in terms of the unperturbed angular drift velocity

$$\mu = \frac{\Omega_0 r_0^2 e}{3}.$$

Equation (4-91) can then be integrated (the second term is integrated by parts) to give

$$\int_{r_0}^{r(t)} dr = r(t) - r_0 = -\frac{5}{7}\frac{r_0^2\Omega_0}{B_0}\int_0^t A(\xi)\sin(\Omega_0\xi + \eta)d\xi - \frac{r_0}{2B_0}[S(t) - S(0)]$$

$$-\frac{8}{21}\frac{r_0^2}{B_0}\{A(t)\cos(\Omega_0 t - \eta) - A(0)\cos\eta\} \qquad (4\text{-}92)$$

With the exception of the integral, all terms on the right of Eq. (4-92) are bounded and are of order b/B_0. Hence, the displacements represented by these terms will not grow beyond this order regardless of how long a time interval is chosen or how the time variation proceeds. On the other hand the integral term can grow without limit for suitable variations of $A(\xi)$. It is therefore apparent that the integral term will be the only important one in describing the radial displacements of the trapped particles.

This result demonstrates rigorously the conclusion mentioned earlier that only the asymmetric part of the perturbation $A(t)$ is effective in producing radial diffusion. Unfortunately, the symmetric part of the disturbance field is usually much larger than the asymmetric part, and the experimental determination of $A(t)$ cannot be done with high precision. A second difficulty in determining $r(t) - r_0$ from magnetometer measurements arises in the estimation of $\mathbf{E}(t)$. This difficulty was alluded to earlier in discussing the *frozen field* assumption. One should note that in passing from Eq. (4-90) to

Eq. (4-92), the electric field term and magnetic field term contribute almost equally to the value of $r(t)-r_0$. Hence, the assumptions leading to the electric field values may be quite important to the final result.

The calculation of $\langle (\Delta r)^2 \rangle$ from Eq. (4-92) requires careful consideration of the type of averaging which is needed. Note that Eq. (4-92) is a deterministic equation which gives the radial coordinate $r(t)$ of a particle of given initial r_0 and longitude after it has experienced a specified magnetic perturbation. The Fokker–Planck coefficient on the other hand is the average value of the square of the particle displacement per unit time, the averages involving the initial longitude of the particle and a statistically complete sample of the magnetic perturbations. The resulting expression for $\langle (\Delta r)^2 \rangle$ will therefore involve statistical properties of the field variation $A(\xi)$ rather than the detailed time variation. After eliminating terms of order b/B_0 Eq. (4-92) becomes

$$[r(t)-r_0] = -\frac{5}{7}\frac{r_0^2\Omega_0}{B_0}\int_0^t A(\xi)\sin(\Omega_0\xi+\eta)d\xi \tag{4-93}$$

The square of the displacement can be expressed as

$$[r(t)-r_0]^2 = \left(\frac{5}{7}\right)^2\left(\frac{r_0^2\Omega_0}{B_0}\right)^2\int_0^t d\xi' A(\xi')\sin(\Omega_0\xi'+\eta)\int_0^t d\xi'' A(\xi'')\sin(\Omega_0\xi''+\eta)$$

$$= \left(\frac{5}{7}\right)^2\left(\frac{r_0^2\Omega_0}{B_0}\right)^2\int_0^t d\xi'\int_0^t d\xi'' A(\xi')A(\xi'')\sin(\Omega_0\xi'+\eta)\sin(\Omega_0\xi''+\eta) \tag{4-94}$$

Let $\zeta = \xi''-\xi'$ replace ξ'' as the inner variable of integration and expand the $\sin(\Omega_0\xi'+\eta+\Omega_0\zeta)$ to give

$$[r(t)-r_0]^2 = \left(\frac{5}{7}\right)^2\left(\frac{r_0^2\Omega_0}{B_0}\right)^2\int_0^t d\xi'\int_{-\xi'}^{t-\xi'} d\zeta A(\xi')A(\xi'+\zeta)\sin(\Omega_0\xi'+\eta).$$

$$\{\sin(\Omega_0\xi'+\eta)\cos\Omega_0\zeta+\cos(\Omega_0\xi'+\eta)\sin\Omega_0\zeta\} \tag{4-95}$$

The averaging over initial longitude η can be performed immediately; the average of $\sin^2(\Omega_0\xi'+\eta)$ is $\frac{1}{2}$ and the average of $\sin(\Omega_0\xi'+\eta)\cos(\Omega_0\xi'+\eta)$ vanishes. The remaining factors of Eq. (4-95) must be averaged over the possible perturbations. As discussed previously the time interval t must be much longer than the particle drift period and include a sufficiently large sample of the perturbations. Specifically this statement means that t is taken to be much greater than the correlation length of $A(\xi)$. Denoting the averages by brackets $\langle \rangle$, the Fokker–Planck coefficient is given by

$$\langle (\Delta r)^2 \rangle = \frac{\langle [r(t) - r_0]^2 \rangle}{t}$$

$$= \frac{1}{2}\left(\frac{5}{7}\right)^2 \left(\frac{r_0^2 \Omega_0}{B_0}\right)^2 \frac{1}{t} \int_0^t d\xi' \int_{-\xi'}^{t-\xi'} d\zeta \langle A(\xi')A(\xi'+\zeta) \rangle \cos \Omega_0 \zeta \quad (4\text{-}96)$$

where $\langle A(\xi')A(\xi'+\zeta) \rangle$ is the autocorrelation function of $A(\xi')$. If $A(\xi')$ is a stochastically stationary function, the autocorrelation function will depend only on ζ, the time difference in the arguments of the two values of A appearing in the integrand. For $\zeta = 0$, the function will be a maximum $\langle A^2(\xi') \rangle$, and for ζ much larger than the correlation length $\langle A(\xi')A(\xi'+\zeta) \rangle = 0$ since $A(\xi'+\zeta)$ is just as likely to have the same as the opposite sign of $A(\xi')$. By interchanging the order of integration in Eq. (4-96), noting that $\langle A(\xi')A(\xi'+\zeta) \rangle = \langle A(\xi')A(\xi'-\zeta) \rangle$, and recognizing that the correlation length is much less than t, Eq. (4-96) reduces to

$$\langle (\Delta r)^2 \rangle = \left(\frac{5}{7}\right)^2 \left(\frac{r_0^2 \Omega_0}{B_0}\right)^2 \frac{1}{4} P_A(\Omega_0) \quad (4\text{-}97)$$

where

$$P_A(\Omega_0) = 4 \int_0^\infty d\zeta \langle A(\xi')A(\xi'+\zeta) \rangle \cos \Omega_0 \zeta$$

is the power spectrum of $A(\xi)$ at angular frequency Ω_0. Since B_0^2 varies as r_0^{-6}, the value of $\langle (\Delta r)^2 \rangle$ varies as r_0^{10}.

For off-equatorial particles the diffusion coefficient is obtained in a similar manner although the derivation is considerably more complex. Equation (4-82) is replaced by the more complete guiding center drift velocity expression containing the curvature drift term

$$\mathbf{v}_\perp = \frac{\mathbf{B}}{eB^2} \times \left\{ e\mathbf{E} - \mu\left(1 + \frac{2w_\parallel^2}{w_\perp^2}\right)\nabla B \right\} \quad (4\text{-}98)$$

where $w_\parallel$ and $w_\perp$ are the parallel and perpendicular velocity components of the particle. Since $v_\parallel = w_\parallel$ the parallel velocity $w_\parallel$ is obtained from the equation of motion of the guiding center along the field line

$$\frac{d}{dt}(mv_\parallel) - \mu\nabla_\parallel B = 0 \quad (4\text{-}99)$$

In calculating $w_\parallel$ from Eq. (4-99) the dipole field is used for B. This approximation is consistent with the overall perturbation approach in which the zero order motion is that of a particle in a dipole field and there is no radial component to the drift velocity. The total field value $\mathbf{B} = \mathbf{B}_0 + \mathbf{b}$ is then substituted into Eq. (4-98) and the component of $\mathbf{v}_\perp$ which lies in the meridional plane is selected. This velocity is then projected back to the equatorial plane to give the rate of change in L value which is produced by the meridional component of $\mathbf{v}_\perp$ at an arbitrary latitude. As was the case for equatorial particles, the only components of $\mathbf{E}$ and $\mathbf{B}$ which lead to radial motion are E_ϕ and $\partial/\partial\varphi(\mathbf{b}\cdot\hat{\mathbf{B}}_0)$ where $\hat{\mathbf{B}}_0$ is a unit vector in the direction of the dipole field. The resulting equation gives the instantaneous value of dL/dt as a function of L, θ, and the perturbation fields. Because only the radial motion averaged over a complete bounce is of interest here, the equation for dL/dt is averaged over latitude, weighting each latitude interval by a factor proportional to the time interval that the guiding center spends in that interval.

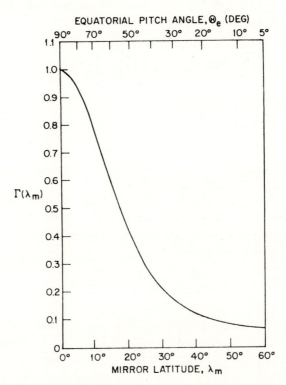

FIGURE 12   Latitude dependent factor of the radial diffusion coefficient for magnetic fluctuations. Note the more rapid diffusion of equatorially trapped particles.

In the final result, one obtains for $\langle (\Delta L)^2 \rangle = \langle (\Delta r)^2 \rangle / R_e^2$ the expression in Eq. (4-97) multiplied by a correction factor $\Gamma(\lambda_m)$ where λ_m is the mirroring latitude. The value of $\Gamma(\lambda_m)$ is given in Figure 12, and it is apparent that particles mirroring in the equatorial plane diffuse much more rapidly than those with small equatorial pitch angles.

The diffusion coefficient can be expressed in more familiar terms by letting $r_0 = R_e L$, $\Omega_0 = 2\pi v_{\mathrm{drift}}$, and $B_0 = \mathcal{M}/R_e^3 L^3$, where v_{drift} is the azimuthal drift frequency of the trapped particle, and $\mathcal{M}$ is the dipole moment of the earth. In these terms the diffusion coefficient for radial motion produced by magnetic fluctuations with power spectrum $P_A(v)$ becomes

$$\tfrac{1}{2}\langle (\Delta L)^2 \rangle = D_M(L, \lambda_m, v_{\mathrm{drift}}) = \Gamma(\lambda_m)\frac{\pi^2}{2}\left(\frac{5}{7}\right)^2\frac{R_e^8}{\mathcal{M}^2}L^{10}[v^2 P_A(v)]_{v=v\mathrm{drift}} \qquad (4\text{-}100)$$

The coefficient is explicitly a function of the L value, the drift frequency v_{drift}, and the mirroring latitude. However, since the drift frequency depends on the magnetic moment as well as the L value, the last factor can introduce an energy dependence as well as an additional variation with L. The non-relativistic drift frequency of a particle varies as μ/L^2. Therefore, the overall L dependence of D_M is influenced by the v dependence of $P_A(v)$. In the special case where $P_A(v)$ is proportional to v^{-2}, which is characteristic of sudden impulse or sudden commencement type disturbances, the v dependence in Eq. (4-100) vanishes. In that special case particles of all energies diffuse at the same rate and D_M varies as L^{10}. For other forms of $P_A(v)$, D_M will be energy dependent and the L^{10} variation will be modified. For a power spectrum $P_A(v)$ which varies as v^{-n}, D_M varies as $L^{6+2n}\mu^{2-n}$.

2. Radial diffusion in electric potential fields

The rate of radial diffusion induced by electric potential fields is obtained in a manner similar to that for magnetic perturbation. However, the origin and nature of the electric potential fields are sufficiently uncertain that a more general representation of the driving fields is desirable.

For equatorial particles and for an applied electric field perpendicular to $\mathbf{B}_0$, the radial drift velocity is obtained from Eq. (4-90)

$$\frac{dr}{dt} = -\frac{E_\phi}{B_0} \qquad (4\text{-}101)$$

It is assumed that the time variations are stochastic in the same sense as the magnetic perturbations of the previous section. Only the ϕ component is involved in Eq. (4-101) and the azimuthal variation of E_ϕ can be expressed generally as a Fourier expansion in ϕ.

$$E_\phi(r_0, \phi, t) = \sum_{n=1}^{N} E_{\phi n}(r_0, t) \cos [n\phi + \gamma_n(r_0, t)] \qquad (4\text{-}102)$$

The upper index N is set by the fineness of the electric field scale to be considered. As a further simplification the phase constants γ_n will be assumed independent of t. This qualification fixes the azimuthal position of the nodes for each field coefficient $E_{\phi n}(r_0, t)$. Such an assumption is reasonable if the solar wind, which has a fixed azimuthal direction with respect to the magnetosphere, is the ultimate source of the electric field variations.

Over a time interval t the displacement of a particle whose initial coordinates are $r = r_0$, $\phi = \eta$ is obtained by integrating Eq. (4-101) to give in analogy to Eq. (4-93)

$$r(t) - r_0 = -\frac{1}{B_0} \int_0^t \sum_{n=1}^{N} E_{\phi n}(r_0, \xi) \sin [n\Omega_0\xi + n\eta + \gamma_n(r_0)]d\xi \qquad (4\text{-}103)$$

where use has been made of the equation for unperturbed motion $\phi = \Omega_0 t + \eta$.

By using Eq. (4-103) the expression for $\langle (\Delta r)^2 \rangle$ is obtained in the same manner as for the magnetic perturbations. Since the algebra is tedious and is given by Fälthammar (1965), it will not be repeated here. The essential points are that in the averaging over initial longitudes of the trapped particles all terms except the ones leading to the power spectrum expressions vanish. Furthermore, only the fluctuating part of the electric potential field has an influence on the time averaged motion. This result is to be expected since a time independent electric field will distort the azimuthal drift path, but the drift orbit will remain closed and no net radial motion will occur.

The second Fokker–Planck coefficient for equatorial particles is

$$\langle (\Delta r)^2 \rangle = \frac{1}{B_0^2} \sum_{n=1}^{N} \int_0^t \langle \tilde{E}_{\phi n}(r_0, \xi') \tilde{E}_{\phi n}(r_0, \xi' + \zeta) \rangle \cos \Omega_0 \zeta d\zeta \qquad (4\text{-}104)$$

where $\tilde{E}_{\phi n}$ is the fluctuating part of $E_{\phi n}$. For particles mirroring off the equator the derivation begins at Eq. (4-101) with L replacing r as the radial coordinate. If the electric field is everywhere normal to $\mathbf{B}_0$ the instantaneous change in the L value of a particle at latitude λ is given by

$$R_e \frac{dL}{dt} = -\frac{E_\phi(L, \lambda)}{B(L, \lambda)} \frac{\sqrt{1 + 3\sin^2\lambda}}{\cos^3\lambda} \qquad (4\text{-}105)$$

The last factor in Eq. (4-105) is the ratio of the field line separation at the equator to the field line separation at latitude λ. Since

$$E_\phi(L, \lambda) = E_\phi(L, 0)/\cos^3\lambda$$

and

$$B(L, \lambda) = B(L, 0) \sqrt{1+3 \sin^2\lambda}/\cos^6\lambda \tag{4-106}$$

the latitude dependent factors in Eq. (4-105) cancel giving

$$\frac{dL}{dt} = -\frac{1}{R_e} \frac{E_\phi(L, 0)}{B(L, 0)} \tag{4-107}$$

which is identical to Eq. (4-101). This surprising result means that Eq. (4-104) which was derived for equatorial particles is also applicable to off-equatorial particles. Expressed in terms of the power spectrum $P_n(L, v)$ of the fluctuating part of the n'th Fourier component of the electric field, the diffusion coefficient for electric fields is

$$D_E(L, v_{\text{drift}}) = \frac{L^6 R_e^6}{8\mathcal{M}^2} \sum_{n=1}^{N} P_n(L, nv)_{v=v_{\text{drift}}} \tag{4-108}$$

Note that the power spectrum is evaluated at nv_{drift} rather than at the actual drift frequency. Thus, if the disturbance electric field has a large number of azimuthal nodes, fluctuations at relatively high frequencies are required to drive the diffusion. This particular result indicates the difficulty commonly encountered in deducing D_E from electric field measurements. The electric field must first be decomposed into its azimuthal Fourier components before computation of the power spectrum of each component. Determination of the azimuthal components in principal requires measurements made simultaneously at many different longitudes. Although D_E is not explicitly a function of particle energy or mirroring latitude, v_{drift} is a function of both energy and mirroring latitude. Hence unless P_n is independent of v, D_E will depend on λ_m and energy. However, v_{drift} is only a weak function of mirroring point, and in contrast to the situation for magnetic variations the radial diffusion rate produced by electric potential fields is almost independent of mirroring point.

Since v_{drift} depends on L the variation of D_E with L is also influenced by the actual form of $P_n(L, v)$. If $P_n \propto L^0 v^{-m}$, then $D_E \propto L^{6+2m}/\mu^m$.

Measurements of the electric field fluctuations throughout the magnetosphere are notably lacking and the use of Eq. (4-108) requires far more detailed information than is available. In view of the limited data the numerical calculations have been carried out with various simplifying approximations such as neglecting all terms with $n > 1$ and assuming the field is independent of L. Birmingham (1969) assumed the autocorrelation function had the form

$$\langle \tilde{E}_{\phi 1}(\xi') \tilde{E}_{\phi 1}(\xi' + \zeta) \rangle = E^2(L) \exp \left(-\frac{\zeta^2}{\tau_c^2} \right) \tag{4-109}$$

where $E^2(L)$ is a constant describing the amplitude of the electric field fluctuations and τ_c, the correlation time, is selected to be the time interval over which the magnetospheric electric field fluctuations retain a memory of their past behavior. With Eq. (4-109) inserted in Eq. (4-104) and neglecting terms for $n > 1$, the diffusion coefficient becomes

$$D_E = \frac{\sqrt{\pi}}{4} \frac{R_e^4}{\mathcal{M}^2} L^6 E^2(L) \tau_c \exp \left(-\Omega_0^2 \tau_c^2 / 4 \right) \tag{4-110}$$

Another form has been used by Cornwall (1968) who took for the auto-correlation function

$$\langle \tilde{E}_{\phi 1}(\xi') \tilde{E}_{\phi 1}(\xi' + \zeta) \rangle = E^2(L) e^{-2\zeta/\tau_c} \tag{4-111}$$

With (4-111) the equation for D_E becomes

$$D_E = \frac{R_e^6}{4\mathcal{M}^2} \frac{L^6 E^2(L) \tau_c}{[1 + (\tau_c \Omega_0 / 2)^2]} \tag{4-112}$$

G. Interactions between particles and waves

Early experiments (see Section V-A-2) showed conclusively that throughout much of the magnetosphere trapped particle loss rates are too rapid to be accounted for on the basis of atmospheric collisions or large scale electric and magnetic field perturbations. Although several mechanisms have been invoked to explain these rapid losses, the most promising theories involve the interaction of particles with electromagnetic waves or turbulence. Wave-particle interactions have been discussed for many years (Dungey, 1963; Brice, 1964; Cornwall, 1964), both as a means of amplifying the various types of electromagnetic noise observed in the magnetosphere and as a mechanism for changing particle pitch angles. Following the classic paper of Kennel and Petschek (1966) the importance of the process in setting upper limits to the allowable particle flux was recognized. In spite of the considerable effort to date the subject is still very much in a developing state. While it does seem clear that for electrons at $L \gtrsim 1 \cdot 25$ and protons at $L \gtrsim 1 \cdot 7$ pitch angle diffusion rates are dominated by the interaction of particles with waves, it is not possible at present to identify which of the possible types of interactions plays a dominant role in the various regions of the magnetosphere.

In this section the basis of the wave-particle interaction processes and their effects on radiation belt particles are described. Nonrelativistic velocities are used throughout and only the simplest cases are considered. In order to provide physical insight even at the cost of rigor an effort is made to avoid the mathematical complexity which is inherent in this subject.

The majority of the work to date has dealt with circularly polarized electromagnetic waves propagating parallel to the geomagnetic field. In these waves the **E** and **b** vectors are perpendicular to each other and to the direction of propagation. A stationary observer would see these vectors rotate about the magnetic field line at the wave frequency ω. The waves may propagate in the left hand or right hand sense, the direction of rotation of the field vectors for right hand waves being the same as that of a gyrating electron. If the right hand were to grasp a field line with the thumb pointing in the direction of B, the direction of right hand rotation would be indicated by the fingers. Right hand waves are commonly called electron whistlers

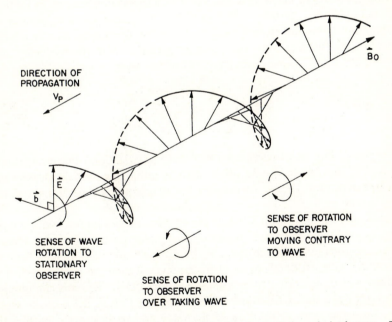

FIGURE 13 Electromagnetic fields of a right hand circularly polarized wave. The frequency as well as the sense of rotation depends on the velocity of the observer.

and have frequencies less than the electron gyration frequency Ω_e while left hand waves, commonly called ion cyclotron waves, have frequencies less than the gyration frequency Ω_p of the ambient ions, which over most of the magnetosphere are protons. The nature of a circularly polarized wave, in this case a right hand wave, is shown in Figure 13. To indicate the sense of rotation mathematically, the frequencies of right hand waves will be negative by convention. Cyclotron frequencies of electrons and protons will retain the sign of the electric charge of the particle. Similarly, if the particle velocity and wave phase velocity are in opposite directions, they will have opposite signs.

To an observer located on a particle moving parallel to the field line at velocity $w_{\parallel}$ the wave frequency is doppler shifted to

$$\omega_{obs} = \omega\left(1 - \frac{w_{\parallel}}{v_{ph}}\right) \tag{4-113}$$

where v_{ph} is the phase velocity of the wave. If the doppler shifted wave frequency ω_{obs} matches the gyration frequency Ω of the particle and is in the same sense of rotation, the electric and magnetic fields of the waves will appear stationary to the particle and will have a strong influence on its motion. For example, the particle will experience a continuous acceleration or deceleration from the electric field of the wave depending on the relative phases of the rotating wave and gyrating particle. With the frequency sign conventions stated above resonance will occur when

$$\Omega_{e,\,p} = \omega\left(1 - \frac{w_{\parallel}}{v_{ph}}\right) \tag{4-114}$$

for any combination of particle type or wave polarization. For convenience the type of interaction will be abbreviated by indicating particle identity and wave polarization, i.e. (e, L) denotes the interaction of electrons and left hand waves.

Since $\omega < \Omega_e$ for a right hand wave, an electron can interact with a right hand wave only if the wave frequency is doppler shifted upwards. Hence, the (e, R) interaction requires $w_{\parallel}$ to be directed opposite to v_{ph}. Protons interacting with left hand waves of frequency less than the proton cyclotron frequency also require $w_{\parallel}$ to be opposite to that of v_{ph}. Protons can interact with right hand waves travelling in the same direction as the proton. However, since the sense of rotation of the wave and particle are opposite, the proton must be overtaking the wave so that the observed rotation of the wave

is in the correct direction to match the proton gyration. A similar interaction can take place between a left hand wave and an overtaking electron.

The effect of the waves on a particle can be calculated from the resonance
condition (Eq. 4-114) and the dispersion equation which gives the relation
between the wave frequency and phase velocity. For left and right hand
waves, respectively, the dispersion relations for most magnetospheric
applications where the ambient plasma density is high can be approximated
by

$$v_{ph} = v_A\left(1-\frac{\omega}{\Omega_p}\right)^{\frac{1}{2}} \qquad (L)$$

$$v_{ph} = v_A\left(1-\frac{\omega}{\Omega_e}\right)^{\frac{1}{2}}\left(1+\frac{m_p}{m_e}\frac{\omega}{\Omega_e}\right)^{\frac{1}{2}} \qquad (R) \qquad (4\text{-}115)$$

where v_A is the Alfén velocity of the ionized medium. In a transverse electro-
magnetic wave the $\mathbf{E}$ and $\mathbf{b}$ vectors are related by $|\mathbf{E}/\mathbf{b}| = |v_{ph}|$. Therefore,
an observer moving with the phase velocity of the wave will see the electric
field of the wave transformed to

$$\mathbf{E}' = \mathbf{E}+\mathbf{v}_{ph}\times(\mathbf{B_0}+\mathbf{b}) = \mathbf{E}+\mathbf{v}_{ph}\times\mathbf{b} = 0 \qquad (4\text{-}116)$$

Hence, in the frame of reference moving with the phase velocity of the wave,
the electric field of the wave vanishes, and the wave-particle interaction in
that frame of reference will conserve energy. This condition is readily
expressed by

$$\tfrac{1}{2}d[w_\perp^2+(w_\parallel-v_{ph})^2] = w_\perp dw_\perp+(w_\parallel-v_{ph})dw_\parallel = 0 \qquad (4\text{-}117)$$

It is apparent that particle energy in the rest frame will not be conserved in
the interaction, and particles will have both their pitch angles and energies
altered. However, for $|w_\parallel| \gg |v_{ph}|$ the change in energy is small. In
velocity space, with axes $w_\perp$ and $w_\parallel$ the differential Eq. (4-117) describes the
element of a circle whose center is located at position $-v_{ph}$ along the $w_\parallel$
axis. As seen in Figure 14, which applies to (p, L) interactions, the particle
trajectory deviates from the constant energy circle, which is centered at the
origin. As the perpendicular velocity decreases (pitch angle decreases) the
particle energy decreases, the energy going into the wave. For (p, R) interactions the sign of v_{ph} is changed and Eq. (4-117) describes the element of a
circle whose center is at $+v_{ph}$. The overall path of a particle in velocity
space under the influence of a wave field can be obtained by integrating

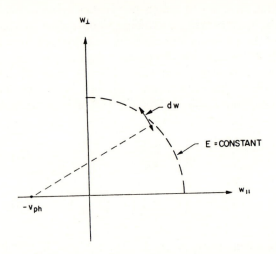

FIGURE 14   Differential element of the trajectory of a proton in velocity space during an interaction with a left-hand wave. The arrows dw, which indicate the proton path, do not follow a constant energy trace.

Eq. (4-117). Since v_{ph} varies with $w_{\parallel}$ in accordance with Eq. (4-114) and (4-115), the overall trajectory will not be a circle.

An individual particle will lose or gain energy depending on the phase relation between the wave and particle gyration. If particles are equally distributed in phase angle as is generally assumed, there will be an equal number accelerated or decelerated to the order of the interaction considered here. However, the question of whether the particles lose or gain energy depends on the particle distribution along the trajectory in velocity space. This feature is illustrated schematically for (p,L) interactions in Figure 15, where the particle distribution is heavily weighted in favor of large pitch angles. In the interaction more particles will move toward the $v_{\parallel}$ axis than towards the $v_{\perp}$ axis and energy will be transferred from the particles to the wave. Such a situation is unstable as long as the gradient persists, and particles will diffuse towards the $v_{\parallel}$ axis, transferring energy to the waves.

The derivation of a diffusion equation which applies for wave particle interactions is rather involved and will not be repeated here (see for example, Gendrin, 1968). However, the final result can be made plausible by an argument developed by Dungey (1963) and Gendrin (1968). Let $f(w_{\perp}, w_{\parallel})$ be the distribution function of trapped particles, namely the number of particles in a unit volume of velocity space. One would expect that if $f(w_{\perp}, w_{\parallel})$ were constant along a diffusion path there would be no net flow of

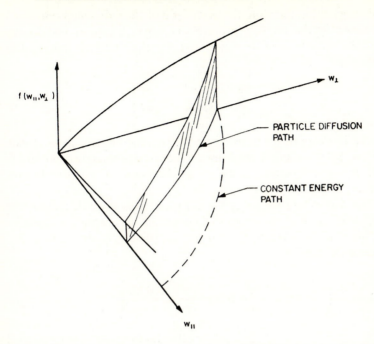

FIGURE 15 Illustration of the velocity space distribution of particles along a diffusion path. In the presence of circularly polarized waves of the proper frequencies particles with large $w_\perp$ will move along the diffusion path towards the $w_{\parallel}$ axis until the distribution is uniform along the diffusion path.

particles. The constancy of f along a path is given by

$$\nabla f \cdot \mathbf{w} = \frac{\partial f}{\partial w_\perp}dw_\perp + \frac{\partial f}{\partial w_{\parallel}}dw_{\parallel} = 0 \qquad (4\text{-}118)$$

while the trajectory equation relating $w_\perp$ and $w_{\parallel}$ is

$$w_\perp\, dw_\perp + (w_{\parallel} - v_{ph})dw_{\parallel} = 0. \qquad (4\text{-}119)$$

Combining Eq. (4-118) and (4-119) gives for the condition of no net flow

$$v_\perp\frac{\partial f}{\partial v_{\parallel}} - (v_{\parallel} - v_{ph})\frac{\partial f}{\partial v_\perp} = \frac{\mathscr{D}}{\mathscr{D}v}f = 0 \qquad (4\text{-}120)$$

In a standard diffusion equation the flow or current of particles is equal to the diffusion coefficient, D, multiplied by some differential operator applied to the distribution function, the exact form of the differential operator being determined by the physical processes at work. Equation (4-120) suggests that $\mathscr{D}/\mathscr{D}v$ is the form of the differential operator, in which case the diffusion equation is

$$\frac{\partial f}{\partial t} = \frac{1}{\mathscr{J}} \frac{\mathscr{D}}{\mathscr{D}v}\left[\mathscr{J}D\frac{\mathscr{D}f}{\mathscr{D}v}\right]. \qquad (4\text{-}121)$$

$\mathscr{J}$ is the Jacobian of the transformation from the velocity space coordinates w_x, w_y, w_z to $w_\perp, w_\parallel$ and ϕ which are the coordinates being used in Eq. (4-121), where ϕ is the azimuthal angle giving the phase of the particle gyration. Since $dw_x\,dw_y\,dw_z = w_\perp\,dw_\perp\,dw_\parallel\,d\phi$, $\mathscr{J} = w_\perp$.

In the case where $w_\parallel \gg v_{ph}$ the diffusion takes place at almost constant energy and Eq. (4-121) reduces to a simple diffusion equation in pitch angle Θ. This result is obtained by changing coordinates from $w_\perp, w_\parallel$ to w, Θ in the standard manner.

$$\frac{\partial}{\partial w_\perp} = \sin\Theta\frac{\partial}{\partial w} + \frac{\cos\Theta}{w}\frac{\partial}{\partial\Theta}$$

$$\frac{\partial}{\partial w_\parallel} = \cos\Theta\frac{\partial}{\partial w} - \frac{\sin\Theta}{w}\frac{\partial}{\partial\Theta} \qquad (4\text{-}122)$$

Inserting these operators into Eq. (4-120) gives

$$\frac{\mathscr{D}}{\mathscr{D}w} = -\frac{\partial}{\partial\Theta} \qquad (4\text{-}123)$$

and Eq. (4-121) becomes

$$\frac{\partial f}{\partial t} = \frac{1}{\sin\Theta}\frac{\partial}{\partial\Theta}\left[\sin\Theta\,D\frac{\partial f}{\partial\Theta}\right]. \qquad (4\text{-}124)$$

The diffusion coefficient D in Eq. (4-121) depends on the intensity of the wave field at the resonant frequency and is (see Gendrin, 1968)

$$D = \frac{1}{2}\frac{\Omega^2}{v_\parallel}\frac{b_k^2}{B^2} \qquad (4\text{-}125)$$

where Ω is the gyrofrequency of the diffusing particles and b_k^2 is the spectral power of the wave magnetic field per unit wave number k at the resonant frequency. Note that Eq. (4-124) as written applies at one position where the instantaneous pitch angle is Θ. To apply Eq. (4-124) to radiation belt problems it is necessary to include contributions to the diffusion from all points along the particle bounce trajectory. If one knows the wave intensity (by measurements for example), it is possible to calculate the diffusion rates from Eq. (4-124) and (4-125). This approach can be used most effectively in studying pitch angle diffusion by waves which are not generated by the diffusing particles themselves.

In the general case, however, the waves are produced by the particle population whose characteristics are being studied, and one wishes to calculate both the particle distribution function and the amplitude and frequency spectrum of the waves. In that situation one must include a conservation equation for wave energy and an equation describing the transfer of energy between particles and waves. Conservation of wave energy can be expressed by

$$\frac{\partial b_k^2}{\partial t} = -v_g \frac{\partial b_k^2}{\partial z} + 2(\gamma - l)b_k^2 \tag{4-126}$$

where v_g is the wave group velocity, z is the coordinate parallel to the field line, γ the growth rate of wave amplitude, and l describes internal losses of wave energy. The first term on the right allows for the propagation of inhomogeneities in wave energy along the field line. The transfer of energy between particles and waves is usually formulated as an equation for the growth rate of the wave amplitude and is schematically given by

$$\gamma(\omega) \simeq g(\omega) \int_0^\infty dv_\perp \, v_\perp^2 \frac{\mathscr{D}}{\mathscr{D}v} f(v_\perp, v_\parallel) \Big|_{v_\parallel = v_{res}} \tag{4-127}$$

where $\mathscr{D}/\mathscr{D}_v$ is the differential operator defined in Eq. (4-120), and $g(\omega)$ is a slowly varying function of ω. The integral in Eq. (4-127) has a simple interpretation. Apart from the factor $v_\perp^2$ the integrand is the gradient of f along the particle diffusion path, and the growth rate (which may be positive or negative) depends primarily on the magnitude and direction of this gradient. For a given wave frequency ω, all particles with the proper $w_\parallel$ will contribute to $\gamma(\omega)$. Therefore the integral is taken over $w_\perp$ at $w_\parallel = w_{res}$, and at each element of $dw_\perp$ the integrand contains the gradient of f along the curve of particle diffusion. If the gradient is negative in the direction of decreasing particle energy, then energy will be transferred from particles to the wave and the wave growth rate γ will be positive. The construction of the integral is schematically shown in Figure 16, which indicates how the overall distribution function contributes to the growth.

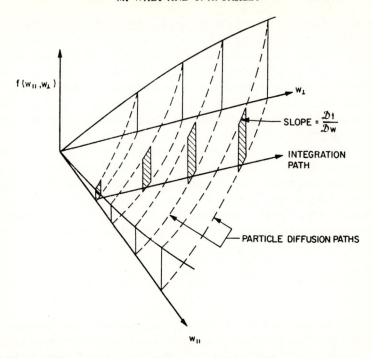

FIGURE 16   Effect of particle distribution on the wave growth rates. The growth rate at a particular-frequency depends on the integral over $w_\perp$ of the slope $\mathscr{D}f/\mathscr{D}w$ of the partical distribution function. Large particle anisotropies lead to larger growth rates. The differential operator $\mathscr{D}f/\mathscr{D}w$ is defined by Eq. (4-120).

For a different wave frequency, the integral would be taken along a different $w_\parallel$ path.

If $v_{ph} \ll w_\parallel$ the diffusion paths are almost circles and particles change pitch angle at nearly constant energy. In that case the integral in Eq. (4-127) will contain only the derivative of f with respect to pitch angle.

A general self-consistent solution of the Eqs. (4-121), (4-125), (4-126), (4-127), and the resonance and dispersion Eqs. (4-114) and (4-115) has not been obtained for realistic magnetospheric geometries. However, for an assumed source equilibrium solutions, i.e. $\partial/\partial t = 0$, have been obtained utilizing simplifying assumptions. These solutions correspond to a situation in which particles are continuously supplied with some arbitrary distribution in energy and pitch angle. Waves will then grow until particle diffusion into the loss cone balances the injection of new particles. The result will give the particle and wave distributions at equilibrium.

In all such self-consistent calculations to date it has been assumed that the interaction takes place over a small distance ΔS near the equator and that upon reaching the ionosphere a fraction $\mathscr{R}$ of the wave is reflected. With these assumptions the wave conservation Eq. (4-126) can be replaced by the much simpler expression

$$\exp\left[\gamma \Delta S / v_{ph}\right] = 1/\mathscr{R} \qquad (4\text{-}128)$$

This equation simply states that the loss in wave energy upon reflection at the ionosphere is compensated for by the transfer of energy from the particles to the wave during its passage across the equator. Such self-consistent calculations have been carried out by Etcheto, *et al.*, (1973) and others for the outer belt region $L \sim 4$ and lead to realistic values of the particle distributions and wave intensities. In the inner belt and slot region similar calculations give disappointing results. The theory predicts far too many electrons, particularly at high energies. This result is independent of the source strength, and although a reflection coefficient $\mathscr{R} \approx 1$ could reduce the flux to acceptable limits, such a high value of $\mathscr{R}$ seems unreasonable. A corollary to this result is that the electron lifetime predicted by the theory is much longer than found experimentally.

The flaws in this approach probably lie in the assumption that the waves propagate only parallel to the geomagnetic field. Calculations of wave trajectories in the inner magnetosphere show that waves of the frequencies which interact with electrons do not follow field lines but propagate more or less freely throughout the plasmasphere. Under these circumstances a trapped electron will interact with waves propagating at all angles to the field line. It is also unlikely that the entire interaction would occur near the equator. Non-parallel wave propagation complicates the problem enormously since the particles on one L shell will be affected by waves generated by a different group of particles on a distant L shell. A self-consistent calculation including these effects would involve simultaneous consideration of L shell diffusion but would give as a solution the entire radiation belt. Such an effort has not been attempted as yet.

A less ambitious step but one which considers non-parallel wave propagation has been accomplished in a series of papers (Lyons, *et al.*, 1972; Lyons and Thorne, 1973). Based on experimental data, these authors made assumptions regarding the characteristics for the wave field and calculated the diffusion coefficients which electrons would experience in that field. Since they included interactions with all harmonics of the waves, the equations for D are much more complicated than Eq. (4-125). The resulting diffusion coefficients give realistic values for the diffusion rates and give impressive agreement with the energy dependence and L-dependence of the trapped electron lifetimes within the inner radiation belt. However, there are still rather large uncertainties in the experimental values of the wave fields and the values of D thus computed have a substantial uncertainty. The next step in this approach is to carry out a self-consistent calculation.

At the present time it is recognized that wave-particle interactions are an extremely important aspect of the radiation belts at all positions except at very low L values where the neutral atmosphere dominates the loss processes. However, the general wave-particle situation is so complex that a quantitative assessment of all aspects of this important subject will not be readily accomplished. One should bear in mind that the outline presented here has considered only the most transparent situations in the interest of illustrating the principal physical mechanisms. A complete theory must include interactions with other types of waves as well as the cases where the interaction is so strong that the particles are bunched in gyration phase angle. If such interactions occur frequently, the entire process becomes non-linear and may require computer simulations to achieve theoretical solutions.

H. Reversible adiabatic effects

The discussion of the effect of the geomagnetic secular variation (Section IV-E) is an example of an adiabatic process which must be included in a theory which predicts the structure of the proton radiation belt. There are also short-term field variations which are reversible, producing reversible particle redistributions. One such process is caused by the temporary existence of a ring current in the outer radiation zone. Such ring currents are formed for periods of days during geomagnetic storms. The current is due to the azimuthal drift of low energy protons and electrons which are added or accelerated during magnetic storms. The current is therefore in a westward direction, reducing the value of B temporarily everywhere within the inner zone.

The general effect of this field perturbation on trapped particles was discussed by Dessler and Karplus (1961). McIlwain (1966a) made specific calculations of the effects on trapped inner zone protons and verified the existence of the effect by comparison with experimental data. The presentation here follows that of McIlwain.

If the build-up time of the ring current is long compared with the drift periods of inner-zone protons, all three adiabatic invariants will be conserved and, as in the secular variation example,

$$\frac{\partial f_1(\mu, J, \Phi)}{\partial t} = 0. \tag{4-129}$$

A time-dependent distribution function in μ, J, L coordinates would require an appropriate coordinate transformation as before, and the time-dependent variation of the magnetic field would have to be specified. However, since all adiabatic invariants are conserved, the final field configuration is sufficient to specify the final distribution function without knowledge of the details of the field variations.

Let the initial state be characterized by a dipole field so that the equatorial field on the fixed spatial shell at L_1 is given by

$$B_1 = \frac{\mathcal{M}}{L_1^3} \tag{4-130}$$

A positive sign will indicate a northward field or flux and a negative sign a southward field or flux.

The net flux within a drift shell at L_1 is given by the integral of the flux outside the drift shell (see discussion of Φ, Section II)

$$\Phi_1 = -2\pi \int_{L_1}^{\infty} \frac{\mathcal{M}}{L^3} L \, dL = -\frac{2\pi \mathcal{M}}{L_1} \tag{4-131}$$

In the final state, with the ring current present, a uniform southward perturbation field ΔB is present inside the ring current. The field value on a shell at L_2 is given by

$$B_2 = \frac{\mathcal{M}}{L_2^3} - \Delta B \tag{4-132}$$

and the flux within the shell is given by

$$\Phi_2 = -2\pi \int_{L_2}^{\infty} \frac{\mathcal{M}}{L^3} L \, dL + 2\pi \int_{0}^{L_2} -\Delta B \, L \, dL$$

$$= \frac{-2\pi \mathcal{M}}{L_2} - \pi L_2^2 \Delta B \tag{4-133}$$

Note that the southward flux within a ring of given L will be increased, and therefore the trapped protons will move outward so that the drift path encloses more northward flux as the field goes from its initial to its final state. If L_1 represents the initial proton drift shell and L_2 the final proton drift shell, conservation of Φ during this adiabatic change requires that

$$-\frac{2\pi \mathcal{M}}{L_1} = -\frac{2\pi \mathcal{M}}{L_2} - \pi \Delta B L_2^2 \tag{4-134}$$

or

$$L_2 = L_1\left(1 + \frac{\Delta B}{2\mathcal{M}}L_2^3\right)$$ (4-135)

and

$$\frac{L_2 - L_1}{L_1} = \frac{1}{2}\frac{\Delta B}{B_2}$$ (4-136)

The fractional change in L is one half the fractional change in B on the final shell. For a ring current field $\Delta B = 100$ gamma at $L = 1 \cdot 5$, $\Delta B/B = \cdot 0108$ so that an equatorially trapped proton moves outward from $L = 1 \cdot 500$ to $L = 1 \cdot 508$.

The change in the equatorial value of B for the proton may be found by using (4-130) and (4-132) to eliminate L_1 and L_2 from (4-134), giving

$$2B_1^{1/3} = 2(B_2 + \Delta B)^{1/3} + \Delta B(B_2 + \Delta B)^{-2/3}$$ (4-137)

If $\Delta B/B \ll 1$, then to first order in $\Delta B/B$

$$B_2 \approx B_1 - 2 \cdot 5\Delta B$$ (4-138

and

$$\frac{B_2 - B_1}{B_1} = -2 \cdot 5\frac{\Delta B}{B_1}$$ (4-139)

The fractional change in the equatorial value of B for the proton is negative and equal to $2 \cdot 5$ times the fractional change in B caused by the ring current. Note that $\Delta B/B_1 \simeq \Delta B/B_2$ as a result of the small ΔB approximation used to obtain (4-138) from (4-137).

The change in proton momentum is given by conservation of the first invariant, $\mu = p_\perp^2/2mB$ where $p_\perp$ is the perpendicular component of the momentum at the equator. Thus

$$p_{\perp 2}^2 = p_{\perp 1}^2\frac{B_2}{B_1} = p_{\perp 1}^2(1 - 2 \cdot 5\Delta B/B)$$ (4-140)

and

$$\frac{p_{\perp 2}^2 - p_{\perp 1}^2}{p_{\perp 1}^2} = -2 \cdot 5\Delta B/B$$ (4-141)

The fractional change in perpendicular momentum squared at the equator is negative and equal to 2·5 times the fractional change in B.

These equations specify the effect of a ring current field on each individual proton; the change in count rate of a directional detector at fixed L may be deduced as follows. Suppose that $j_i(p, L)$ is a known function specifying the initial variation of the directional intensity j of equatorially trapped protons with momentum p and position L in the unperturbed field. For a detector located on the equator at L_2 responding to equatorially trapped protons of momentum p_2, the initial count rate R_i will be

$$R_i \sim j_i(p_2, L_2) \tag{4-142}$$

After the addition of the perturbation field ΔB, the detector will respond to protons which initially had momentum p_1 at L_1.

As outlined in Section III, the Liouville equation requires the distribution function to be unchanged ($df/dt = 0$) along the dynamic path of the protons from $p_1 L_1$ to $p_2 L_2$. Since F is proportional to j/p^2 (see Section III), it follows that

$$\frac{j_i(p_1, L_1)}{p_1^2} = \frac{j_f(p_2, L_2)}{p_2^2} \tag{4-143}$$

and the final count rate will be

$$R_f \sim j_f(p_2, L_2) = \frac{p_2^2}{p_1^2} j_i(p_1, L_1) \tag{4-144}$$

The ratio of final count rate R_f to initial count rate R_i will be

$$\frac{R_f}{R_i} = \frac{p_2^2}{p_1^2} \frac{j_i(p_1, L_1)}{j_i(p_2, L_2)} \tag{4-145}$$

Since, by assumption, j_i is a known function of p and L, the ratio can be determined from ΔB, using (4-136), (4-140) and (4-145).

Using this theory, McIlwain (1966a) found reasonable agreement with experimental data for 40-110 Mev protons during a magnetic storm for which $\Delta B = 70$ gamma.

Since this physical process is adiabatic, the initial conditions will be restored during the gradual decay of the ring current.

V. APPLICATION OF THEORETICAL FORMULATION TO INNER RADIATION BELT OBSERVATIONS

The theoretical developments described in the previous sections are being applied with mixed success to experimental data obtained in the inner radiation belt. In general, the objective of this work is to understand the basic processes taking place in the radiation belt. Hence, the approach is to make quantitative comparisons of observational data with theoretical predictions based on the assumed processes. Agreement between theory and experiment is interpreted as an indication that the important processes have been identified. Alternatively, a lack of agreement shows that the theoretical basis is unsatisfactory.

In this chapter a number of attempts to compare theory and experiment will be described for both electrons and protons in the inner belt region. With a few exceptions the presentation follows the historical sequence of the original research. In only a few cases is a clear cut evaluation of the theory possible. Most of the theoretical results involve parameters whose values are not known, such as the strength and power spectrum of the electric and magnetic field fluctuations. These quantities can be varied over a wide range of uncertainty. If agreement between theory and experiment is reached through varying the unknown quantities, it is not clear whether the theory is correct or whether one has compensated for an inappropriate theory by choosing incorrect values for some of the parameters. This ambiguity in concert with the substantial experimental errors has led to the conflicting conclusions offered by different authors. In many cases these difficulties cannot be resolved until more and better experimental data are available.

A. Electrons in the inner zone

One of the first sources postulated for radiation belt electrons was the albedo neutron decay source. The character of this source is described in Section IV-C, and the interaction of these electrons with the atmosphere is well understood and described in Section IV-A. Consequently the neutron decay source and atmospheric loss processes by themselves lead to an easily calculable radiation belt.

A very rough estimate of the intensity of such a belt can be obtained by combining the known neutron albedo decay density (Hess, et al., 1961) with the experimental observation that artificially injected electrons disappear with a characteristic time of about one year at $L = 1 \cdot 5$. The electron flux near the maximum of the neutron decay electron spectrum (say 300 kev) will be given by the product of source strength and apparent lifetime τ_a.

$$j_e \approx \frac{nw}{4\pi} \frac{\tau_a}{300} \cong \frac{10^{-11} \times 2 \times 10^{10} \times 3 \cdot 14 \times 10^7}{4 \times 3 \cdot 14 \times 3 \times 10^2}$$

$$= 1 \cdot 7 \times 10^3 \text{ electrons cm}^{-2}\text{sec}^{-1}\text{ster}^{-1}\text{kev}^{-1} \quad (5\text{-}1)$$

The one year disappearance rate is not necessarily due to atmospheric collisions and may well result from the combined effect of several loss and transport mechanisms.

A proper calculation including the neutron decay source and atmospheric collision losses proceeds from the equation for the distribution function

$$0 = \frac{\partial f(x, L, E; t)}{\partial t} = \frac{4\pi e^4}{m_0 c} \frac{\partial}{\partial E}\left[\frac{(E + m_0 c^2)}{E^{\frac{1}{2}}(E + m_0 c^2)^{\frac{1}{2}}} \mathfrak{M} f \right]$$

$$+ \frac{2\pi e^4 c (E + m_0 c^2)}{E^{3/2}(E + 2m_0 c^2)^{3/2}} \frac{\partial}{\partial x}\left[\mathscr{L} S(1 - x^2) \times \frac{\partial}{\partial x}\left(\frac{f}{xS} \right) \right] + Q(x, E, L) \quad (5\text{-}2)$$

where the neutron source function $Q(x, E, L)$ is given in terms of the neutron decay density $n(L, s)$ (See Section IV-C) by

$$Q(x, E, L) =$$

$$\frac{1}{2}\frac{\oint n(L, s)ds}{\oint ds} \frac{0 \cdot 614}{m_0 c^2}\left[1 + \frac{E}{m_0 c^2} \right]\left[\frac{E}{m_0 c^2} \right]^{\frac{1}{2}}\left[2 + \frac{E}{m_0 c^2} \right]^{\frac{1}{2}}\left(1 \cdot 53 - \frac{E}{m_0 c^2} \right)^{\frac{1}{2}} \quad (5\text{-}3)$$

Other symbols in Eq. (5-2) are as defined for Eq. (4-34). The integrals are taken along the spiral path of an electron whose equatorial pitch angle is arccos x. It is readily seen that the first term in Eq. (5-2) accounts for energy loss in collisions while the second term describes pitch angle diffusion.

Solution of Eq. (5-2) for f with realistic values of Q, $\mathscr{L}$ and $\mathfrak{M}$ could be accomplished by available numerical techniques, but the computation has not been actually carried out. Estimates of the energy spectrum of f have been made by successively neglecting either the pitch angle diffusion term or the energy loss term, and a solution including both terms but imposing approximations on the atmospheric functions $\mathfrak{M}$ and $\mathscr{L}$ has been obtained (Walt and MacDonald, 1961).

If pitch angle diffusion is neglected the solution of Eq. (5-2) becomes by direct integration

$$f(x, L, E) = \frac{m_0 c}{4\pi e^4} \frac{E^{\frac{1}{2}}(E+m_0 c^2)^{\frac{3}{2}}}{\mathfrak{M}(E+m_0 c^2)} \int_E^\infty dE' Q(x, E', L) \qquad (5\text{-}4)$$

Since the neutron decay source Q is separable into a function of energy and a function of position and pitch angle, the energy spectrum of f is independent of both x and L.

If the energy loss term is neglected, the equation can be solved by successive integrations, noting that $\partial/\partial x \, (f/xS) = 0$ at $x = 0$ and that $f/xS = 0$ at x_c, where the atmosphere becomes too dense for trapping to occur. The result is

$$f(x, E) = \frac{xS}{2\pi e^4 c} \frac{E^{3/2}(E+2m_0 c^2)^{3/2}}{(E+m_0 c^2)} \int_x^{x_c} \frac{dx'}{\mathscr{L}S(1-x'^2)x'} \int_0^{x'} Q(E, L, x'')dx'' \quad (5\text{-}5)$$

It is apparent from Eq. (5-5) that diffusion will cause the equilibrium spectrum to have a larger fraction of higher energy electrons. The physical reason for this result is that the lower energy electrons scatter more rapidly and diffuse into the loss cone more quickly than the higher energy electrons. For a neutron decay source of electrons, the equilibrium flux obtained from Eq. (5-4) and (5-5) are illustrated in Figure 17. Note that while the diffusion process increases the average energy above that of the source spectrum, the energy loss term reduces the average energy. One would expect that the inclusion of both processes would result in an energy spectrum intermediate between the two shown.

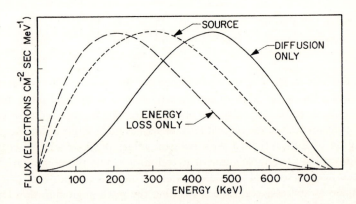

FIGURE 17 Spectra of trapped electrons produced by the decay of slow neutrons. The source curve indicates the expected spectrum of trapped electrons in the absence of loss processes; the other curves show the effects of atmospheric collisions. In the *diffusion only* curve, the angular deflections in scattering collisions are included. In the *energy loss only* curve, electrons are assumed to lose energy in the atmospheric collisions but are not deflected.

Recent experimental measurements on inner zone trapped electrons, uncontaminated by the artificially-injected Starfish electrons, are inconsistent with the solutions of Eq. (5-4) and (5-5). In general there is no well defined maximum near 300 kev, the differential flux increasing with decreasing energy. Even at 200 kev, the measured flux is more than an order of magnitude higher than the rough estimate of Eq. (5-1). There is also a considerable flux of electrons with energy far in excess of the neutron β-decay endpoint of 780 kev. From these comparisons, one concludes that the majority of inner zone trapped electrons do not result from a neutron decay source and atmospheric losses.

1.   Atmospheric scattering of electrons

Injections of intense fluxes of trapped electrons by high altitude nuclear detonations furnished some of the first and most definitive tests of trapped electron theory. These injections made it possible to separate the source processes from the loss processes and allowed an examination of the loss and transport phenomena without confusion from the injection of new particles. While there have been nine high-altitude nuclear explosions which resulted in the injection of electron fluxes, only the Starfish event on 9 July 1962 is of interest for comparison with the atmospheric scattering theory. The other detonations at low L values were so poorly instrumented that little experimental data on the trapped fluxes were obtained, and as will be seen in the following sections, the electrons from the detonations at higher L are not appreciably influenced by atmospheric collisions.

Following the Starfish event, an extensive radiation belt was formed. The initial pitch angle distributions differed appreciably from those of the natural radiation belts; however, over a period of time the fluxes gradually evolved into distributions similar to the natural ones. The flux decay was rather rapid at very low L values ($L < 1 \cdot 2$) and at L values near the slot of the natural radiation belt. However, at $L \approx 1 \cdot 5$ the decay time, defined as the period in which the flux decreased by a factor e, was on the order of a year. This behavior suggested that at low L values the electrons were being removed by the atmosphere, while at higher L values some other loss mechanism was responsible. A quantitative evaluation of this hypothesis was made by Walt (1964) and Walt and Newkirk (1966).

The formulas derived in Section IV-A were applied directly to the artificial radiation belt situation. In this case the time dependent form was used, and Eq. (4-34) was solved as an initial value problem. Because of the complexity of the coefficients $\mathcal{L}$ and $\mathfrak{M}$ at low altitude analytic solutions were not possible and finite difference techniques were required. With these techniques it was possible to integrate Eq. (4-34) over the time variable and from a given initial distribution obtain the flux and spectrum as a function of time. This evolving distribution was then compared with the observations to see how well the theory accounted for the data. In this theory it was assumed that electrons remained on the same L shell throughout their lifetimes.

The available experimental data on the decay of the Starfish electrons consisted of omnidirectional fluxes from detectors sensitive to all electrons above a given threshold. The theoretical results gave the complete distribution in energy and pitch angle, and from these values the omnidirectional flux above any arbitrary energy could be calculated for comparison with the data. The initial conditions for pitch angle distributions were taken from experimental data. Unfortunately, during the first few months after the injection energy spectrum measurements were not made. Hence it was assumed that the electrons injected by the nuclear explosion had an energy distribution corresponding to the beta particles emitted by fission products.

A comparison of the atmospheric scattering calculations and the experimental data for the first forty days following the detonation is shown in Figures 18 and 19 for L values near 1·185 and 1·25. The figures give the counting rates as a function of time for various values of total magnetic field B. For a given L shell, the B value denotes position along a field line, higher B values indicating position at higher latitudes. The curves are the theoretical results, while the symbols denote the experimental data of Van Allen (1966). In these calculations the only adjustable parameters are the

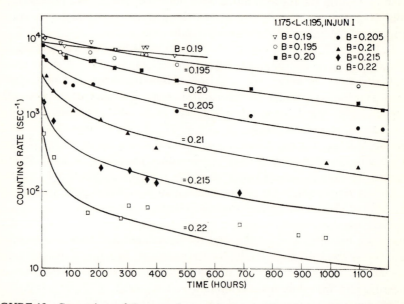

FIGURE 18 Comparison of theory and experiment at $1\cdot175 < L < 1\cdot195$ for the decay by atmospheric collisions of electron fluxes produced by the Starfish nuclear detonation.

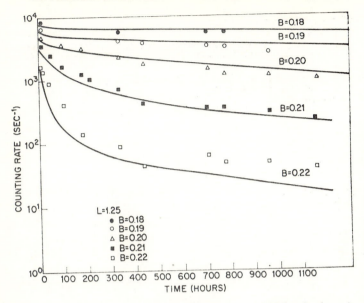

FIGURE 19 Comparison of theory and experiment at $L = 1\cdot25$ for the decay by **atmospheric** collisions of electron fluxes produced by the Starfish nuclear detonation.

initial conditions. Therefore the agreement at $t = 0$ is forced, but the close correspondence between theory and experiment as time progresses indicates a successful test of the hypothesis that atmospheric collisions played a dominant role in the decay of the electron fluxes. Even such detailed behavior as the crossing of the $B = 0\cdot19$ and the $B = 0\cdot195$ curves at $L \approx 1\cdot185$ is reproduced by the theory, giving considerable confidence in the above conclusions.

There have been additional opportunities to test the atmospheric loss mechanisms at low L-values using natural electrons. On at least two occasions the trapped electron flux at low L values was sufficiently perturbed by major magnetic storms that easily identifiable features in the particle population were produced (Imhof and Smith, 1965). In October and November of 1963 magnetic activity caused a selective redistribution of trapped electrons such that the energy spectrum at $L < 1\cdot18$ exhibited well defined isolated peaks. In one instance the peak was at $0\cdot74$ Mev while in the other case the peak occurred at $1\cdot34$ Mev. In the quiescent period following the magnetic storms the decay of these peaks could be followed and the rate of electron loss compared to that predicted by the atmospheric collision process.

The nature of the experimental data for the October 30 perturbation is illustrated in Figure 20 which gives the energy spectrum measured at several L values below $1\cdot25$. The isolated maximum at about $1\cdot3$ Mev is present for

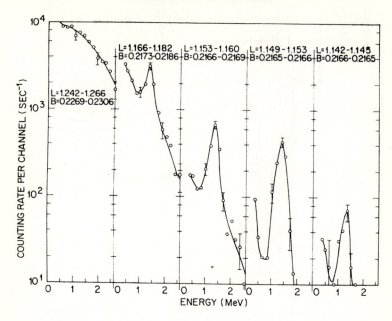

FIGURE 20 Observations of isolated peaks in the energy spectra of trapped electrons at low L values.

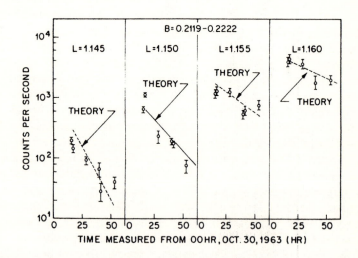

FIGURE 21 Comparison of observed decay rates of the electron flux peaks with the decay rates expected from atmospheric collisions.

all L values below about 1·18. By assuming that the electrons in the energy peak were injected during the magnetic storm and were later removed by atmospheric collisions, it is possible to treat this situation as an initial value problem similar to the artificial radiation belts produced by nuclear detonations. The comparison of theoretical and experimental values of the time dependence is given in Figure 21. It is apparent that the rate of decay is described to within experimental error by the atmospheric collision process and that no other processes need be invoked to explain the decay rates.

At higher L values, the atmospheric loss mechanism is inadequate. This conclusion was reached during the study of Starfish electrons when it was found that the actual decay of the electron fluxes was more rapid than could be accounted for by collisions alone. After the first few days the curves of

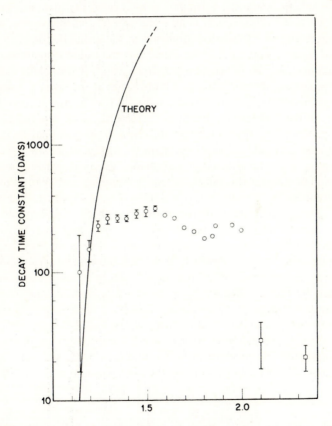

FIGURE 22   Decay rates of electron fluxes injected by nuclear detonations. Points are experimental data (McIlwain, 1966b; Van Allen, 1966), and the curve denotes the calculated decay rates expected from atmospheric collisions.

Figures 18 and 19 can be approximated by exponentials, particularly for data near the equatorial plane, and the time required for the counting rate to decrease by a factor of e can be interpreted as a crude "lifetime". Values of this "lifetime" for omnidirectional fluxes in the equatorial plane are shown in Figure 22. The theoretical values increase monotonically with increasing L as would be expected from the rapid decrease in atmospheric density with increasing altitude. In contrast the experimental "lifetimes" reach a broad plateau at about $L = 1\cdot3$. It is evident therefore that some other process was important in removing electrons at $L > 1\cdot25$, and that the region in which the atmosphere is the only significant loss mechanism is the very small region below that L value.

2.   *Radial diffusion of inner zone electrons*

The atmospheric scattering formulation applied in the previous section did not include radial diffusion. It is now recognized that this neglect was justified in these cases only because gradients in the radial distribution over the region of interest were initially so small that the radial displacement effects were unnoticeable. However, as is apparent from Figure 22, the exponential decay from atmospheric collisions is more rapid at the lower L values where the atmospheric densities are greater, and with the passage of time the radial gradient of the distribution function increased. Since the radial electron current produced by third invariant diffusion is proportional to this radial gradient, the diffusion transport mechanism, negligible at low L in the first few months after the detonation, gradually assumed greater importance. The first indication of this effect, noted by Newkirk and Walt (1968a), was an apparent electron decay rate for $1\cdot15 < L < 1\cdot21$ which was less than the asymptotic decay rate predicted from atmospheric collisions and actually observed soon after the detonation. For example, at $L = 1\cdot15$, some two years after Starfish the observed decay time was 110 days whereas the expected decay time was 3 days. Newkirk and Walt interpreted this experimental observation as evidence of a continuous supply of electrons transported downward by radial diffusion from the more intense regions of the inner zone near $L = 1\cdot5$.

Newkirk and Walt (1968a) and Farley (1969a) used the available data on equatorially trapped electrons with $E > 0\cdot5$ Mev to compute the magnitude of the radial diffusion coefficient which would provide the necessary transport of electrons. The equation governing the distribution function in this case is

$$\frac{\partial f}{\partial t} = \frac{\partial}{\partial L}\left[\frac{D_{LL}}{L^2}\frac{\partial (fL^2)}{\partial L}\right] - \frac{f}{\tau_{col}} = -\frac{f}{\tau_a} \tag{5-6}$$

where $\tau_{col}(L)$ is the asymptotic atmospheric collision lifetime and $\tau_a(L)$ is the apparent lifetime as determined from several experiments extending up to three years after the detonation.

The solution of Eq. (5-6) for D_{LL} may be found by integrating with respect to L from a boundary at L_B up to L to obtain

$$D_{LL}(L) = \frac{L^2}{(\partial/\partial L)(fL^2)}\left\{\int_{L_B}^{L}\left(\frac{1}{\tau_{col}}-\frac{1}{\tau_a}\right)fdL+\frac{D_{LL}(L_B)}{L_B}\left[\frac{\partial(fL^2)}{\partial L}\right]_{L=L_B}\right\} \quad (5\text{-}7)$$

The second term in the braces is the arbitrary constant which must be specified by a boundary condition in order to obtain the appropriate solution of this first order differential equation. The constant is readily recognized as the radial current through the chosen boundary at L_B.

Unfortunately there is no L_B for which $D_{LL}(L_B)$ was known or could be estimated for $1\cdot15 < L_B < 1\cdot21$, and Newkirk and Walt (1968) as well as Farley (1969a) were only able to exhibit members of the family of possible solutions. An estimate of the minimum possible member of this family is

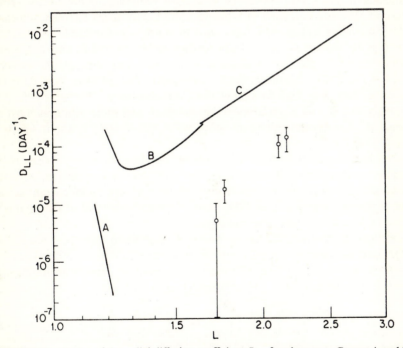

FIGURE 23   Values of the radial diffusion coefficient D_{LL} for electrons. Curves A and B are minimum and maximum values derived from Starfish electrons. Curve C gives the best fit to the redistribution of electrons in L following a magnetic storm, and the points give values of D_{LL} obtained directly from the radial spreading of electron bands produced by nuclear detonations.

shown in Figure 23 as Curve A. Family members below this curve have negative values somewhere in the interval.

In spite of this limitation, two significant conclusions could be reached. First, all reasonable solutions decreased by at least an order of magnitude with increasing L from $L = 1 \cdot 15$ to $L = 1 \cdot 21$. This behavior is in contrast with the evidence obtained with various particle populations at larger L, all of which indicate an increase in D with an increase in L. The increase in D_{LL} with L is a feature of all theories (See Section IV-B) which ascribe the diffusion to time-varying electric or magnetic fields whose scale is that of the magnetosphere. This anomalous result suggests that radial diffusion at low L is dominated by local field variations due perhaps to ionospheric currents or other phenomena more pronounced at low altitude. No satisfactory quantitative theory of such local field variations has yet been advanced, and perhaps other mechanisms are involved.

The minimum possible diffusion coefficient shown in Figure 23 is obtained by systematically reducing $D_{LL}(L = 1 \cdot 21)$, thereby reducing the arbitrary constant in Eq. (5-7). As $D_{LL}(L = 1 \cdot 21)$ is reduced, the downward radial current through this shell is reduced until finally there are just sufficient electrons being transported downward to supply the Coulomb losses. Any attempt to reduce $D_{LL}(L = 1 \cdot 21)$ further results in a diffusion coefficient which suddenly plunges to negative values part way through the interval $1 \cdot 15 < L < 1 \cdot 21$. Mathematically this means that particles are diffusing upward from $L = 1 \cdot 15$ to supply the Coulomb losses. Negative diffusion coefficients have no physical meaning, however, and these solutions must be rejected. The critical minimum value of $D_{LL}(L = 1 \cdot 21)$ is about 4×10^{-7} day^{-1}.

No member of the family of solutions can be selected as a reasonable maximum and arbitrarily large diffusion coefficients are consistent with the data in this L range. Large coefficients result in a large radial current across the region boundary at $L = 1 \cdot 21$, most of which passes out of the region at $L = 1 \cdot 15$ and into the dense atmosphere.

Farley (1969b) noted that a boundary condition can be obtained if $\partial(fL^2)/\partial L$ vanishes at some L_0 within the L region under consideration, for then the numerator of Eq. (5-7) must also vanish to prevent an unphysical infinite divergence of D_{LL}. Consequently

$$\frac{D_{LL}(L_B)}{L_B}\left[\frac{\partial(fL^2)}{\partial L}\right]_{L=L_B} = -\int_{L_B}^{L_0}\left(\frac{1}{\tau_{\text{col}}} - \frac{1}{\tau_a}\right)f\,\mathrm{d}L \qquad (5\text{-}8)$$

and the solution shown in equation (5-7) may be written in the concise form

$$D_{LL}(L) = \frac{L^2}{(\partial/\partial L)(fL^2)}\int_{L_0}^{L}\left[\frac{1}{\tau_{\text{col}}} - \frac{1}{\tau_a}\right]f\,\mathrm{d}L \qquad (5\text{-}9)$$

With more extensive Starfish electron data for the period from January 1963 to November 1965, Farley (1969b) used Eq. (5-9) to compute the diffusion coefficient for electrons having first invariant values of 14·4 Mev gauss^{-1} for $1·20 < L < 1·65$. A plot of fL^2 showed a peak at $L = 1·38$, thereby establishing the value of L_0 and making a unique solution for D_{LL} possible. This solution is also shown in Figure 23 as Curve B. It shows the same upward trend at low L, with a minimum occurring near $L = 1·3$. Evidently it would join one of the larger members (not shown) of the family computed earlier for the region $1·15 < L < 1·21$. Electrons having first invariant values of 14·4 Mev gauss^{-1} have energies from 2·5–3·0 Mev in the L range $1·15 < L < 1·21$, and therefore the new curve might not necessarily join one of the earlier family members because of the variation of the diffusion coefficient with energy. In Eq. (5-9) Farley used for τ_{col} a value based on Coulomb collisions only. If appreciable wave-particle scattering also occurs, the actual value of τ_{col} would be larger, and the true value of D_{LL} less. Hence, curve B represents an upper limit to the value of D_{LL} obtained from Starfish data.

The Starfish nuclear detonation was not the only one to inject electrons with which radial diffusion studies could be made. The three Argus explosions in 1958 and the USSR test of 1 November 1962 provided distinct, if temporary, additions to the natural electron belts at L values ranging from $L = 1·76$ to $L = 2·2$. During the injection process these electrons were confined to a thin shell about the L value of the burst point, appearing as a very sharply defined peak in the distribution function at that L.

At these L values pitch angle diffusion lifetimes are relatively short, and the electrons disappeared within several weeks. Some spreading of the distribution function across L shells was observed during the gradual disappearance, and from this spreading estimates of the radial diffusion coefficient were made (Newkirk and Walt, 1968b; Walt, 1971). In these four cases Eq. (5-6) was employed and D_{LL} was varied to obtain the observed rate of spreading in L. Boundary conditions were established by fixing $f = 0$ on L shells sufficiently distant from the electron shells to have no effect on the evolution of the distribution function. Since the distributions were so narrowly confined in L, the radial diffusion coefficient D_{LL} was assumed to have no significant variation across the shell, and a single value of D_{LL} was obtained in each of the four cases. These values are shown as open circles in Figure 23.

The disappearance of the Starfish electrons took years instead of days or weeks because so many of them were injected into the L range ($1·25 < L < 1·6$) in which Coulomb collisions are relatively rare and pitch angle diffusion lifetimes are long. While some of these electrons were radially diffusing downward to be lost in the atmosphere, the maximum in the function fL^2

at $L = 1.38$ establishes that electrons at L values above 1.38 are diffusing outward toward regions in which pitch angle diffusion is an important loss mechanism. Several years were required for the removal of the electrons by these mechanisms because of the small size of the radial diffusion coefficient in this L range.

During and after geomagnetic storms large additions of energetic electrons occur in the outer radiation zone. The mechanisms for these additions are not presently well established, but apparently they do not operate in the inner zone, for impulsive injection or local acceleration have not been observed there. As the Starfish electron fluxes decayed away year by year, these outer zone additions become relatively larger in relation to the Starfish electron remnants. Finally, in September 1966, large outer zone electron increases took place during a geomagnetic storm and reversed the sign of the radial gradient of fL^2 at constant μ from negative to positive. For the first time since the Starfish detonation, a radial diffusion current was directed inward as far as the inner zone, and important electron additions were made to the inner zone, particularly in the energy range of a few hundred kev.

Tomassian, *et al.* (1972) made a detailed study of this event in the crucial L range $1.7 < L < 2.8$ in order to extract a value for the diffusion coefficient required to produce such diffusion. Again, Eq. (5-6) was assumed to govern the distribution function. For such an event it is not possible to find a constant value of the decay constant τ_a so that $\partial f/\partial t$ may not be represented as $(-f/\tau_n)$. Indeed, $\partial f/\partial t$ is positive through much of the eight week period under study as fresh electrons are transported into this L range more rapidly than they are removed by pitch angle diffusion.

Tomassian, *et al.* subdivided the eight-week study period into ten time intervals. The required diffusion coefficient D_{LL} was assumed to vary as $L^m \mu^n$. Integration of Eq. (5-6) over one time period then yielded

$$\int_{t_1}^{t_2} \frac{\partial f}{\partial t} dt = D_{LL} \int_{t_1}^{t_2} \left[\frac{\partial}{\partial L} \frac{1}{L^2} \frac{\partial(fL^2)}{\partial L} + \frac{m}{L^3} \frac{\partial(fL^2)}{\partial L} \right] dt - \frac{1}{\tau_0'} \int_{t_1}^{t_2} f dt \quad (5\text{-}10)$$

where τ_0' now includes losses from all forms of pitch angle scattering. In the above equation it has been assumed that D_{LL}, the diffusion coefficient, and τ_0', the pitch angle lifetime, do not vary between t_1 and t_2 so that they may be factored from the integrals. Since the integrands are known from the data, the integrals can be evaluated numerically. Equation (5-10) is no longer a differential equation, but an algebraic equation in the two unknowns D_{LL} and m. Using estimates based on experiments for the pitch angle lifetime τ_0', Tomassian et al. (1972) used Eq. (5-10) to make 157 determinations of D_{LL} as a function of t, L and μ. The exponent m was varied until the calculated result for D_{LL} agreed with the assumed value of m, which occurred for $m = 7 \cdot 9$.

The values obtained by Tomassian *et al.*, shown in Figure 23 as Curve C, are consistent with the values obtained at lower L from Starfish electron decay (Curve B). They are also consistent with electric potential field diffusion (See Section IV-E) as applied by Cornwall (1968) to electrostatic fields across the magnetosphere. Cornwall's result,

$$D_E = \frac{R_e^6 L^6}{4\mathcal{M}^2} \frac{E^2(L)\tau_c}{[1 + (\tau_c \Omega_0/2)^2]} \tag{5-11}$$

was found to fit the values of Tomassian *et al.* (1972) provided that $E(L)$, the rms value of the fluctuating field, was chosen to be $0 \cdot 28$ mv/m, and τ_c, the characteristic period of the fluctuations, was 1600 seconds. Some direct experimental field measurements exist to support these values. In the above expression $\mathcal{M}$ is the magnetic moment of the earth, Ω_0 is the angular drift frequency, and τ_c is the correlation time of electric field fluctuations.

The radial diffusion coefficients in Figure 23, derived from observations, show a very wide range in value. The theory of Cornwall suggests one possible reason: the coefficients depend on the electron drift period τ_d, and the electrons from which the coefficients in Figure 23 have been obtained have a wide range of drift periods.

Tomassian *et al.* (1972) reviewed these reported diffusion coefficients in an effort to reconcile them using the Cornwall estimate of the electric diffusion coefficient. The reconciliation was modestly successful in the inner zone, but left important discrepancies in outer zone results.

The role of $\partial/\partial L(fL^2)$ at constant first and second invariant as a key indicator of the direction of the diffusion current needs to be emphasized. The function fL^2 is easily obtained from experimental equatorial directional intensities, and simple inspection of this function plotted for $1 \cdot 1 < L < 6$ will reveal whether the inner zone is being repopulated by radial diffusion at any given time. If $\partial/\partial L(fL^2)$ is anywhere negative, the gate is closed, and no electrons are diffusing inward across the negative region. Wherever it is positive, inward diffusion is occurring. These conclusions are true without regard to whatever other source, loss, or transport mechanisms may be acting simultaneously.

According to the studies described here, the gate between inner and outer zone was closed from the time of the Starfish detonation in July 1962 until the additions made in September 1966. Pfitzer and Winckler (1968) and Tomassian *et al.* (1972) suggested that the gate might be open only after important magnetic storms, and then closed by the depletion of electrons in the so-called slot region by pitch angle diffusion.

More recently Lyon and Thorne (1973) have suggested that the gate remains open (i.e. $\partial/\partial L\ fL^2 > 0$) all of the time in the absence of Starfish electrons, providing a continuous transport of electrons into the inner zone. Their assumption was tested by obtaining a time stationary solution of equation (5-6) ($\partial f/\partial t = 0$) using an experimentally determined electron spectrum at $L = 5\cdot5$ and $f = 0$ at $L = 1$ as boundary conditions. They used Eq. (5-11) to estimate D_{LL}, but reduced the value of E from $0\cdot28$ mv m^{-1} as found by Tomassian et al. (1972) to $0\cdot1$ mv m^{-1} to simulate quiet time conditions. They also used pitch angle lifetimes estimated on the basis of wave-particle interactions with plasmaspheric hiss amplitude levels appropriate to quiet periods. Their solution reproduces the observed equatorial electron distribution rather well over a wide range of energies. Continued experimental observation will be required to determine just what fraction of the time the gate is open allowing electrons to diffuse into the inner zone.

B. Protons in the inner zone

The distribution of energetic inner zone protons has been studied longer and more intensively by the methods described in this work than any other geomagnetically trapped particle distribution. Partly as a result, a proper description of the inner zone protons must include a large number of mechanisms which have been shown to be significant. Unfortunately, a completely satisfactory description does not yet exist, due either to the omission of relevant mechanisms, unsatisfactory or incomplete data for reconciliation of theory and experiment, or perhaps both. Nonetheless, much progress on the problem has been made in the last fourteen years.

The inner zone of trapped protons, unlike most other particle distributions in the magnetosphere, may be treated as time stationary to a reasonable first approximation. The first attempt at such a treatment was made by Singer (1958) even before the particles had been identified and their energy spectrum measured. Singer chose to balance an albedo neutron source with atmospheric ionization losses, using $\beta = w/c$ as the independent coordinate, so that the distribution function $f(\beta)$ is the number of protons per unit increment of β. In equilibrium, the equation for the distribution function is

$$\frac{\partial f(\beta)}{\partial t} = 0 = -\frac{\partial}{\partial \beta}\left[\frac{d\beta}{dt}f\right] + Q(\beta) \qquad (5\text{-}12)$$

Q, the neutron source term, was chosen to be proportional to $E^{-1\cdot8}$ per Mev corresponding to $\beta^{-2\cdot6}$ per unit β when β is substituted for E. This calculation and several others which followed, predicted only the β (or energy) dependence of f, since information on the absolute albedo neutron energy spectrum, the orbit-averaged atmospheric densities encountered by the protons, and the value of the injection coefficients were not then available.

Singer approximated the ionization energy loss $d\beta/dt$ as a power law in β, and integrated the equation directly to predict that f would have an energy dependence given by $E^{-0.8}$ over the energy range from 20 to 500 Mev. This energy dependence was in rough agreement with the results of a subsequent nuclear emulsion spectral measurement made by Freden and White (1959), and numerous refinements made in subsequent calculations by others have not resulted in very different theoretical predictions of the spectral shape in the region of space in which the nuclear emulsions were exposed.

Hess (1959) made a rather similar calculation using the energy coordinate E, assuming a neutron differential spectrum $Q(E)$ varying as E^{-2}, and approximating dE/dt as a power law in E.

$$\frac{\partial f(E)}{\partial t} = 0 = -\frac{\partial}{\partial E}\left[\frac{dE}{dt}f\right]+Q(E) \qquad (5\text{-}13)$$

The result of solving Eq. (5-13) indicated that f should vary approximately as $E^{-1.3}$ for 20 Mev $< E <$ 700 Mev, in reasonable agreement with Singer (1958).

Freden and White (1960) added a term to represent the loss of protons by collisions with atmospheric nuclei obtaining

$$\frac{\partial f(E)}{\partial t} = 0 = -\frac{\partial}{\partial E}\left[\frac{dE}{dt}f(E)\right]+Q(E)-\rho\sigma\beta cf(E) \qquad (5\text{-}14)$$

They used an absolute estimate of the neutron albedo flux (Hess *et al.*, 1959) so that the source term became

$$Q(E) = (0\cdot8E^{-2}/\beta c\gamma\tau_n)\,(R_e/R)^4 \qquad (5\text{-}15)$$

The last factor in Eq. (5-15) was intended to compensate for the reduction in neutron flux with distance R away from the earth, and is in effect the first estimate of an injection coefficient for the neutrons.

The use of an absolute neutron flux estimate left only the atmospheric density as an adjustable parameter for normalization to the data. Equation (5-14) was solved essentially by an iterative estimation technique to give $f(E)$ for the interval 80 to 700 Mev. The single unknown parameter, the orbit-averaged atmospheric density, was determined by normalizing $f(E)$ to the nuclear emulsion proton measurements at 100 Mev (Freden and White, 1959). The atmospheric density obtained, 2×10^5 atoms cm^{-3}, was re-inserted in Eq. (5-14) to give a solution in the energy region $10 < E < 80$

Mev, in which nuclear interactions are negligible. The complete solution provided somewhat better agreement with the nuclear emulsion data than previous calculations, principally because the newly added nuclear interactions remove some of the high energy protons which were in excess in the earlier calculations. The fit to the experimental data is shown in Figure 24 where the solid line represents the normalized solution of Eq. (5-14) and the dashed curve is drawn to best fit the experimental data. In spite of the nuclear interaction term, the theory still predicted too many protons at the high energy end of the curve.

The first absolute intensity calculations of the trapped proton flux were made by Lenchek and Singer (1962) and later extended by Lenchek and

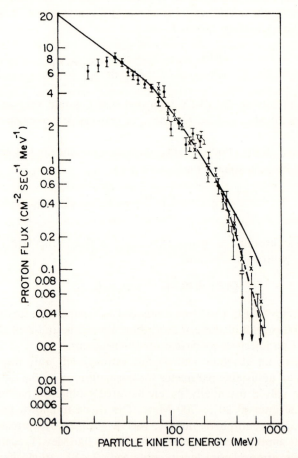

FIGURE 24 Energy spectrum of inner belt protons. Points and dashed line represent experimental values, and the solid curve is the normalized theoretical result from Eq. (5-14).

Singer (1963). They employed Eq. (5-14), made neutron albedo flux estimates from a very simple model, and calculated orbit-averaged atmospheric densities from an atmospheric model. The computed spectral shape was in satisfactory agreement with experiment, and the computed intensity, compared at 100 Mev, was generally 2 to 3 times larger than the experimentally determined intensity. This result was very encouraging, since such discrepancies might easily be accounted for by approximations and uncertainties in the calculations.

These approximations and uncertainties were reviewed, refined, and reduced by several authors in subsequent papers. Lingenfelter (1963) made a study of all earlier neutron measurements and made extensive calculations to obtain not only the globally averaged neutron leakage intensity, but estimates of its latitude and solar cycle variation as well. There were no available data above 10 Mev on which to base any estimates, and Lingenfelter simply extrapolated his intensity curves from 10 to 50 Mev. In the absence of any experimental data, Lingenfelter's leakage rates were used in many subsequent studies, usually with extrapolation far beyond the already extrapolated 50 Mev values. The extrapolated rates on a globally-averaged basis are a factor of 2 or 3 lower than the estimates of Lenchek and Singer (1962, 1963) and are about an order of magnitude lower at the geomagnetic equator.

Further calculations of atmospheric densities, properly averaged over the proton trajectories, were obtained by Cornwall et al. (1965) for two different magnetic field models at two different periods of the solar cycle. The rate of energy loss dE/dt may be written as a product of the orbit-averaged atmospheric density $\bar{\rho}$ and a factor $F(E)$ which depends only on the proton energy. Cornwall et al. noted that, for energies at which nuclear interactions can be ignored, the equation for the distribution function may be written

$$\frac{Q(E)}{\bar{\rho}} = \frac{\partial}{\partial E}[F(E)f(E)] \qquad (5\text{-}16)$$

By direct integration

$$\frac{\int_E^\infty Q(E)dE}{\bar{\rho}f(E)} = F(E) \qquad (5\text{-}17)$$

which is independent of position. Cornwall et al. used this relation as a test of the validity of Eq. (5-16). The ratio actually varied by a factor of at least eight and sometimes much more depending on the particular proton data used and the particular magnetic field model. As a test of Eq. (5-16) the result was inconclusive because of the rather large uncertainties in all three factors Q, $\bar{p}$ and f.

Subsequently Dragt et al. (1966) computed injection coefficients as outlined earlier in this paper. They used the neutron leakage estimates of Lingenfelter (1963), the averaged atmospheric densities of Cornwall et al. (1965), and the proton flux values of Freden and White (1959, 1962) for further testing of the neutron albedo source-atmospheric collision loss model of the high-energy trapped proton belt. The comparison between the calculated and observed distribution functions showed reasonable agreement on the spectral shape, as had all previous calculations, but the calculated function was about a factor of fifty too small at $L = 1\cdot3$. This comparison, much less favorable than that obtained by Lenchek and Singer (1962, 1963) was in part due to the lower neutron leakage rates of Lingenfelter (1963), and in part due to numerous minor differences in neutron angular distributions, injection coefficients, and atmospheric densities. Since the neutron leakage rate at high energy had never been measured, but merely extrapolated from low energy experiments, it seemed reasonable to believe that an erroneous neutron source function was the most likely reason for the discrepancy.

When broader regions of space were considered other discrepancies became apparent. The factor of 50 deficiency of the theoretical calculation decreases with radial distance until a match is found at about $L = 1\cdot7$ on the equator. At larger distances, there are too many protons in the calculated distribution function. Furthermore, the proton energy spectrum varies systematically with radial distance and also with pitch angle (Farley et al., 1969). Such variations are not predicted by the steady-state neutron albedo source-atmospheric collision loss model.

Lenchek and Singer (1963) studied the possibility that neutron albedo produced by solar cosmic rays might also contribute to the proton population of the inner zone, particularly at the higher L values and smaller pitch angles. Since solar cosmic rays produce a different albedo neutron energy spectrum having relatively more low-energy neutrons, a combination of both the solar and galactic cosmic ray neutron sources might produce a systematically varying proton energy spectrum. Dragt et al. (1966) determined that the solar cosmic ray contribution was unimportant, basing their conclusion on the low injection coefficients which they calculated and the estimates of solar cosmic ray-induced neutron albedo given by Lingenfelter and Flamm (1964).

Spatially-dependent time variations of the trapped proton spectrum can and do occur as a result of atmospheric density variations related to the eleven year solar activity cycle. During the active portion of the cycle increased amounts of ultraviolet radiation reach the earth and are absorbed high in the atmosphere, increasing the local temperature and the scale heights of the various constituents except for hydrogen. (Hydrogen is an exception since the increased temperature increases the escape rate and decreases the scale height.) Overall this effect produces an atmospheric density increase with solar activity over the lower part of the inner radiation zone. Following earlier estimates by Blanchard and Hess (1964) of the effect on trapped protons, Dragt (1971) undertook an exhaustive analysis designed to predict the spectrum of trapped protons at any location at any stage of the solar cycle.

A time-varying atmosphere causes time-varying losses since both the ionization energy loss rate dE/dt and the nuclear interaction loss rate are proportional to the atmospheric density ρ. After a series of transformations of both the dependent and independent variables in equation (5-14), Dragt succeeded in reducing the calculation of the time-dependent solution $f(E, t)$ to the numerical evaluation of two definite integrals.

There are two time scales which enter the problem: T_s, the eleven year solar cycle period, and $T_p = -\partial/\partial t\,(\log j)$, the characteristic time for atmospheric collisions to remove protons in the absence of any source. Dragt obtained asymptotic solutions for the two extreme cases $T_p \gg T_s$ and $T_p \ll T_s$ keeping first order terms in the ratio of the characteristic periods in each case. The solutions demonstrated the expected results that the proton flux has little variation and depends on the average atmospheric density when $T_p \gg T_s$ and that the proton flux varies inversely as $\bar{\rho}(t)$ when $T_p \ll T_s$.

For the intermediate cases it is necessary to know the orbit-averaged, time-dependent atmospheric density for each proton orbit under consideration. The level of extreme ultraviolet solar radiation, which determines the atmospheric density variations, can be obtained reasonably well from measurements of the solar radio noise at 2800 MHz. The level of EUV is characterized by an intensity index $\mathscr{S}$ and model atmospheres for various values of $\mathscr{S}$ are available from which orbit-averaged densities $\bar{\rho}$ may be calculated. Cornwall et al. (1965) calculated $\bar{\rho}$ for many orbits at various values of $\mathscr{S}$, using several different magnetic field models. Examining these results, Dragt (1971) concluded that the orbit-averaged densities of interest could always be represented, within an error of 50 per cent, by an expression of the form

$$\bar{\rho}(t) = c_1 \exp\left\{c_2[\mathscr{S}(t)-100]\right\} \qquad (5\text{-}18)$$

where c_1 and c_2 are constants which depend on the particular proton orbit and on the particular magnetic field model utilized. Dragt provided plots of c_1 and c_2 for various proton mirror points and L values. Once c_1 and c_2 have been selected, $\bar{\rho}(t)$ is determined from equation (5-18) using experimental values of $\mathscr{S}(t)$. The values used by Dragt were quarterly averages for the solar cycle period from 1956 through 1966. The proton flux as a function of time within the solar cycle can then be computed numerically by evaluating the integrals which Dragt obtained as a solution to Eq. (5-13). A separate numerical calculation must be made for each set of the input parameters c_1, c_2, t, and proton energy E. Dragt provided extensive tabulations of the proton flux for various combinations of these four input parameters at sufficiently small intervals to permit interpolation for parameter sets not in the tables. The proton flux values were tabulated as a ratio to the equilibrium flux which would be present in that orbit at that energy if the atmospheric density were constant at its solar cycle-averaged value.

Selected ratios were plotted versus time during the solar cycle, showing the interesting lag effects between atmospheric density variations and the resulting proton flux variations. One of these plots is shown in Figure 25.

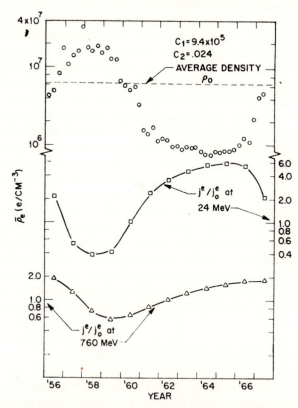

FIGURE 25   Solar cycle variations in atmospheric density and the resulting variation in equatorially trapped proton fluxes. The solar cycle variations are largest at the lower (24 MeV) proton energy since the lifetimes of these protons are comparable to the period of a solar cycle.

Experimental verification of predicted solar cycle variations has been rather difficult because a long and continuous series of observations is required, preferably with the same instrument in order to eliminate errors due to uncertainties in the geometrical factors of different instruments. Nevertheless, substantial evidence for such variations in reasonable agreement with the theory has been presented (Heckman *et al.*, 1969; Macy *et al.*, 1970) for the low altitude regions in which $T_p \ll T_s$.

The solar cycle variations, while they are undoubtedly important at the lower altitudes and lower energies, are not of the right character to account for the apparent overall deficiency of the neutron source. In a series of papers Farley *et al.* (1970), Farley and Walt (1971), and Farley *et al.* (1972) explored the possibility that inward radial diffusion of protons from the outer zone might make up the deficiencies of the neutron albedo source at lower energies while simultaneously redistributing the albedo neutron-derived higher energy protons in a manner more consistent with the available data.

When it is assumed that third invariant diffusion occurs without violation of the first and second adiabatic invariants, it is usually most convenient to work in μ, J, Φ or μ, J, L coordinates in order to reduce the radial diffusion effects to a single term in the distribution function equation. All previous work on inner zone protons had been carried out in E, Θ_e, L coordinates, or equivalently β, Θ_e, L coordinates. In these coordinates the atmospheric collision energy loss is a single term, since the process conserves Θ_e and L. In μ, J, L coordinates both μ and J are affected, and the collision term becomes

$$\frac{\partial f}{\partial t}\bigg|_{coll} = -\frac{\partial}{\partial \mu}\bigg[\langle \Delta\mu \rangle f \bigg] - \frac{\partial}{\partial J}\bigg[\langle \Delta J \rangle f \bigg] \qquad (5\text{-}19)$$

Since μ is given by $p^2 \sin^2\Theta / 2mB$ and J by $p \oint ds/\cos \Theta$ it is easy to calculate that

$$\langle \Delta J \rangle = \frac{dJ}{dt} = \frac{J}{2\mu}\frac{d\mu}{dt} = \frac{J}{2\mu}\langle \Delta\mu \rangle \qquad (5\text{-}20)$$

keeping in mind that $d\Theta/dt = 0$ in the atmospheric collision loss process. Inserting Eq. (5-20) into Eq. (5-19), the loss terms become

$$\frac{\partial f}{\partial t}\bigg|_{coll} = -\frac{\partial}{\partial \mu}[\langle \Delta\mu \rangle f] - \frac{\langle \Delta\mu \rangle f}{2\mu} - \frac{J}{2\mu}\frac{\partial}{\partial J}\langle \Delta\mu \rangle f \qquad (5\text{-}21)$$

Since Farley *et al.* (1970) and Farley and Walt (1971) used the distribution function equation for equatorially trapped ($J = 0$) particles only, the last term in Eq. (5-21) was discarded. With the addition of the secular dipole variation term by Farley, *et al.* (1972) the complete equation for equatorially trapped particles became

$$\frac{\partial f(\mu, J, L)}{\partial t} = 0 = Q(\mu, J = 0, L) + \frac{\partial}{\partial L} \frac{D_{LL}}{L^2} \frac{\partial f L^2}{\partial L} - \frac{\partial}{\partial \mu}[\langle \Delta \mu \rangle f] - \frac{\langle \Delta \mu \rangle f}{2\mu}$$

$$- \frac{\partial}{\partial \mu}[a\mu f] - \frac{\partial}{\partial L}[a L f] \quad (5\text{-}22)$$

Boundary values were chosen by using the experimentally determined values of f at $L = 1 \cdot 7$, by setting $f = 0$ at $L = 1 \cdot 10$ for all μ and by setting $f = 0$ at $\mu = 4000$ Mev gauss^{-1} for all L. This last condition is equivalent to the assumption that neutron albedo decay contributes insignificant numbers of protons at extremely high energies.

Inserting the neutron flux values of Lingenfelter (1963), the injection coefficients of Dragt *et al.* (1966) and atmospheric composition and density estimates from current model tabulations, Farley and Walt (1971) and Farley *et al.* (1972) obtained solutions of Eq. (5-22) with and without the dipole variation terms, respectively. The diffusion coefficient was varied to obtain the best fit to experimentally determined values of f.

A comparison between the calculated and experimental values of f appears in Figure 26. While the agreement is quite reasonable overall, there are significant deficiencies in the calculated number of protons at the lowest L values, particularly at $L = 1 \cdot 15$. This solution employs a solar-averaged atmosphere, since numerical solutions with a time varying atmosphere could not be readily obtained. Farley and Walt (1971) also computed solutions for solar-minimum and solar-maximum atmospheres and speculated that a complete time dependent solution of Eq. (5-22) might produce better agreement.

Recently, however, new neutron flux measurements (Heidbreder *et al.*, 1970; Eyles *et al.*, 1971; White *et al.*, 1972; Preszler *et al.*, 1972) have redirected attention to the neutron source term. These new measurements indicate that the widely used extrapolations of Lingenfelter's calculations are too low by perhaps a factor of 25 in the important region above about 30 Mev.

At first glance, these new results might appear to solve the outstanding problems by providing more protons in the deficient regions at low L. The diffusion coefficient, for whose value there is little direct evidence as yet, might then be reduced to prevent an overabundance of protons from appearing in the solution for $1 \cdot 3 < L < 1 \cdot 7$.

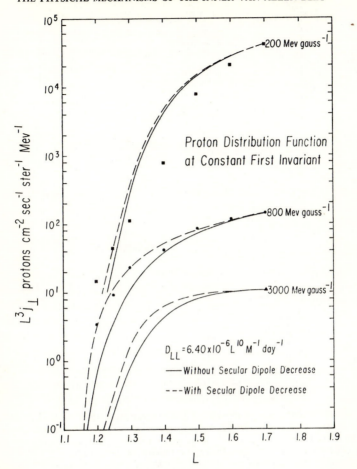

FIGURE 26   Comparison of experimental values (points) of equatorially trapped proton fluxes with the theoretical curves. The theory includes neutron decay injection with neutron fluxes of Lingenfelter (1963), atmospheric collisions, secular variation of the geomagnetic field, and radial diffusion with boundary conditions specified at $L = 1\cdot15$ and at $L = 1\cdot7$.

Unfortunately this approach is not completely successful. **Figure 27** illustrates the solution (plotted as $L^3 j_\perp \sim L^2 f$ to show the direction of the diffusion current) using the most recent neutron data. This figure, from Claflin and White (1973) shows the result at 800 Mev gauss[-1] computed on a basis identical to that of Farley and Walt (1972) except for the larger neutron source. Curve A was computed using an atmospheric density averaged over the solar cycle, while curve B was computed using the atmospheric density corresponding to the maximum period of the solar cycle

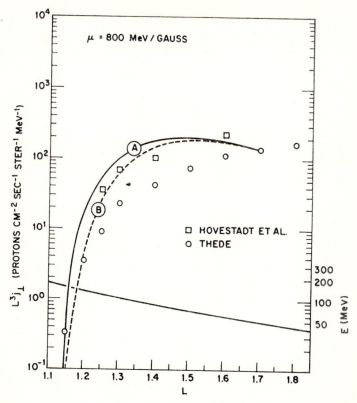

FIGURE 27   Comparison of theoretical and experimental values of equatorially trapped proton fluxes. Theory includes the same processes as in Figure 25 but utilizes increased neutron fluxes observed by Preszler, *et al.* 1972. Curve A is for an atmosphere averaged over a solar cycle, and curve B is for a solar maximum atmosphere.

Data from another measurement (Hovestadt *et al.*, 1972) has been added to the figure.

Since $L^3 j$ at this first invariant value now has a peak in the inner zone, the protons are not diffusing inward from the outer zone, but are diffusing in both directions away from this maximum. The figure indicates proton intensities somewhat in excess of measured values. Better agreement would be produced at 800 Mev gauss^{-1} by increasing the diffusion coefficient to get rid of them more rapidly. Such improvement would be at the expense of agreement at, say, 200 Mev gauss^{-1}, since the larger inward diffusion currents from the outer zone would produce excessive proton intensities at these lower first invariant values.

Claflin and White (1973) actually found that there is little freedom available to vary the diffusion coefficient in either direction without causing deterioration in the agreement between theory and experiment.

In short, the extrapolation of Lingenfelter's neutron values produced a distribution function deficient in protons at low L values; the much larger values indicated by the recent experimental measurements produce excesses in the middle of the inner zone particularly at the higher energies.

It may be that there are still unknown mechanisms such as proton pitch angle diffusion which remain to be taken into account, or it may be that there is still sufficient error among the many input parameters or in the proton data itself to account for the remaining discrepancies.

VI. SUMMARY AND CONCLUSIONS

In the previous sections the various mechanisms which have been identified as important for particles trapped in the inner radiation zone were reviewed, and the quantitative theories for these processes were developed. The processes include 1) collisions of the particles with the ambient atmosphere, 2) radioactive decay of earth albedo and solar neutrons, 3) radial diffusion by fluctuating electric and magnetic fields, 4) secular variation of the geomagnetic field, 5) wave-particle interactions, and 6) reversible adiabatic variations.

The theories of the inner zone were developed with the philosophy that mechanisms influencing the inner belt particles could be isolated from a broader range of processes which make the outer belt much more chaotic and unpredictable. To some extent this separation has been possible. For example, the albedo neutron source and atmospheric collisions are local phenomena, unrelated to outer zone processes. Also, radial diffusion coefficients for inner belt particles can be expressed in terms of local electric and magnetic field variations. Of course these variations arise from currents on the surface of the magnetosphere, or distributed within it, and thus the theory is not entirely satisfactory until the entire magnetosphere is better understood. Also many of the inner zone particles enter the region via the outer zone, and the particle population does depend indirectly on conditions in the outer zone. This problem has generally been handled by establishing an outer boundary for the inner zone and treating quantitatively the boundary conditions which determine the particle movement across the boundary.

Although there is good reason to believe that the major mechanisms affecting inner-zone particles have been identified and are understood at least qualitatively, and in many cases quantitatively, there are a number of areas where the theory needs further development. In particular the radial diffusion theories are incomplete. It is not clear whether electric or magnetic field variations are the more important in the inner zone, and perhaps both cccur to a comparable extent. Also, the theories have been developed thus far for simplified models of the disturbance fields. In the magnetic field perturbations, for example, the azimuthal variation of the perturbation field was assumed to vary as cos φ. Substorm field variations very likely have a higher spatial harmonic content than a simple cos φ variation. The radial diffusion theory also suffers from the limitations of a perturbation theory. The perpendicular drift velocity resulting from the disturbance field is assumed to be very much smaller than the normal gradient and curvature drifts. Appreciable changes in L can take place only over many drift periods. This perturbation approach may not always be valid since even in the inner radiation zone some of the changes induced by severe magnetic storms take place within a few hours or less. The wave-particle interactions processes outlined in section 4-G are probably in the least satisfactory state. In addition to the electromagnetic cyclotron wave interactions described in that section, there are a variety of other modes whose importance remains to be assessed. Also, as was the case for radial diffusion, the waves responsible for the particle diffusion may originate outside the inner radiation zone so that the wave-particle interaction phenomena may be a total magnetospheric one, and the inner belt cannot be entirely divorced from the rest of the magnetosphere.

Some comments on the experimental data are also in order. Experimental uncertainties in such input quantities as the albedo neutron flux and energy spectrum, the atmospheric density and composition, and solar cycle variations of these quantities contribute appreciable error to the theoretical estimates of the expected radiation belts. There are also uncertainties in the radiation belt fluxes. In spite of the fact that measurements on inner belt particle populations have been made for a decade, the data base is not as comprehensive as one would like. For comparison with the theory one needs to construct the particle distribution as a function of the adiabatic invariants. Most of the experimental data give omnidirectional fluxes above a given threshold; only occasionally is it possible to obtain the differential energy spectrum, and even fewer measurements give the energy and pitch angle variations needed to derive the particle distribution function., In order to refine the theoretical values of the radial diffusion coefficients even for the restrictive theory, much better experimental information of the fluctuations of magnetospheric electric and magnetic fields are required. In section 4-F the difficulty in measuring the asymmetric component of the magnetic field variations in the presence of large symmetric components was mentioned. Better data will also be needed to see whether the field models assumed in the present theories are adequate.

Even with the reservations enumerated above the inner radiation zone is the best understood region of particle trapping in the magnetosphere. The stationary or relatively slow time dependence of most of its characteristics has allowed the synthesis of an overall picture of that region from a number of individual measurements taken at different times.   From the theoretical point of view the stability or slow time rate of change has frequently allowed the use of time independent equations to describe the phenomena. Also of importance to the theory, the low magnitude of distrubances in the inner zone has made it possible to treat many of the mechanisms by perturbation formulations.

Although many inner zone mechanisms are not of importance to the outer zone and the theoretical treatments derived for the inner zone are unsuitable at large L, the knowledge of inner zone mechanisms does give a basis for recognizing similar processes in the outer zone. The mathematics is much more involved there, and it is already apparant that the relative importance of the various mechanisms is different at high L. Nevertheless, information derived from the studies of the inner zone is useful in outer zone investigations and will be needed in the coming investigations of the radiation belts of other planets.

List of symbols

The following is a list of symbols used throughout the article. Infrequently used symbols which appear in the text but which are not listed below are defined where used.

$A(t)$	azimuthally asymmetric term in perturbation magnetic field
a	$(= -5 \times 10^4 \, \mathrm{yr}^{-1})$, decay constant of dipole term in geomagnetic field
$\mathbf{B}$	magnetic field value
$\mathbf{B}_0$	dipole magnetic field value
b_K^2	spectral power density of wave magnetic field
$\mathbf{b}(t)$	disturbance or perturbation magnetic field value
c	velocity of light in a vacuum
D_E	radial diffusion coefficient produced by electric potential field fluctuations
D_{ij}	diffusion coefficient for i, j coordinates
D_M	radial diffusion coefficient produced by magnetic field disturbances
E	particle energy
$\mathbf{E}$	electric field
$E_{\varphi n}$	nth Fourier expansion coefficient of the azimuthal component of the electric field perturbation

e	electron charge
$\hat{e}_r, \hat{e}_\theta, \hat{e}_\varphi$	unit vectors in the r, θ, and φ directions
F	particle distribution in phase space
$f(y, t)$	particle distribution functions in coordinates y at time t
g_n^m, h_n^m	coefficients of spherical harmonic expansion of geomagnetic field.
H	Hamiltonian function
h	Planck's constant
I_i	geometric mean ionization potential of ith atomic species
J	second adiabatic invariant
$\mathscr{J}(\mathbf{x}:\mathbf{y})$	Jacobian function giving ratio of volume elements in x coordinate system to y system
j	particle directional flux, subscripts used to denote type of particle
j	current density
k	wave number of electromagnetic wave (Boltzman constant)
L	magnetic shell coordinate of McIlwain
$\mathscr{L}(x, E)$	atmospheric average scattering function, defined by Eq. (4-33)
$\mathfrak{M}(x, E)$	atmospheric average energy loss function, defined by Eq. (4-33)
$\mathscr{M}$	dipole moment of the earth
$\mathscr{M}_0$	dipole moment of the earth at present time
m	particle mass, subscripts used to denote species
m_0	particle rest mass
n	number density, subscripts used to denote species
P_n^m	Legendre Polynomials
$\mathbf{p}$	particle momentum
$p_\perp, p_\parallel$	momentum components parallel and perpendicular to the magnetic field
Q	source function for trapped particles
q	electric charge of a charged particle
$\mathbf{q}$	position vector
R_e	radius of the earth
R_0	equatorial crossing position of magnetic dipole field line
r	($= LR_e$) radial coordinate in equatorial plane
$\mathbf{r}$	radial vector in spherical coordinates
r_0	value of r at $t = 0$

S	spiral distance between particle mirroring points
$S(t)$	azimuthally symmetric term in magnetic field perturbation
s	distance measured along a magnetic field line
s_n	neutron leakage flux
T	period of electromagnetic field fluctuations, temperature
t	time
v	velocity of guiding center of a trapped particle
$v_\perp, v_\parallel$	components of guiding center velocity perpendicular and parallel to the magnetic field
v_A	Alfven velocity
v_g	group velocity of a wave
v_p	velocity of plasma motion
v_{ph}	phase velocity of a wave
w	particle velocity
$w_\perp, w_\parallel$	components of particle velocity perpendicular and parallel to the magnetic field
x	cosine of equatorial pitch angle
$\mathbf{x}$	position vector in phase space
$Y(y)$	distribution function in y coordinate system
y	arbitrary coordinate system
Z	electric charge number of particle
$Z(\mu, J, \Phi, \varphi_1, \varphi_2, \varphi_3)$	distribution function in a space defined by the adiabatic invariants and their associated angle variables
$Z'(\mu, J, \Phi)$	distribution function in adiabatic invariant space integrated over angle variables
$\mathbf{z}$	coordinate vector in adiabatic invariant space
α	cosine of the local pitch angle
α_e	cosine of equatorial pitch angle
β	ratio of particle velocity to velocity of light
γ	$(= [1-\beta^2]^{-\frac{1}{2}})$ relativistic mass increase factor
γ	growth rate of electromagnetic wave
γ_n	phase constant of the nth Fourier component of the azimuthal electric potential field. Defined in Eq. (4-102)
ΔS	length of wave-particle interaction region
ζ	zenith angle
η	scattering angle in center of mass frame of reference
η	phase constant giving longitude of particle at $t = 0$
Θ	local pitch angle of particle
θ	spherical polar coordinate

θ_s	scattering angle in laboratory frame of reference
λ	geomagnetic latitude
λ_D	$(= [kT/4\pi m_e e^2]^{\frac{1}{2}})$ Debye length
μ	first adiabatic invariant
v_{drift}	azimuthal drift frequency of particle
$\bar{\rho}$	atmospheric density averaged over trapped particle orbit
$\sigma(\eta)$	differential scattering cross section
τ_a	apparent lifetime of trapped particles
τ_b	full bounce period of particle
τ_c	correlation time of electric and magnetic field fluctuations
τ_{col}	mean time for a trapped particle to be removed by a non-elastic collision
τ_d	azimuthal drift period of trapped particles
τ_E	time required for trapped proton to lose $\frac{1}{2}$ of its energy by collisions with atmosphere
τ_n	mean lifetime of neutron
Φ	third adiabatic invariant
φ	azimuthal angle in spherical polar coordinates
$\varphi_1, \varphi_2, \varphi_3$	angle variables associated with adiabatic invariants
χ	injection coefficient for introducing trapped particles into the geomagnetic field
Ψ	transition probability
ψ	azimuthal scattering angle
Ω	solid angle
Ω	gyration frequency of charged particle, subscript denotes particle type
Ω_0	angular drift velocity of trapped particle in a dipole field
ω	wave frequency

References

Alfvén, H., and C.-G. Fälthammar (1963). *Cosmical Electrodynamics*, 2nd Edition, Oxford University Press.

Birmingham, T. J., and F. C. Jones (1968). *J. Geophys. Res.*, **73**, 5505.

Birmingham, T. J. (1969). *J. Geophys. Res.*, **74**, 2169.

Blanchard, R. C., and W. N. Hess (1964). *J. Geophys. Res.*, **69**, 3927.

Brice, N. (1964). *J. Geophys. Res.*, **69**, 4515.

Claflin, E. S., and R. S. White (1970). *J. Geophys. Res.*, **75**, 1257.

Claflin, E. S., and R. S. White (1973). University of California, Riverside, *Institute of Geophysics and Planetary Physics Rep. IGPP-UCR-73-2* (unpublished).

Cornwall, J. M. (1964). *J. Geophys. Res.*, **69**, 1251.

Cornwall, J. M., A. R. Sims and R. S. White (1965). *J. Geophys. Res.*, **70**, 3099.

Cornwall, J. M. (1968). *Radio Sci.*, **3**, 740.

Davis, L., and D. B. Chang (1962). *J. Geophys. Res.*, **67**, 2169.

Dessler, A. J., and R. Karplus (1961). *J. Geophys. Res.*, **66**, 2289.

Dragt, A. J., M. M. Austin and R. S. White (1966). *J. Geophys. Res.*, **71**, 1293.

Dragt, A. J. (1971). *J. Geophys. Res.*, **76**, 2313.

Dungey, J. W. (1963). *Planetary Space Sci.*, **11**, 591.

Etcheto, J., R. Gendrin, J. Solomon and A.Roux, (1973). *J. Geophys. Res.*, **78**, 8150.

Eyles, C. J., A. D. Luney and G. K. Rochester (1971). Int. Conference on Cosmic Rays, *Conf. Papers*, **2**, 462.

Fälthammar, C.-G. (1965). *J. Geophys. Res.*, **70**, 2503.

Fälthammar, C.-G. (1968). In B. M. McCormac (Ed), *Earth's Particles and Fields*, Reinhold, New York.

Farley, T. A. (1969a). *J. Geophys. Res.*, **74**, 377.

Farley, T. A. (1969b). *J. Geophys. Res.*, **74**, 3591.

Farley, T. A., A. D. Tomassian and M. C. Chapman (1969). *J. Geophys. Res.*, **74**, 4721.

Farley, T. A., A. D. Tomassian and M. Walt (1970). *Phys. Rev. Letters*, **25**, 47.

Farley, T. A., and M. Walt (1971). *J. Geophys. Res.*, **76**, 8223.

Farley, T. A., M. G. Kivelson and M. Walt (1972). *J. Geophys. Res.*, **77**, 6087.

Freden, S. C., and R. S. White (1959). *Phys. Rev. Letters*, **3**, 9.

Freden, S. C., and R. S. White (1960). *J. Geophys. Res.*, **65**, 1377.

Freden, S. C., and R. S. White (1962). *J. Geophys. Res.*, **67**, 25.

Gendrin, R. (1968). *J. Atmos. Terr. Phys.*, **30**, 1313.

Goldstein, H. (1950). *Classical Mechanics*, Addison-Wesley, Cambridge.

Haerendel, G. (1968). In B. M. McCormac (Ed), *Earth's Particles and Fields*, Reinhold, New York.

Heckman, H. H., P. J. Lindstrom and G. H. Nakano (October 1969). University of California, *Lawrence Radiation Lab. Rep. UCRL-19309*.

Heckman, H. H., and P. J. Lindstrom (1972). *J. Geophys. Res.*, **77**, 740.

Heidbreder, E., K. Pinkau, C. Reppin and V. Schonfelder (1970). *J. Geophys. Res.*, **75**, 6347.

Hess, W. N. (1959). *Phys. Rev. Letters*, **3**, 11.

Hess, W. N., H. W. Patterson, R. Wallace and E. L. Chupp (1959). *Phys. Rev.*, **116**, 445.

Hess, W. N., E. H. Canfield and R. E. Lingenfelter (1961). *J. Geophys. Res.*, **66**, 665.

Hovestadt, D., E.Achtermann, B. Ebel, B. Hausler and G. Paschmann (1972). In B. M. McCormac (Ed) *Earth's Magnetospheric Processes*, D. Reidel Publishing Company.

Imhof, W. L., and R. V. Smith (1965). *Phys. Rev. Letters*, **14**, 885.

Kennel, C. F., and H.E Petschek (1966). *J. Geophys. Res.*, **71**, 1.

Kruskal, M. (1962). *J. Math. Phys.*, **3**, 806.

Lenchek, A. M., and S. F. Singer (1962). *J. Geophys. Res.*, **67**, 1263.

Lenchek, A. M., and S. F. Singer (1963). *Planet. Space Sci.*, **11**, 1151.

Lingenfelter, R. E. (1963). *J. Geophys. Res.*, **68**, 5633.

Lingenfelter, R. E., and E. J. Flamm (1964). *Science*, **144**, 292.

Lingenfelter, R. E., E. J. Flamm, E. H. Canfield and S. Kellman (1965a). *J. Geophys. Res.*, **70**, 4077.

Lingenfelter, R. E., E. J. Flamm, E. H. Canfield and S. Kellman (1965b). *J. Geophys. Res.*, **70**, 4087.

Lockwood, J. A., and L. A. Friling (1968). *J. Geophys. Res.*, **73**, 6649.

Lyons, L. R., R. M. Thorne and C. F. Kennel (1972). *J. Geophys. Res.*, **77**, 3455.

Lyons, L. R., and R. M. Thorne (1973). *J. Geophys. Res.* (to be published).

MacDonald, W. M., and M. Walt (1961). *Ann. Phys.*, **15**, 44.

Macy, W. W., R. S. White, R. C. Filz and E. Holeman (1970). *J. Geophys. Res.*, **75**, 4322.

McIlwain, C. E. (1961). *J, Geophys. Res.*, **66**, 3681.

McIlwain, C. E. (1966a). *J. Geophys. Res.*, **71**, 3623.

McIlwain, C. E. (1966b). In B. M. McCormac (Ed), *Radiation Trapped in the Earth's Magnetic Field*, D. Reidel Publishing Company.

Møller, C. (1932). *Ann. Physik*, **14**, 531.

Morrison, P. (1953). In E. Segre (Ed), *Experimental Nuclear Physics*, Vol. 2, John Wiley and Sons, N.Y.

Mott, N. F., and H. S. W. Massey (1950). *The Theory of Atomic Collisions*, 2nd Edition, Oxford University Press.

Nakada, M. P. (1963). *J. Geophys. Res.*, **68**, 47.

Nakada, M. P., and G. D. Mead (1965). *J. Geophys. Res.*, **70**, 4777.

Newkirk, L. L., and M. Walt (1968a). *J. Geophys. Res.*, **73**, 1013.

Newkirk, L. L., and M. Walt (1968b). *J. Geophys. Res.*, **73**, 7231.

Northrop, T. G., and E. Teller (1960). *Phys. Rev.*, **117**, 215.

Northrop, T .G. (1963). *The Adiabatic Motion of Charged Particles*, Interscience, New York.

Pfitzer, K. A., and J. R. Winckler (1968). *J. Geophys. Res.*, **73**, 5792.

Preszler, A. M., G. M. Simnett and R. S. White (1972). *Phys. Rev. Letters*, **28**, 982.

Roederer, J. G. (1970). *Dynamics of Geomagnetically Trapped Radiation*, Springer-Verlag, New York.

Schulz, M., and G. A. Paulikas (1972). *J. Geophys. Res.*, **77**, 744.

Singer, S. F. (1958). *Phys. Rev. Letters*, **1**, 181.

Stern, D. P. (1966). *Space Sci. Rev.*, **6**, 147.

Tomassian, A. D., T. A. Farley and A. L. Vampola (1972). *J. Geophys. Res.*, **77**, 3441.

Tverskoy, B. A. (1965). *Geomagnetizm i Aeronomiya*, **5**, 617.

Van Allen, J. A. (1966). In B. M. McCormac (Ed), *Radiation Trapped in the Earth's Magnetic Field*, D. Reidel Publishing Co.

Vasyliunas, V. M. (1972). *J. Geophys. Res.*, **77**, 6271.

Vernov, S. N., N. L. Grigoriv, I. P. Ivanenko, A. I. Lebedinskii, V. S. Murzin and A. E. Chudakov (1959). *Dokl. Akad, Nauk, SSSR*, **124**, 1022.

Walt, M., and W. M. MacDonald (1961). *J. Geophys. Res.*, **66**, 2047.

Walt, M. (1964). *J. Geophys. Res.*, **69**, 3947.

Walt, M., and L. L. Newkirk (1966). *J. Geophys. Res.*, **71**, 3265.

Walt, M. (1970). In B. M. McCormac (Ed), *Particles and Fields in the Magnetosphere*, D. Reidel Publishing Co.

Walt, M. (1971). *Rev. Geophys. and Space Phys.*, **9**, 11.

White, R. S., S. Moon, A. M. Preszler and G. M. Simnett (May 1972). University of California, Riverside, *Institute of Geophysics and Planetary Physics Rep. IGPP-UCR-72-16*.